Optoelectronics
An Introduction

J.C.A. Chaimowicz

BUTTERWORTH
HEINEMANN

Butterworth-Heinemann Ltd
Linacre House, Jordan Hill, Oxford OX2 8DP

PART OF REED INTERNATIONAL BOOKS

OXFORD   LONDON   BOSTON
MUNICH   NEW DELHI   SINGAPORE   SYDNEY
TOKYO   TORONTO   WELLINGTON

First published as *Lightwave Technology: An Introduction* 1989
Reprinted 1992

British Library Cataloguing in Publication Data
Chaimowicz, J. C. A. (John C. A.)
  Optoelectronics: an introduction
  1. Electro-optics
  I. Title
  535
ISBN 0 7506 0803 X

Library of Congress Cataloguing in Publication Data
Chaimowicz, J. C. A.
  Optoelectronics: an introduction
  Bibliography: p.
  Includes index.
  ISBN: 7506 0803 X
  1. Electro-optics   2. Electro-optical devices
  I. Title
  TA1750.C47   1989   621.36'9   88–29590

Printed and bound in Great Britain at the University Press, Cambridge

Preface

Until about 1963 the only two light-producing devices the electronics engineer came into contact with in the course of his or her work were the CRT and the panel indicator. The CRT's use was limited to the oscilloscope, the TV set and radar. The panel indicator was simply a small incandescent bulb or a miniature neon. The electronics engineer could get away with very little optics indeed!

The almost simultaneous appearance of the LED and the laser in the early 1960s, followed only slightly later by the optical fibre, has changed all this. More recent intrusions of electro-optic devices and techniques makes the situation change so rapidly that one feels almost entitled to speak of an opto-electronic 'revolution'. Suddenly, the electronics engineer stands face to face with the Intrusive Photon. If the term revolution, so much used and abused by the media in relation to the microchip, is too strong here, one can state with no fear of exaggeration that the intrusive photon already is, and certainly will go on, producing far-reaching changes in the world of communications, information storage and display, metrology, metal machining, energy conversion, printing, medicine, navigation and, alas, warfare.

Lightwave communications, laser machining, compact disc audio, fibre optic sensing, light guide image transmission and holography are all rapidly becoming part of modern engineering, with integrated optics and optical computing likely to follow quite soon. The impact is enormous. A further penetration of lightwave technology into engineering, not only electronics engineering, will be so extensive that many practising engineers of today run the risk of finding themselves handicapped tomorrow, when faced with new equipment, confronted with fresh articles in the press or forced to do design work with the help of data sheets on 'opto' devices. The use of visible or invisible, coherent or non-coherent light will become so widespread that familiarity with electro-optics will become a must.

So vigorous is the growth of the electro-optics based industries (and expert survey projections for the years to come are breathtaking) that it is bound to create, before long, a strong demand for engineers and technicians. They will be wanted in development, production, quality assurance, field engineering and technical sales. Many will be drawn from those with an electronics background. It is mostly for them, the newcomers to lightwave technology, that this book has been written. Without being a textbook, this volume will also be of assistance in the electronics departments of polytechnics and universities, both to final year undergraduates and to new postgraduates. Finally, most of the text should provide enjoyable reading to the more inquisitive general reader wishing to gain a basic understanding of some of the most fascinating applications of lightwave technology.

The book is structured around a course of lectures first delivered at the Gloucester College of Technology in 1982, the contents of which have been both broadened and updated to take into account the progress made during the intervening years.

The treatment is descriptive rather than scholarly. Profusely illustrated, the explanations rest more on the physical aspects of things than on mathematics, and encourage intuitive perception. Whatever residual mathematics there is has been relegated to the appendices. Optical subjects have been tackled by electrical analogy wherever possible. Some chapters (and all appendices) are only loosely bonded to the remainder of the book, so that they can be selectively picked (or skipped!) by particular categories of readers without loss of comprehension. Thus, Chapters 4 and 11 and most of the appendices could be bypassed by the general reader while Chapters 1, 13 and 14 and large sections of others which lie outside the reader's own immediate needs could be omitted by the practically minded technician or engineer.

The writing of this book was greatly helped by the stimulating atmosphere of the Department of Electronic and Electrical Engineering of University College London. I am indebted to Professor D.E.N. Davies and Professor B.Culshaw for enabling me to benefit from it. Many seemingly casual conversations held there on various aspects of electro-optics proved, with hindsight, to have been not only of an informative but also of a quest-impelling nature. Intentional, as opposed to casual, assistance has also been received. In this respect I wish to thank Ms D. Vickers for setting up the fibre optic gyroscope experiment photographed for Section 12.1.2 and Dr P. Sturges for reviewing Section 13.3 on integrated optics. Thanks are also due to Ms K. Mallalieu and Dr S. Venkatesh for the information on resonant optical fibre sensors (Section 14.2.2) and to Mr A.P. Overbury for that on the IO spectrum analyser (Section 13.3.4). I also wish to thank Dr G. Parry, UCL, for drawing my attention to the moon landing exhibit at the Science Museum (Section 12.2.1), and Christine Lindey of the Courtauld Institute for the research related to the way of illustrating the concept of a shaft of light (Section 4.1). My sincere thanks go to Mrs Betty Smith for her cooperation and patience shown in the preparation of an excellent typescript of the book.

J.C.A. Chaimowicz

Contents

Preface   v

Notation   xi

Abbreviations and acronyms   xv

1 Introduction   1
   1.1 Light, lightwave technology and electronics   2

2 How lenses work and how light is guided   5
   2.1 How lenses work   5
   2.2 Guiding light with and without the help of lenses   17

3 Working with photons: the generalised P–N junction   21
   3.1 The physics of the P–N junction   21
   3.2 The usefulness of P–N junctions to the electronics engineer and the electro-optics specialist   27
   3.3 The three quadrants and the six operating modes: interaction with light   30

4 Photometric and radiometric quantities   36
   4.1 Luminous flux   36
   4.2 Illuminance–irradiance   39
   4.3 Intensity   40
   4.4 Luminance–radiance   42

5 The LED: heart of the light transmitter   44
   5.1 Wavelength varieties: IREDs, VLEDs and LWLEDs   44
   5.2 Packaging the wafer   45
   5.3 Wafer geometries   47
   5.4 The LED as a circuit element   50
   5.5 Driver circuits   52
   5.6 Making full use of data sheets   55
   5.7 LEDs – the exotic breeds   58
   5.8 The attractiveness of the LED as a light source   60

6 The photodiode: core of the light receiver   62
   6.1 The non-amplifying light-receiving junction   62
   6.2 The amplifying light-receiving wafer   81
   6.3 The photoreceiver – core of the RX antenna   85

7 The captive ray: fibre optics communications   89
   7.1 Insulators compete successfully with conductors   89
   7.2 The fundamentals: fibre optic physics   90
   7.3 Fibre optic hardware. Phase velocity   96
   7.4 Why fibres?   102
   7.5 Other components of a fibre optic link   103
   7.6 Fibre optics communication systems   110
   7.7 State of the art   115

8 The liberated ray or fibreless optical communications   119
   8.1 Free space optical communications from 3500 years ago to the
       present   119
   8.2 The nature of free space optical communications   120
   8.3 The hardware of a simplex FSOC link   122
   8.4 The FSOC link as a whole   127
   8.5 Examples of FSOC equipment   132
   8.6 FSOCs and their competitors   134

9 The dangerous and the non-dangerous laser   135
   9.1 Why the laser is such a special light source   135
   9.2 The way lasers work   135
   9.3 Helium–neon (HeNe) lasers   140
   9.4 Gallium arsenide (GaAs) lasers   143
   9.5 Other lasers   150
   9.6 The power range of lasers   158
   9.7 The jobs lasers do   160

10 Laser beam engineering   166
   10.1 Exploiting the laser beam   166
   10.2 The untreated laser beam   166
   10.3 Beam treatment   177

11 Especially for electronics engineers   212
   11.1 Non-electrical transmission of electrical signals by means of
       insulated signal couplers   212
   11.2 Basic structure and operation   213
   11.3 Variations on a theme   214
   11.4 Some applications of ISCs   216
   11.5 Optofollower – the supercoupler   217
   11.6 Digital transmission of analogue signals   220

12 Red and not-so-red rays for engineering   222
   12.1 The sensing fibre   222
   12.2 Optical barriers and laser 'chalk lines'   231
   12.3 Illuminating and imaging FO bundles   241
   12.4 Optical data storage on disc   244
   12.5 The supermarket's bar code, the librarian's magic wand   248
   12.6 Laser Doppler velocimetry   254

13 Holography, Fourier Transforms and integrated optics   259
   13.1 Three-dimensional imaging by holography   259
   13.2 Frequencies and filters both temporal and spatial. Fourier
       Transforms by lens   271
   13.3 Integrated optics   282

14 The electro-optics curiosity shop   296
  14.1 Lenses by the length   296
  14.2 Optically activated levitation, rotation and flexural
  vibration   298
  14.3 Changing colours, reversing time and flip-flopping through non-
    linearities   302

Epilogue   309

Appendix 1 Decibels and optical density   310

Appendix 2 Essential lens formulae   313

Appendix 3 Radiation flux within a solid angle of an emitter, calculated
    from its polar diagram   316

Appendix 4 Lambertian radiators and re-radiators   321

Appendix 5 On noise in semiconductor diode detectors   324

Appendix 6 Determination of the maximum angle of acceptance   327

Appendix 7 The influence of fibre (core) diameter on the number of
    probable propagating modes   329

Appendix 8 The two velocities   330

Appendix 9 A four-step method for range calculations   331

Appendix 10 Range calculations for Section 8.4.2(a)   334

Appendix 11 The derivation of coherence length from spectrum
    width   336

Appendix 12 Circular and elliptical polarisation: retardation plates   338

Appendix 13 List of principal periodicals containing information on EO
    developments   345

Bibliography   346

Index   355

Notation

A	Area
A_{RXA}	Area of receiver antenna
c	Speed of light in vacuum ($= 2.998 \times 10^8$ m/s)
C	Capacitance
C_{D}	Dynamic capacitance
C_{j}	Junction capacitance
C_{p}	Package capacitance
d	Pitch (of grating)
d	Diameter
d_{s}	Diameter of source or of received spot
D	Distance, optical density, detectivity, diameter
D_{L}	Diameter of lens
D^*	Specific detectivity
e	Base of natural logarithms ($= 2.718281828$)
e	Electron charge ($= 1.602 \times 10^{-19}$ C)
E	Extraordinary ray
E	Energy, electric field
E	Vector of electric field
E_{c}	Conduction band edge
E_f	Fermi level
E_{g}	Energy gap
E_{v}	Valence band edge
f	Function, transfer function, frequency
$f_{\#}$	f number of a lens
f_{mod}	Modulation frequency
F	Frequency
$f_{\mathrm{L}}, F_{\mathrm{L}}$	Focal length
G	Gain
h	Planck's constant ($= 6.626 \times 10^{-34}$ J s)
I	Current, irradiance
I_{F}	Forward current
I_{N}	Noise current
I_{o}	Output or saturation current
I_{R}	Reverse current
I_{RX}	Receiver irradiance

k	Boltzmann's constant ($= 1.381 \times 10^{-23}$ J/K)
l	Length
l_c	Critical length
L	Length, inductance
L_c	Coherence length
L_s	Static or package inductance
m	Mass, integer number
m_0	Mass at rest
M	Magnification
n	Refractive index, ratio
n_{cl}	Refractive index of cladding
n_{core}	Refractive index of core
n_e	Refractive index of extraordinary ray
n_o	Refractive index of ordinary ray
n_s	Refractive index of slow medium
N	Luminous sensitivity
O	Ordinary ray
p, P	Power
P	Pressure
$\mathscr{P}$	Radiation pressure
P_i	Input power
P_o	Output power
P_{TX}	Transmitter power
P_{TOT}	Total power
q	Shape factor (of lens)
r	Reflectance factor
r, R	Radius
R	Range, resistance
R_D	Dynamic resistance
R_F	Feedback resistance
R_{FS}	Range of free space link
R_L	Load resistance
R_o	Back distance of free space link
R_S	Series resistance
$\mathscr{R}$	Responsivity
$\mathscr{R}_\lambda$	Spectral responsivity
S_i	Distance of image from lens
S_o	Distance of object from lens
S^*	Complex conjugate of S
t	Time
t_f	Fall time
t_r	Rise time
T	Transmittance, transfer factor, temperature
T_a	Ambient temperature
T_j	Junction temperature
v	Velocity
V	Visibility factor, voltage
V_{BR}	Breakdown voltage
V_F	Forward voltage
v_g	Group velocity

v_p	Phase velocity
v_s	Signal velocity
w	Beamwidth, waist width of laser beam
X_i	Distance of image from focal point
X_o	Distance of object from focal point
α	Angle
α_c	Critical angle
α_d	Deflection angle
α_R	Reflection angle
$\beta/2$	Divergence half-angle
$\gamma/2$	Half-angle of admittance
η	Efficiency
η_Q	Quantum efficiency
θ	Space angle, angle
λ	Wavelength
ν	Frequency
ϕ	Luminous flux
ω	Angular frequency, angular velocity
$\propto$	Proportional to
$\simeq$	Approximately equal to

Abbreviations and acronyms

A/D	Analogue-to-digital (converter)
AM	Amplitude modulation
APD	Avalanche photodiode
AVC	Automatic volume control
$BaTiO_3$	Barium titanate
BER	Bit error rate
BRH	Bureau of radiological health
BSI	British Standards Institution
BWD	Bandwidth
$CaCO_3$	Calcium carbonate (calcite)
CATV	Communal antenna television
CCTV	Closed circuit television
CD	Compact disc
CEGB	Central Electricity Generating Board
CGH	Computer-generated HOE (*see* HOE)
CIE	Commission Internationale d'Eclairage
CMR	Common mode rejection
CRT	Cathode ray tube
CTR	Current transfer ratio
CW	Continuous wave
D/A	Digital-to-analogue (converter)
DBR	Distance between repeaters
DC	Direct current
DHJ	Double heterojunction
DHLD	Double heterojunction laser diode
DIL	Dual in line
DOR	Digital optical storage
DRAW	Direct read after write
e.e.b.e.	Everything else being equal
EMI	Electromagnetic immunity
EO	Electro-optic
FEL	Free electron laser
FEP	Teflon
FET	Field effect transistor
FM	Frequency modulation
FMCW	Frequency modulated continuous wave

FO	Fibre optic
FOCs	Fibre optic communications
FOPS	Fibre optic pressure sensor
FP	Fourier plan
FPE	Fabry–Perot etalon
FQM	First quarter mode
FSOCs	Free space optical communications
FT	Fourier Transform
FT^{-1}	Inverse Fourier Transform
F/V	Frequency-to-voltage (converter)
FWHM	Full width at half maximum
GRIN	Graded index
HeNe	Helium–neon
HJ	Homojunction
HOE	Holographic optical element
Hologon	Holographic polygon
HT	High tension
IC	Integrated circuit
IO	Integrated optics
IR	Infrared
IRED	Infrared emitting diode
ISC	Insulated signal coupler
KDP	Potassium dihydrogen phosphate, KH_2PO_4
LASCR	Light activated silicon controlled rectifier
Laser	Light amplification by stimulated emission of radiation
LD	Laser diode
LDV	Laser Doppler velocimetry
LED	Light emitting diode
LIDAR	Light radar
$LiNbO_3$	Lithium niobate
LNP	Lithium neodymium phosphate, $LiNdP_4O_{12}$
LWLED	Long-wave light emitting diode
Maser	Microwave amplification by stimulated emission of radiation
MTF	Modulation transfer function
NA	Numerical aperture
Nd-YAG	Neodymium-YAG (see YAG)
NEP	Noise equivalent power
NLP	Neodymium lanthanum phosphate
OE	Opto-electronic
OFS	Optic fibre sensor
OO	Opto-optic, opto-optical
$PbMoO_4$	Lead molybdate
PCM	Phase conjugate mirror
PIN	Positive–intrinsic–negative
P–N	P-doped–N-doped (junction)
PSU	Power supply unit
PWM	Pulse width modulation
PZT	Piezoelectric transducer
RAPD	Reach through avalanche photodiode
ROM	Read-only memory

RXA	Receiver antenna
SAW	Surface acoustic wave
Selfoc	Self-focussing
SHJ	Single heterojunction
SHLD	Single heterojunction laser diode
SiO_2	Silicon dioxide
SNR	Signal-to-noise ratio
SQM	Second quarter mode
SQM(I)	SQM current
SQM(P)	SQM power
SQM(V)	SQM voltage
TAT	TransAtlantic
TCL	Total conjugate length
TEA	Transversely excited atmosphere (laser)
TEM	Transverse electromagnetic
TIR	Total internal reflection
TQM	Third quarter mode
TQM(I)	TQM current
TQM(VP)	TQM voltage/power
TTL	Transistor–transistor logic
TX	Transmitter
TXA	Transmitter antenna
UCL	University College London
UPC	Universal Product Code
UV	Ultraviolet
V/F	Voltage-to-frequency (converter)
VLED	Visible light emitting diode
VLP	Video long playing
YAG	Yttrium aluminium garnet

Chapter 1

Introduction

All light rays are invisible. Astounding as it sounds, this is absolutely true of all light rays. Unless a bundle of rays hits our eye – be it directly, by reflection or by diffusion – its passing cannot be noticed. Objects need light to be seen; light needs objects to be sensed. In an empty matt black painted room, a light beam entering through a hole in the wall and leaving it by another hole in the opposite wall (and I am not talking about infrared or ultraviolet, but about 'ordinary' red, green, yellow or white radiation) remains unseen unless there is dust or smoke in the atmosphere to scatter it about.

Invisible or not, light is totally and utterly indispensable to life, and not only to our life, but to all life on earth – to all plants and animals. All life on this planet runs on sunlight. This was recognised early in the history of man. The oldest human religions worshipped light (usually the Sun). Light is indispensable not only to enable us to see around us, but to our very existence. We even have a physiobiological unit of light, the einstein (a tribute to Einstein in recognition of his contribution to photoelectricity), to account for the relevance of light to the life of plants. Even sightless moles couldn't exist without light on earth. Our own individual existence may well start in darkness but it could not continue without light, even in the comfort of a mother's womb. Why? Because we need food, much more than the lighting and the warmth luminous radiation provides us with. Plants produce the hydrocarbons, proteins and vitamins we need, thanks to photosynthesis. Primitive man knew instinctively how vital light was to him – instinctively he felt a veneration for this Primeval Force, which even today the Mystic, the Spiritual and the Poet treat as 'the nearest thing to God'. The World's energy too (apart from nuclear energy) comes from fossil deposits, coal, oil and gas, which are the result of geological remains of plants and their transformations.

Photons help to make our food by providing the energy for photosynthesis (Figure 1.1). Water and carbon dioxide, under the influence of light (photons), become sugars and oxygen.

Proteins are most important. Albumen (egg white) is one of them. '*Green* plants are the food factories of the world', some books state. (I would add 'photons are the workers'.) Note the word 'green' in the sentence above. Chlorophyll is the fourth agent of photosynthesis; it produces the right kind of absorption (absent in fungi and some parasitic

1

$$\Phi \ (h\nu)$$

Water	+	Carbon dioxide	$\longrightarrow$	Glucose	+ Oxygen to breathe
$6H_2O$		$6CO_2$		$C_6H_{12}O_6$	$6O_2$

Proteins (food)   Starch (food)

Energy in glucose (food)

Vitamins (food)   Fats (food)

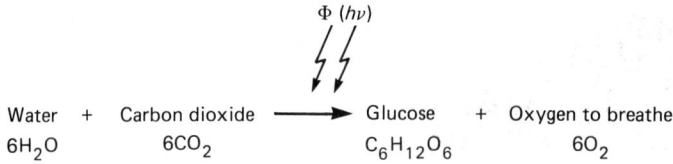

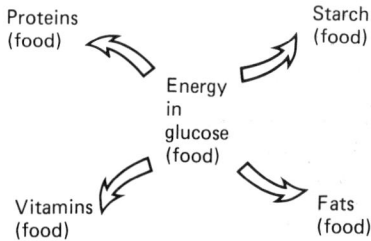

Figure 1.1 How photons help to make our food. Energy trapped by photosynthesis is used in producing all the essential foodstuffs

plants). A most welcome sub-product is oxygen for us to breathe (although some bacteria produce sulphur instead of oxygen). (Experiments have shown that a mouse can survive in a sealed jar, provided it has green plants in it!) Photosynthesis converts 400×10^9 tonnes of carbon dioxide into oxygen each year (10% on land). So when we speak of the intrusive photon, photosynthesis is certainly where its intrusion is most welcome [12, 15].

Perhaps I ought to apologise for using the term 'photon' without explaining what it is. Well, I am sure that most of you 'felt' that the photon is 'a particle of light', more precisely, the elementary particle of luminous energy. We shall return to this subject; let it suffice for now to say that light – even the weakest light from the nearest star – can be thought of as a cloud of an unimaginably large number of photons. Not only to man's stomach, but also to man's soul and to man's heart is *light* necessary. Before we get absorbed in technical matters I would like to invite you now to stop for a moment and reflect, quietly, very quietly, on Figures 1.2 and 1.3.

1.1   Light, lightwave technology and electronics

Luminous energy is contained in and carried by electromagnetic (EM) waves. Like radiowaves, light consists of *harmonic* EM oscillations and thus it is only natural that they should be linked by a large amount of concept sharing. Pick at random an issue of *Optical Engineering, Applied Optics* or any similar magazine of the last decade, and you will see at once that the language of modern optics has much in common with that of electronics. Concepts and terms such as frequency, wavelength and phase, stemming from the harmonic nature of both types of radiation, appear

Figure 1.2 15000-year-old worked hollow stone oil lamp from the Shetlands (Reproduced by permission of the Trustees of the Science Museum (London))

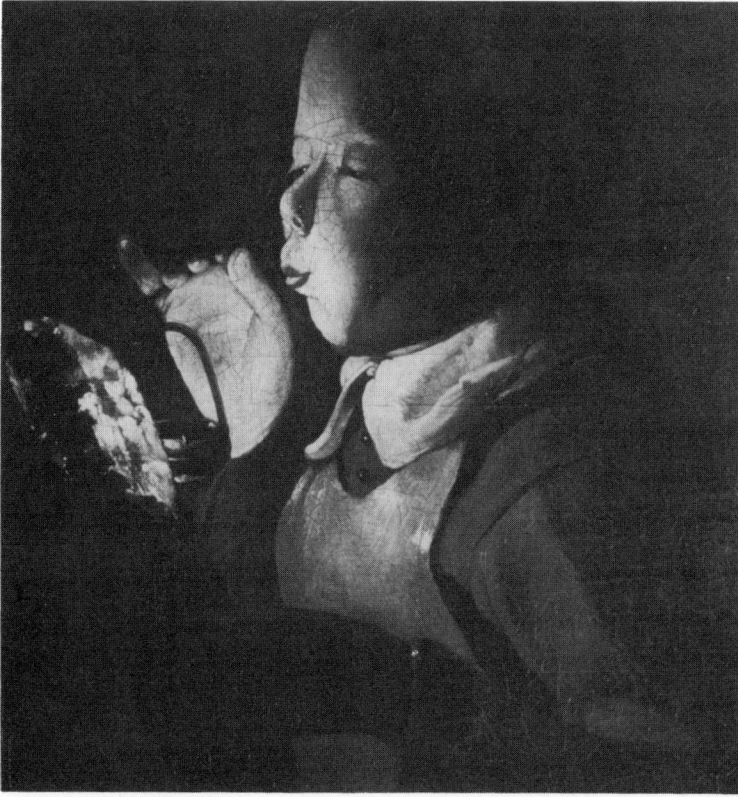

Figure 1.3 *Le Souffleur à la lampe*, Georges de la Tour. The interplay of light and shade, always strongly felt by the painter, is often magnificently rendered by him (Courtesy of the Musée des Beaux Arts, Dijon)

again and again. The same holds for dispersion, velocity (phase and group) or signal-to-noise ratio, to mention but a few. This was not so, however, at the time many optics manuals were written, leaning strongly either on 'glassware optics' or on mathematical analysis as they did. Since this book is destined mainly for engineers with a strong electronics background, it seems appropriate to use in it electronic concepts, the electrical way of argument and the language familiar to engineers whenever possible. For instance, the term 'frequency' will be used not only in connection with time-variable phenomena, but also in connection with the new independent variable – space (spatial frequency).

Phenomena such as interference or non-linearity can lead, in optics as in radio, to heterodyning, harmonics generation and even more recently to 'optical' rectification [17, p.503]. Modulation of a carrier transfer function (MTF), dispersion, wave propagation modes, can all be paired up in opto and radio, and so can the selective frequency attenuation – whether temporal or spatial, as for example with the notch filter. The concept of the decibel too has its opposite number in optics in the shape of the optical density, D. Both represent a log (base 10) function of a ratio of two values of a physical quantity: $\log_{10} P$ and $\log_{10} E$. Consider this worked example, based on Tables A1.1 and A1.2 of Appendix 1.

Worked example
(a) A power loss of 9.1 dB is read from Table A1.1 as −9.1 dB. This corresponds to a power transfer ratio of 0.123 = 12.3%:

$$-9.1 \text{ dB} = \log_{10} 0.123 = \log_{10} \frac{P_o}{P_i}$$

In bels we have −0.91 bels.
(b) The optical density, 'degree of blackness', $D = 0.91$, is read from Table A1.2 as corresponding to:

$$T = \frac{\Phi_{out}}{\Phi_{in}} 100 = 12.3\%$$

Perhaps the most staggering aspect of the kinship between the two technologies lies in their common use of the Fourier Transform: in the temporal frequency domain for the one and the spatial frequency domain the other.

Little did Maxwell see of the cross-fertilisation of techniques that was to take place in electromagnetic radiation, an area *he* had declared to be common to electrodynamics and optics. While Maxwell's equations, this cathedral of science, majestically dominate both technologies, deep differences between them should temper our enthusiasm for the pairing up process. Of these, two come to mind:

1. No movement of electrical charges is involved, generally speaking, in optics. The often praised electromagnetic immunity of fibre optic transmission lines probably stems from this. The concepts of 'capacitance' (charge storage), 'inductance' (inertia) and 'resistance' (friction) have no equivalent here.
2. The power levels involved in optics are generally much lower than those of electrodynamics.

Chapter 2

How lenses work and how light is guided

2.1 How lenses work

In older books we simply find statements of what lenses do, usually supported by geometrical drawings of 'light rays'. I shall explain lens action by considering the *wave* nature of light. Such a treatment will be more in tune with an electronics engineer's way of thinking, and prepare us, from this early stage, for the whole field of interference optics, to be dealt with in later chapters.

2.1.1 Plane waves instead of rays

Figure 2.1 shows that a plane-wave approach to an explanation is permissible. (Most light waves are either spherical or plane and the spherical one becomes planar not very far from the source of radiation, though in theory

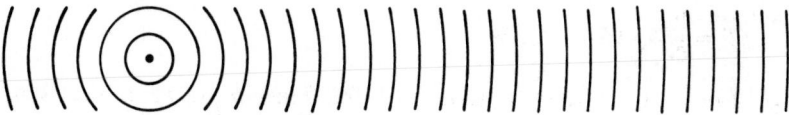

Figure 2.1 Spherical waves flattened with distance

at infinity.) Consider the case of such waves falling onto a glass plate, parallel and parallel to the wavefronts (see Figure 2.2). The surrounding medium is air. The closing up of wavefronts inside the plate is a consequence of a *lower* speed of light in the new, denser medium – glass. Note that the wavelengths λ_0 in air and λ_1 in glass are directly proportional to the velocity of light in these media.

Let n be the factor relating these velocities v_0 and v_1:

$$v_1 = \frac{v_0}{n}$$

n is called the *refractive index*: the reason for this becomes apparent from Figure 2.3, in which the plate of Figure 2.2 has been replaced by a prism.

$n_0 = 1.0$ $n_1 = 2.0$ $n_0 = 1.0$

$\lambda_1 = \frac{1}{2}\lambda_0$

λ_0

Air   Glass   Air

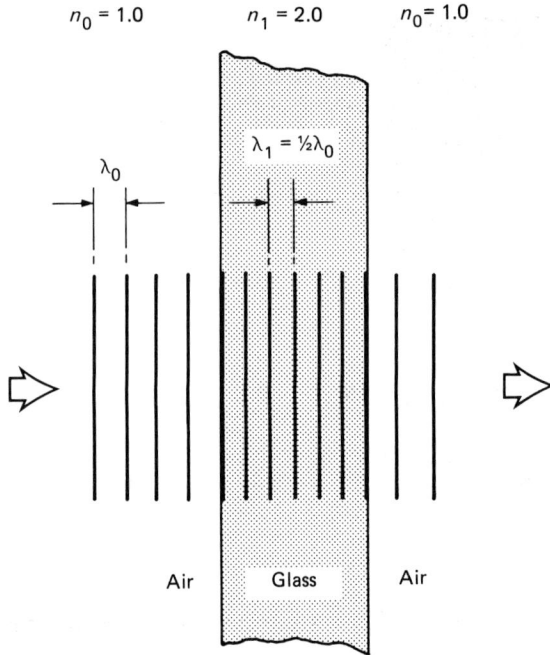

Figure 2.2 Plane wavefronts traversing a glass plate. The closing up of wavefronts in the plate is the consequence of a lower velocity of light in glass

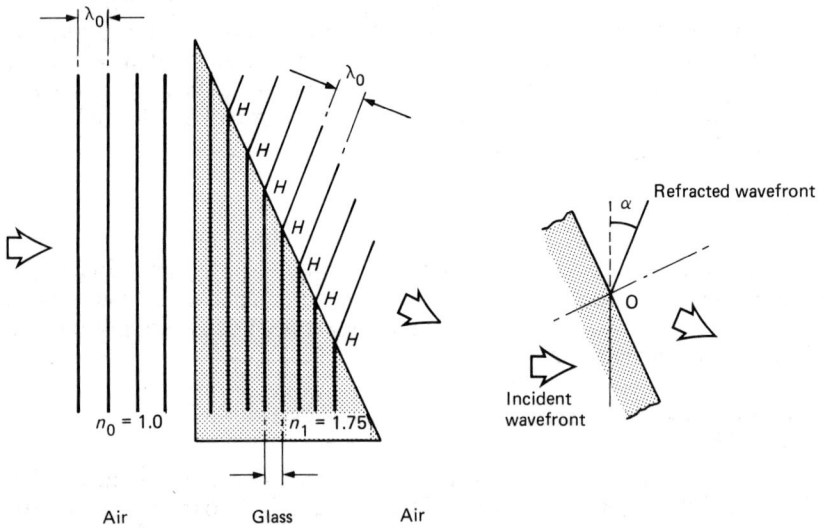

λ_0

λ_0

H
H
H
H
H
H
H

$n_0 = 1.0$ $n_1 = 1.75$

Air   Glass   Air

Refracted wavefront

α

O

Incident
wavefront

Figure 2.3 Plane wavefronts traversing a glass prism

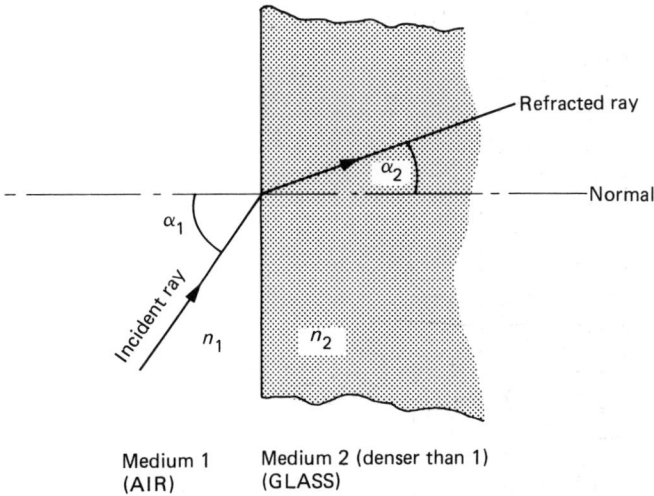

Figure 2.4 Interface refraction

As in Figure 2.2, $n = 2$ and wavelength λ_0 is halved upon entering the glass:

$$\lambda_1 = \tfrac{1}{2}\lambda_0$$

Upon leaving the glass the wavefronts, once again in the air, can only resume the original spacing λ_0, the speed of light having reverted to its original value. The consequence of this is that they *must* tilt at the interface. The amount of slant the wavefronts undergo is then strictly and uniquely defined by the ratio λ_0/λ_1. The waves are refracted and hence the name of refractive index for n, the λ_0/λ_1 ratio. (If we try, mentally, to swivel the whole comb of the refracted wavefronts at the hinging points H, simultaneously, we find that there is one and only one incident angle, α, for which the spacing between them assumes the right value, i.e. λ_1 given by λ_0/n. Thus, the amount of tilt α is uniquely defined by the refractive index n.) Simple geometry shows that:

$$\sin \alpha = \frac{1}{n}$$

Were the outer medium other than air or vacuum (strictly speaking, v_{air} is 0.03% smaller than v_{vacuum}) and the incident wavefronts other than parallel to the input face of the glass, as in Figure 2.4, we would use the generalised relation for an interface refraction:

$$\frac{n_1}{n_2} = \frac{\sin \alpha_2}{\sin \alpha_1} \tag{2.1}$$

n_1 and n_2 relate the light velocity in media 1 and 2 to that in vacuum which is 3×10^{10} cm/s (300000 km/s). Equation 2.1 is known as Snell's (or Descartes's) Law. It is usually stated for the fictitious 'ray of light', a very useful teaching and design aid. The 'ray' represents the direction of

8

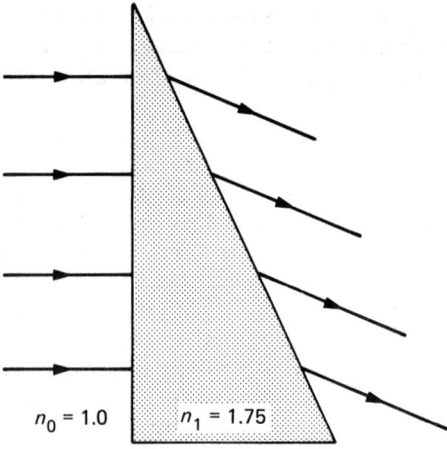

Figure 2.5 A translation into the ray language of Figure 2.3

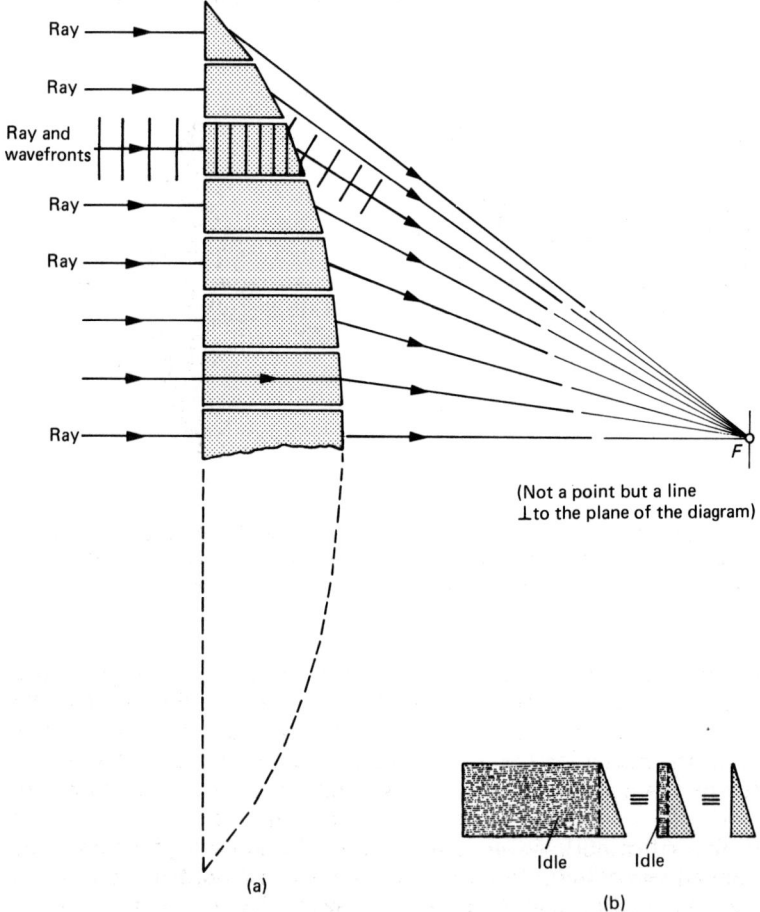

Figure 2.6 Rays corresponding to wavefronts incident upon a succession of small prisms

propagation of the wavefronts. As such, it is everywhere perpendicular to them. The heavy lines of Figure 2.4 represent the conventional (ray) way of illustrating refraction and Snell's Law. Figure 2.5 is a translation of Figure 2.3 into the ray language.

Refraction is of fundamental importance in the study of propagation of light. Snell's Law is to optics what Ohm's Law is to electronics.

Returning to prismatic refraction, let us now imagine a group of prisms with progressively decreasing apex angles, arranged as depicted in Figure 2.6(a). The incident plane wavefronts are broken up into individual 'frontlets', travelling in the direction of a common perpendicular to the drawing line F. In ray language, all the rays are refracted so that they travel towards the common 'point' F. It is easily seen that the apex angles of the prismlets can be chosen so that the intersection of the refracted wavefronts takes place along the line F and all the rays come together, or *focus*, as we say, at point F. Increasing the number of prismlets while reducing their height indefinitely creates the situation of Figure 2.7(a). Assuming now that all prismlets have very little depth and rotating Figure 2.7(a) on its axis of symmetry, we produce the *plano-convex lens* of Figure 2.7(b).

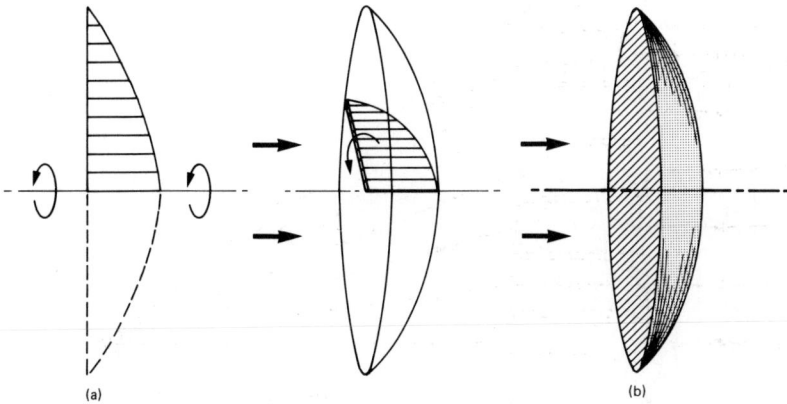

(a)   (b)

Figure 2.7 Increasing the number of prismlets while reducing their height indefinitely creates the situation shown in (a). Assuming that all the prismlets have very little depth, and rotating (a) on its axis of symmetry, we produce the plano-convex lens shown in (b)

Figure 2.6(b) shows that there is in each prismlet a portion of glasswork which does not participate in refraction. Removal or addition of such idle or 'dead' zones is then permissible. The Fresnel lens can be thought of as the result of first removing the idle glass bulk from Figure 2.8(a) and then adding to the result (Figure 2.8(b)) a strengthening, common substrate (Figure 2.8(c)). Fresnel lenses with small incremental steps have an effect similar to that of smooth plano-convex lenses. In many applications they have the advantage of being lighter (especially when moulded in plastic, as is often the case) and cheaper than smooth lenses. This is mostly of interest for large diameters. They can also be more readily manufactured in the

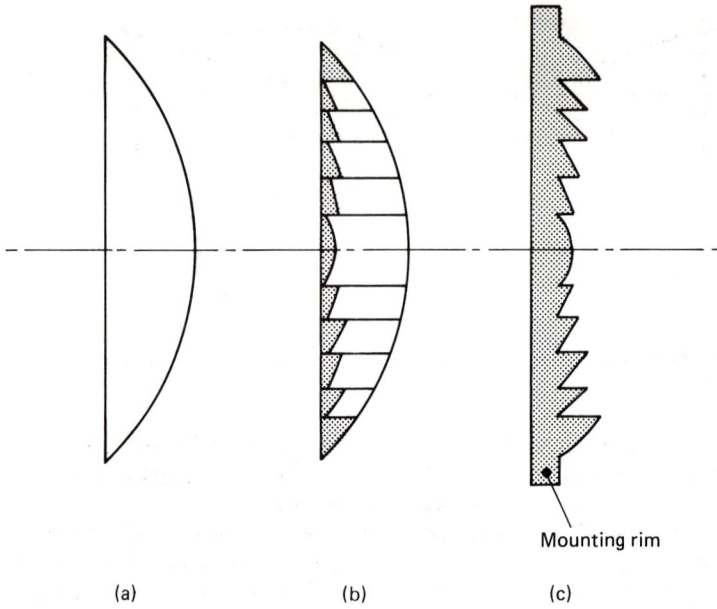

Mounting rim

(a) (b) (c)

Figure 2.8 Metamorphosis of a succession of prismlets into a Fresnel lens

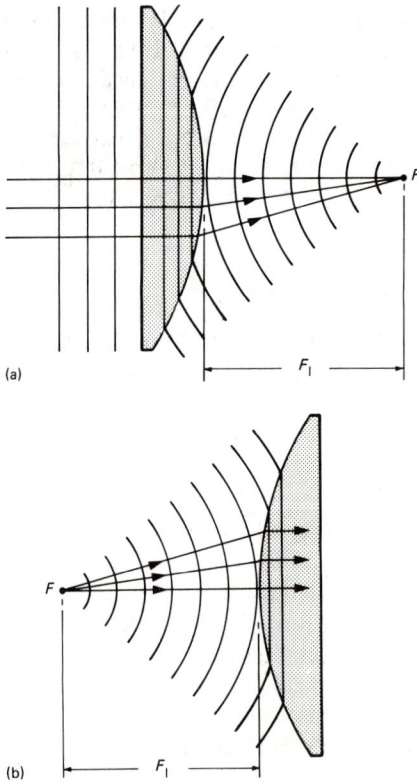

Figure 2.9 (a) The lens converts plane wavefronts into spherical ones. F is the centre of the concentric spheres. (b) Conversion of spherical wavefronts emerging from point F into parallel ones

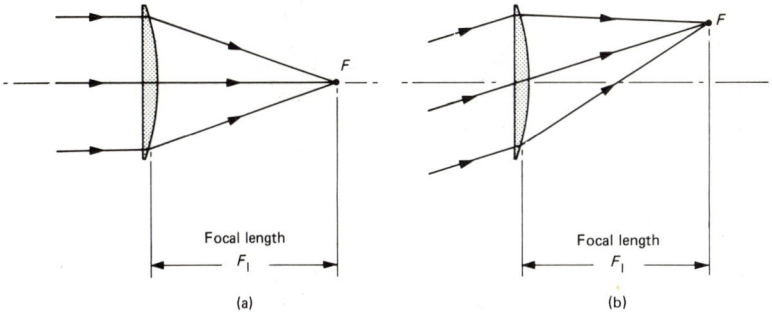

Figure 2.10 The distance at which a lens makes all the rays of an incident parallel bundle converge is the focal length

aspheric version than smooth lenses and this also has its attractions [3, (d)]. Generally speaking, however, Fresnel grooved elements are inferior substitutes for their conventional, smooth counterparts.

The distance at which a lens (Fresnel or smooth) makes all the portions of an incident plane parallel wavefront converge, or focuses all the rays of an incident parallel bundle (Figures 2.9 and 2.10(a)) is the *focal length*. This is F_L, the most important single parameter characterising a lens. For smooth lenses, the plane wavefront need not be parallel (or the rays perpendicular) to the input face, providing it is understood that F_L is measured as the on-axis distance between the lens and its *focal* plane (Figure 2.10(b)). (Strictly speaking, F_L is not measured from the lens vertex, but from one of its two 'principal points', which only occasionally coincides with the vertex, as in Figure 2.10 [17, p.167]. In cases of our concern the difference is negligible.) In most catalogues and some books, F_L is referred to as the EFL (effective focal length) to distinguish it from some other, similar but not identical, engineering parameters [3(b), p.5].

2.1.2   The plano-convex species and lens bending

Figure 2.11 shows seven lenses of four different shapes. They all have the same focal length. The reason for this is that they all have the same

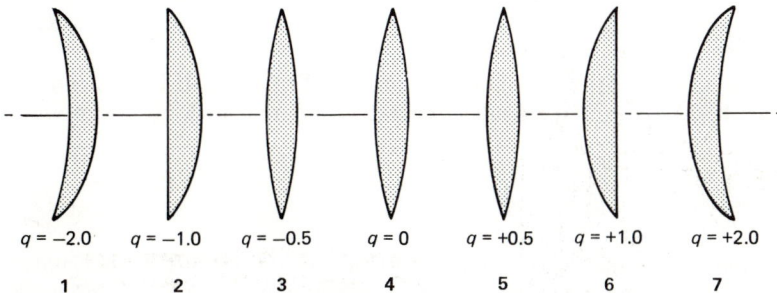

Figure 2.11 Seven lenses of four different shapes, obtained by lens bending. All have the same focal length

thickness. For relatively thin lenses of a given diameter all made of the same material, F_L is uniquely determined by their thickness. Thus, the behaviour of the symmetrical biconvex lens 4 will be very similar to that of the plano-convex lens 2 used in this text so far. Making the transition from one variety to another is called, imaginatively, *lens bending*. The numbers under each of the lenses of Figure 2.11 characterise the degree of bending, conveniently expressed by the ratio of the sum of its radii of curvature to their difference, the shape factor q:

$$q = \frac{r_2 + r_1}{r_2 - r_1} \tag{2.2}$$

The q values of both members of the pairs 1 and 7, 2 and 6, and 3 and 5 vary in sign only, e.g. -2.00 and 2.00. This difference is reflected in some of the more subtle aspects of lens behaviour, e.g. the spherical aberration [1, p.132] which depends on which of the lens faces is chosen as the input port for light. In the remainder of this work, lenses will be drawn in their preferred orientation, whenever relevent, as for example in optical communications in free space (Chapter 8).

Through lens bending we can make an easy transition from the plano-convex lenses of Figures 2.7–2.10 to the biconvex variety shown in the centre of Figure 2.11. This is the oldest and the most common representative of the species, the lens *par excellence*. The shape reveals the etymology of the word 'lens', through its resemblance to the edible brown lentil. (This is most evident in some languages other than English, e.g. French, *lentille = lentille*; Italian and Spanish, *lenticchia* $\simeq$ *lente*; Polish, *soczewica* $\simeq$ *soczewka*; German, *Linse = Linse*; Arabic, *adas* $\simeq$ *adasè*.) It is this variety that we shall use in the rest of the book as the generalised symbol of a lens, except in cases in which such a simplification could be detrimental to the argument.

2.1.3   Lenses used to form images

In the case of optical communications and most other electro-optics applications, lenses are used to form images of light sources in the sensitive planes of light receivers. The knowledge of the focal length, F_L, of a lens will enable us to calculate the lens–image distance, S_i, from a known value of the object–lens distance, S_o, and vice versa:

$$\frac{1}{F_L} = \frac{1}{S_o} + \frac{1}{S_i} \tag{2.3}$$

This is the mathematically innocent Gaussian Equation. It is illustrated in Figure 2.12(a). To an electronics engineer it bears a striking resemblance to:

$$\frac{1}{R} = \frac{1}{r_1} + \frac{1}{r_2} \tag{2.4}$$

which makes it easier to remember. Both equations are rather naive, neglectful as they are of secondary effects, e.g. contacts, connecting leads, high frequencies and transient behaviour for Equation 2.4; lens thickness, spherical, astigmatism or colour aberrations for Equation 2.3. Both are,

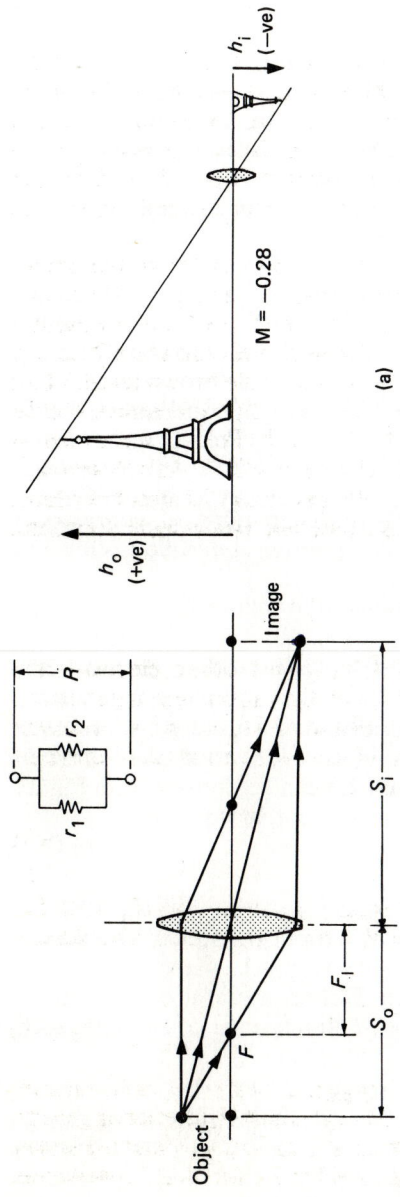

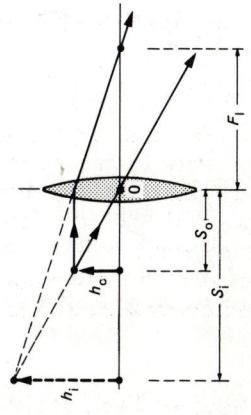

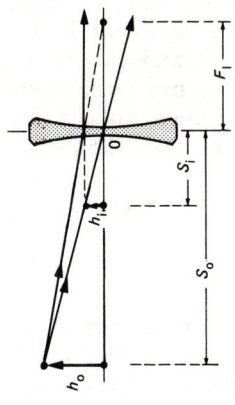

Figure 2.12 How real and virtual images are formed by 'thin' lenses. (a) Positive lens forming a real image and the electrical analogue of the Gaussian equation. (b) Positive lens forming a virtual image. (c) Negative lens forming a virtual image

nevertheless, useful tools. For *positive* lenses, the image will be *real* when $S_o > F_L$. (A lens is said to be positive when it focuses a parallel bundle of rays on the side opposite to the input port.) With a *negative* lens (e.g. a biconcave one) no real image is ever formed. Its images are, under all circumstances, *virtual*, i.e. only *appearing* to be there but, in fact, absent, unable to impress either a photographic film or a photoreceiver, like the image of a candle *behind* a plane mirror, where it appears to be coming from but where no candle rays can be found.

Table 2.1 The sign convention (Reproduced from Jenkins and White [1] by permission of McGraw-Hill Inc.)

1. All figures are drawn with the light travelling from left to right

2. Object and image dimensions are positive when measured upward from the axis and negative when measured downward

3. All convex surfaces *encountered* are taken as having a positive radius, and all concave surfaces *encountered* are taken as having a negative radius

4. Both focal lengths are positive for a converging system, and negative for a diverging system

5. All object distances (S_o or S) are considered as positive when they are measured to the left of the vertex, and negative when they are measured to the right

6. All image distances (S_i or S') are positive when they are measured to the right of the vertex and negative when to the left

While we shall deal in this book with real images only (with the one and only exception of the Galilean beam expander, Chapter 10), note that in Equation 2.3, F_L, S_o and S_i may assume negative values, and so can the object and image height h_o and h_i. To get your sums right under all circumstances from the Gaussian and other equations of geometrical optics, you must respect a set of rules called the *sign convention*, shown in Table 2.1.

Of the above mentioned 'other equations', the following:

$$M = \frac{h_i}{h_o} = -\frac{S_i}{S_o} \tag{2.5}$$

is very useful, as it gives the linear magnification of a lens for a set value of the S_i/S_o ratio. If M is negative, the image is inverted, i.e. directed downward for an upright object. Equation 2.5 can be derived from Figure 2.12 using elementary triangle relationships. Another equation is:

$$\text{TCL} = -F_L\frac{(1 - M)^2}{M} \tag{2.6}$$

Its solution for M gives the magnification of an optical system with a known F_L and an imposed TCL (Total Conjugate Length), which is no other than the sum $S_o + S_i$ (Figure 2.13). The TCL is a useful concept when designing optical systems in the presence of dimensional constraints, such as a set maximum for the total length of a machine.

$$M_{1,2} = \frac{-\left(\dfrac{\text{TCL}}{F_L} - 2\right) \pm \sqrt{\left[\left(\dfrac{\text{TCL}}{F_L} - 2\right)^2 - 4\right]}}{2} \tag{2.7}$$

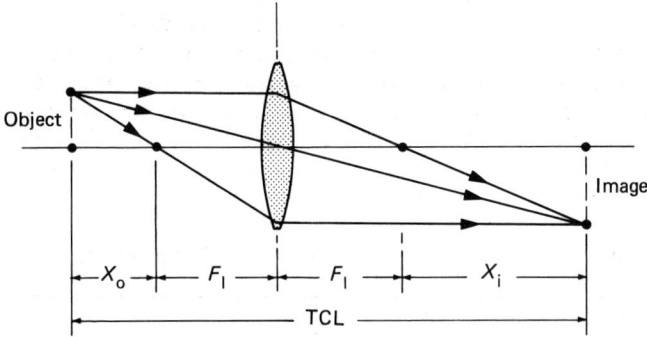

Figure 2.13 The total conjugate length (TCL) is a useful concept when designing optical systems in the presence of dimensional constraints

with minification (size reduction) given by the smaller of the roots of the quadratic equation 2.6, M_2. Several interesting results can be obtained by studying the conditions of existence, of value and of sign of the roots. Note two of them:

1. The TCL can never be smaller than $4F_L$ (which occurs from $M_1 = M_2 = -1$, the case of the reversed image of the same size as the object, produced by a positive lens bang in the middle of TCL).
2. The product of both roots always equals unity ($M_1M_2 = 1.0$), meaning that the minification is always the reciprocal of the magnification.

Result 1 is confirmed by the Newtonian lens formula, Equation 2.8:

$$X_oX_i = F_L^2 \qquad (2.8)$$

in which the x values represent the object and image distances from their focal points, respectively (Figure 2.13). For the minimum TCL, we simply have:

$$X_o = X_i = F_L \qquad \text{and} \qquad S = 2X$$

giving

$$\text{TCL} = X_o + F_{L1} + X_1 + F_{L2} = 4F_L$$

The Newtonian lens formula (Equation 2.8) is sometimes easier to handle than the Gaussian Equation 2.3.

Returning for a moment to the electrical analogy between Equations 2.3 and 2.4, we see that finding the image distance S_i from Equation 2.3:

$$S_i = \frac{S_oF_L}{S_o - F_L} \qquad (2.9)$$

is performed in a way not unfamiliar to the electronics engineer who trims down a resistor r_1 to a lower value R by adding in parallel to r_1 a resistor r_2 calculated from:

$$r_2 = \frac{Rr_1}{R - r_1}$$

as, for example, by adding 110 kΩ in parallel to 11 kΩ to obtain 10 kΩ.

An important concept of geometrical optics is that of Numerical Aperture (NA). It applies to lenses, mirrors, optical fibres, etc. Its cousin, the *f*-number, is highly relevant to photography, where it has been used for a very long time. The NA (shown in Figure 2.14 for a lens and discussed in Chapter 7 for an optical fibre) characterises the angle of acceptance of light rays:

$$NA = \sin\frac{\alpha}{2} \tag{2.10}$$

α being the widest angle of a bundle of rays that the lens will accept to form a real image of their source. (For media other than air or vacuum,

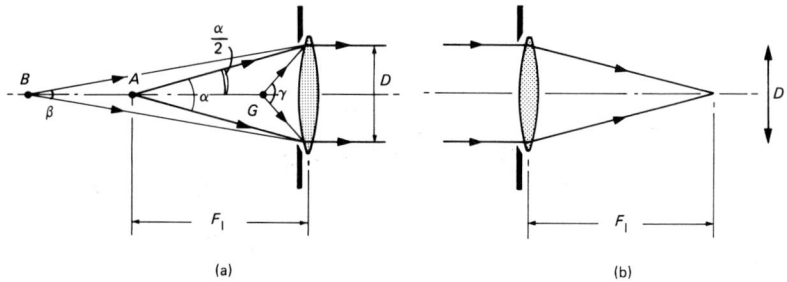

(a)   (b)

Figure 2.14 (a) The numerical aperture, NA, is given by sin $\alpha/2$, α being the absolute maximum limit of the angle of acceptance of a lens for incident rays necessary for the formation of a real image of the source. Angle β is smaller than α. Angle γ, while larger than α, does not lead to the formation of a real image of the source, thus corresponding to the case in Figure 2.12(c). (b) The *f*-number is the numerical ratio f_l/D

sin ($\alpha/2$) has to be multiplied by the refractive index *n*.) It will be seen from Figure 2.14 that

$$NA = \frac{D}{2F_L} \tag{2.11}$$

The *f*-number is the numerical ratio of the focal length of a lens to its clear or effective diameter (Figure 2.14). (See Chapter 10 for laser work.) The usual way of writing the *f*-number is $f_\# = 2$ or $f/2$ for a lens with, for example, $F_L = 50$ mm and $D = 25$ mm. We say then that the lens has an aperture of $f/2$ (pronounced 'eff two'). In photography and in optical linkages the irradiance produced on the sensitive element, film or photoreceiver respectively is usually proportional to $1/(f_\#)^2$. So is the so-called 'speed' of a camera.

Appendix 2 contains a list of formulae published by an American manufacturer of optical systems. It applies to all single-element thin as well as to their multi-element thick lenses. Useful as it is, this document ought to be used with a little caution as it does not use exactly the same notation and sign convention adopted elsewhere in this book. A simple sketch on graph paper supporting the calculations should remove all doubt regarding image reversal.

2.2 Guiding light with and without the help of lenses

2.2.1 Lens ducts

When an image has to carry over distances several times their own size, an array of so-called field and relay lenses can be used, arranged as in Figures 2.15(a) and (b), which are self-explanatory.

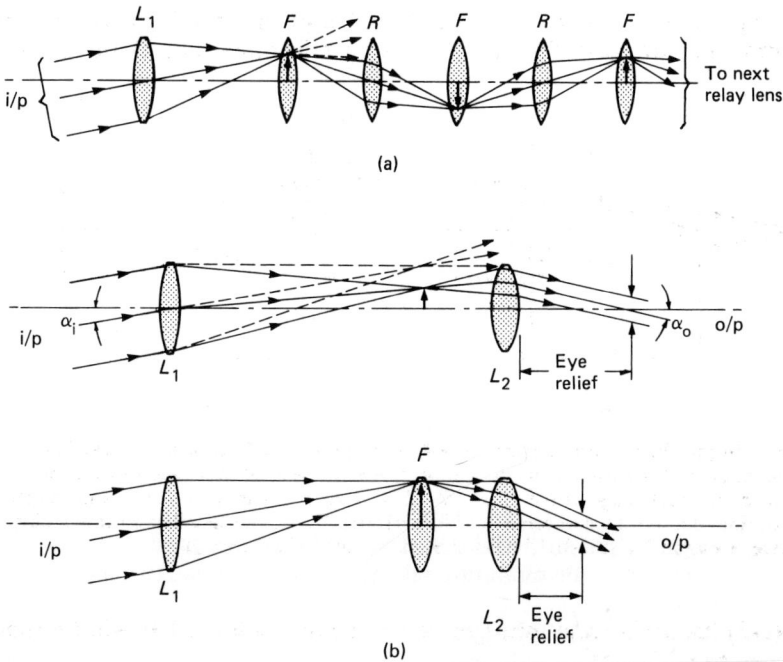

Figure 2.15 (a) Image ducting by field, F, and relay, R, lenses. (b) Field lens, F (bottom), brings in the dashed rays (top) which would otherwise be lost

In the early days of *endoscopy*, doctors and surgeons used tubes containing such arrays when they wanted to see down a patient's throat, gullet and into the stomach. A reasonable degree of flexibility had been attained before the mid-20th century.

The 1960s saw some proposals for long-range optical communications based on field-and-relay lens ducting. Developments along these lines were short-lived, lensed light guides having been superseded by optical fibres.

2.2.2 Flat mirrors

The *periscope* (Figure 2.16), made up of just two ordinary mirrors and a box is one of the simplest and, at the same time, one of the most impressive luminous path-bending instruments. The simple periscope will enable a train guard to keep an eye on the rooftops of all his carriages (and/or see

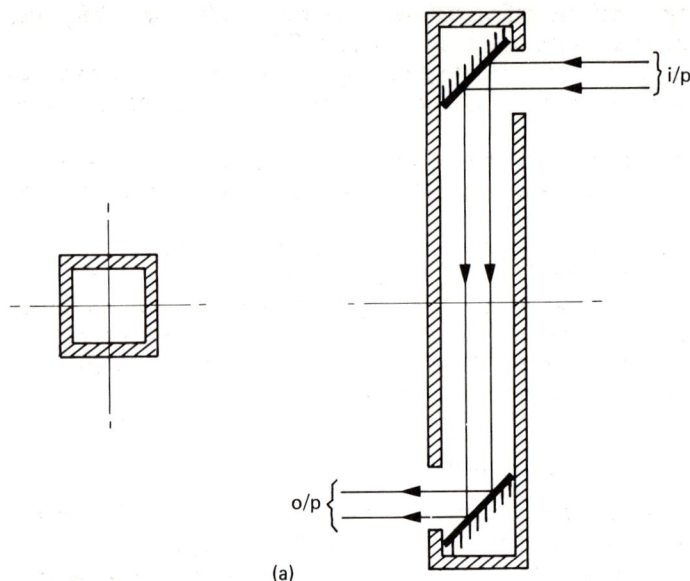

Figure 2.16 A simple two-mirror periscope

the track), a short person in a big crowd situation (coronation, royal wedding, papal visit, etc.) to see the 'superstars' of the day, and a child to view the roads and houses of a miniature village as its imaginary inhabitants would. This can be experienced in Bourton-on-the-Water, Gloucestershire, England, where such a looking *down* periscope is available to the visitors of the miniature village. And, of course, to peep inside a walled garden!

On a more serious tone, however, periscopes are used by submarines and tanks. While still based on the 'two mirrors in a box' principle, such periscopes can be complex and expensive instruments, containing telescopes, gimbals, night-seeing equipment, electronic controls etc. (A beautiful submarine telescope can be seen (and operated) in the Science Museum, Kensington, London, England.)

2.2.3 Convex mirrors

These can be used to increase one's field of view, e.g. in factory yards or to see incoming traffic around the corner.

2.2.4 Prisms

In the periscopes shown in Figure 2.16, beam bending is accomplished by prisms – not mirrors. (The expression 'two mirrors in a box' is used rather freely above, extending the word 'mirror' to prisms as, like mirrors, prisms can bend light beams, and because all periscopes are based on the same principle.) As beam benders, prisms have the additional advantages of superior rigidity, ease of mounting and freedom from 'ghost' images.

(Beam bending in prisms relies on Total Internal Reflection (TIR), the very basis of optical cables, treated in Chapter 7.)

In binoculars, prisms are used to guide light rays along rather tortuous paths to reduce the physical length of an instrument that needs long focal length lenses.

A 90° prismatic beam bender can be used by photographers wishing to remain unnoticed: the operator shoots sideways, while appearing to be aiming ahead.

Two more of the many interesting applications of beam bending by prism are the pentaprism [2, p.93] and the cube corner retroreflector. (See Chapter 12 on optical barriers.) Both prisms are remarkable inasmuch as they deflect the incident beam by a *constant* angle, i.e. 90° for the pentaprism and 180° for the cube corner. These angles remain constant almost regardless of the prism's own position.

2.2.5 Cone channel condensers

Figure 2.17 shows a device sometimes used to increase the field of view of a photoreceiver of small size. It is a kind of a *light funnel*, capable of trapping a wide bundle of rays and channelling it onto a sensitive area of a receiver which, for some valid reason (e.g. frequency response), has to remain small [2, p.234]. The term *light pipe*, sometimes given to this light

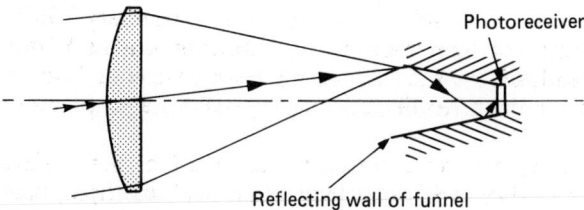

Figure 2.17 The light funnel

channel, is a misnomer, the geometry of this device being that of a truncated hollow cone, rather than that of a pipe. (The device can be turned into a genuine *light trap*, useful in laser work, by giving its interior a matt black finish, turning the cone back to front and shutting off its wide end.)

2.2.6 Streaks of running liquid

Light can create the illusion of 'flowing', when a thin water stream is illuminated at its origin. The famous colour fountains of the Paris Trocadero of the late 1930s used to splash luminous jets and streaks of blue, red and white water high into the air with much panache – or rather seemed to, as in fact clear mains water was being used. The effects are due, once again, to TIR.

2.2.7   Perspex (Lucite, Plexiglass) light guides

Here, as in the above case, light seems to be following a curved path. TIR makes this possible. A doctor will illuminate his patient's throat by a light-guiding Perspex spatula, replacing the old fashioned spoon and the torch. An engineer will design and use a strangely, almost spider-shaped multibranch light guide of moulded acrylic to take light into or out of a

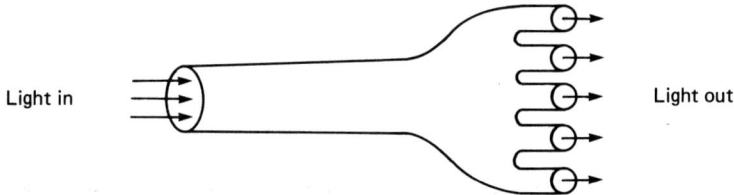

Light in   Light out

Figure 2.18 Multibranch light guide of moulded acrylic

number of inaccessible or closely packed areas (Figure 2.18). Mark sensing (pencil strokes on IBM cards) and punched tape reading are examples of such fan-out and fan-in of light applications.

An essential difference between the first five types of light guide (Sections 2.2.1–2.2.5) and the last two (Section 2.2.6–2.2.7) is that each member of the first groups carries images. The last two do not. The differences between carrying static pictorial *intelligence*, 'in parallel' so to speak, and carrying simply a light flux, is an important one from the viewpoint of communications.

2.2.8   Optical fibres

To crown the set of examples of lensless light guides comes the optical fibre. Hair-thin strands of glass or plastic, batched into orderly bundles, sometimes remaining supple and sometimes fused together into rigid assemblies, will carry light *and images* out of a patient's lung, stomach or bladder to the surgeon's eye, or out of an inaccessible location of a machine through a multibend path, to the place where it is wanted. Better still, properly clad and strengthened, a *single* fibre will become the core of an optical cable, capable of transmitting speech, pictures or data with a considerable information-carrying capacity and amazingly little attenuation over considerable distances. This, however, is another story, the story of the captive ray, which deserves – and gets – a chapter of its own.

Chapter 3

Working with photons: the generalised P–N junction

To an electronics engineer involved with electro-optics, working with photons means controlling the photon/electron interplay in devices, which requires a good understanding of this interplay.

In modern electro-optics all conversions of optical into electrical and most conversions of electrical into optical signals take place in the P–N junction of a semiconductor material. The P–N junction acts as a *Bureau de Change*, swapping photons for electrons and vice versa whenever converting energy of one kind into the other. Hence, having clearly in mind, at all times, the ways in which a P–N junction works is fundamental to this course of study. This chapter therefore revises, and perhaps clarifies, these fundamentals.

3.1 The physics of the P–N junction

First of all, the so-called P–N 'junction' is not a junction at all. Once again we face a misnomer: you don't take a piece of P material and *join* it to a piece of N material to make a diode. 'Transition' would be a better name. Unfortunately, the term *junction* is universally used. Figure 3.1 shows a modern silicon P–N junction.

It is the *addition* of adequate impurities that decides whether the material is of the P or N type. Remember that:

1. The material must be crystalline.
2. There must be lattice continuity throughout and uppermost at the junction.
3. The impurities (**P** – acce**P**tor, **N** – do**N**or) are in fact very, very few (e.g. 10^{-8}) and their atoms fit into the regular crystal lattice structure (like a few yellow bricks into a red brick wall – maybe a little longer or shorter than the red ones but still fitting).
4. When conduction takes place the *impurities* themselves don't move, only their outer electrons (or holes) do. They are the *carriers*.

The concept of the hole, the 'non-electron' carrier, becomes less abstract when likened to an air bubble in water ('no-water'). Its motion, too, 'up' the potential hill in situations in which electrons roll down it, is easily grasped and well remembered with the help of the analogy of a glass of

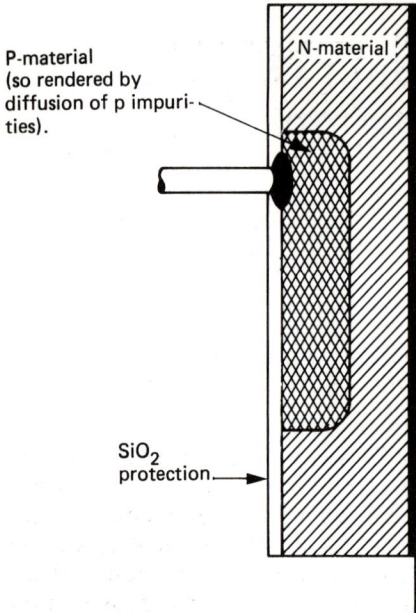

P-material
(so rendered by
diffusion of p impuri-
ties).

N-material

SiO$_2$
protection

Figure 3.1 A modern silicon P–N junction

water with both air bubbles and fine grit in it, where the same gravity field
forces the bubbles *up* and the grit *down*.

3.1.1   Carrier energy diagrams

Carrier energy diagrams are very important for the understanding of the
junction mechanism. Figures 3.2–3.5 are such diagrams, all drawn with the
energy level of the *electrons* increasing towards the top of the figure (as
shown in Figure 3.3). They are a useful abstraction and should never be
(but often are!) confused with the geometrical representation of situations
in semiconductors. It must be stressed, at the risk of offending some
readers, that, unlike Figure 3.1, Figure 3.2 is *not* the picture of a junction.
It is not even a pictorial representation of energy levels in a P–N junction.
The same has to be said about Figure 3.3. All such figures are only
representations of energy levels in junctionless portions of semiconductor
material, Figure 3.2 being that of *intrinisic* (very pure, undoped) material
at three different temperatures.

Figures 3.4 and 3.5 *do* show situations in junctions, but again do so from
the viewpoint of energy levels, the only association with device geometry
being that of 'probing' the energy situation step by step along a line
perpendicular to the transition from the N-side to the P-side of the
material. This is done from left to right in the drawings. It would be wrong
to view such drawings as cross-sections of junctions.

3.1.2   The forbidden band

At very low temperatures (achievable in laboratories only) practically all
orbital electrons of the crystal's atoms occupy valence energy levels, the

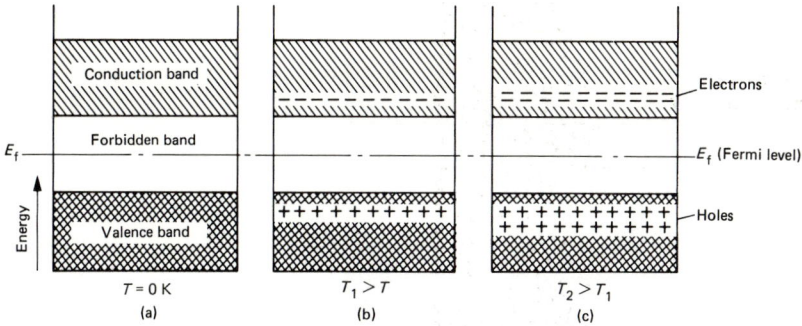

Figure 3.2 Carrier concentration in an intrinsic semiconductor at three different temperatures. (a) corresponds to the 'absolute zero' (0 K = −273.15 °C) case

lowest possible ones. The term 'band' is used here because of the enormous numbers of levels present. These large numbers result from countless couplings of discrete levels in neighbouring atoms. At the theoretical absolute zero, all the 'vacancies' (allowable states) of the valence band are filled (Figure 3.2(a)). As soon as the temperature rises, some electrons acquire additional energy and become sufficiently excited to leave their orbits and to move 'up' to the conduction band (Figure 3.2(b)). No longer atom-bound, they become *free* electrons, capable of taking part in the conduction of electric current. However, they are not the only carriers the temperature rise has created; each electron 'promoted' to the conduction band leaves behind it a 'hole', which is also a carrier, this time a positive one, also capable of drifting under the influence of an electric field, albeit in a direction opposite to that of an electron, and also capable of contributing to conduction.

The 'forbidden band' represents the non-allowable states in the crystal and has, in common semiconductor materials, e.g. germanium (Ge), silicon (Si) and gallium arsenide (GaAs), a width of the order of an electron-volt (1 eV). If electrons from the valence band are to become conduction electrons, they must acquire sufficient extra energy to jump the forbidden band. Figure 3.2 summarises what happens when the material is subjected to progressive temperature rises. Thermal agitation provides 'the little extra' energy needed.

3.1.3   The role impurities play

An enormous change takes place in a piece of pure crystalline germanium or silicon when a small, in fact an extremely small, amount of a suitable impurity is introduced into it. The rise in conductivity, at a given temperature, is dramatic and totally disproportionate to the quantity of the additive. A few parts in 100 million turn an almost-insulator into an almost-conductor: the material becomes an *extrinsic* semiconductor. (It is worth remembering that apart from having lower conductivities than conductors, semiconductors differ in yet another way: should the temperature rise, the conductivity of an intrinisc or moderately doped extrinsic semiconductor will increase, while that of a conductor will decrease.)

(a) Semiconductors with do⃞N⃞ors.

(b) Semiconductors with acce⃞P⃞tors.

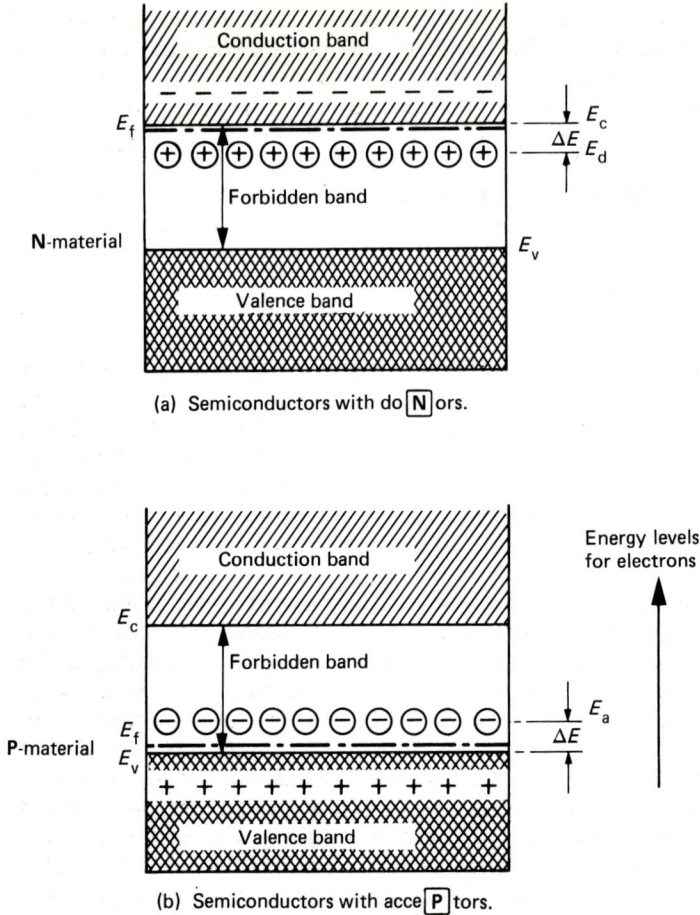

Figure 3.3 Band structure in a semiconductor. (a) doNors, (b) accePtors

Take the case of a *donor* impurity, such as arsenic (As, five valence electrons) added to the host germanium (Ge, four valence electrons). Being more loosely bound to the four-bond architecture of the host lattice, the fifth electron of the impurity can quit the valence bond more easily, jump to the conduction band and, in doing so, leave a positively charged atom behind. Only a few tens of meV of boost are needed and thermal agitation provides sufficient energy. This is shown in Figure 3.3(a), where the ringed pluses representing the fixed charges of the remaining, now ionised, arsenic atoms. In this case, carriers are not produced in pairs, as no hole in the valence band is generated by the departure of an electron and because the average spacing between arsenic atoms is too great to make atom-to-atom electron jumps very likely. With this impurity, then, current will be conducted in germanium very predominantly by electrons, *negative* carriers. (For further study see Ref. 28, p.23 and Ref. 29, p.66.) A handy means for remembering this is that doNors contribute Negative charges and accePtors Positive ones.

Examine the cases of *acceptor* impurities in the same host material, germanium. Aluminium (Al), boron (B) and indium (In) have only three valence electrons and are often used. Germanium valence electrons are strongly solicited by the atoms of such impurities to join and form the fourth bond. The fact that this impurity is so keen on accepting electrons from the host germanium atoms is reflected by the smallness of ΔE, its separation from the valence band (Figure 3.3(b)). A mere 10 meV push, easily provided by thermal agitation, suffices to produce the trans-bonding. The large statistical probability of the event stems from the shattering numerical superiority of germanium atoms in the material.

When an acceptor atoms succeeds in capturing a germanium electron it becomes negatively charged. This does *not* produce a negative carrier, however: the scarcity of impurity atoms in the material is such that atom-to-atom charge 'hops' under the influence of an external field are very unlikely. The captivator remains a *fixed* negative charge and this is why the minuses in Figure 3.3(b) are ringed. The event does, however, create additional mobile charges. The breakaway electron leaves behind a positively charged atom. While the atom itself is not mobile, its charge, now a hole, is. This *no-electron* emplacement becomes a positive carrier capable of drifting under the influence of an electric field, on average, and thus of contributing to the conduction mechanism. It's a little like empty seats in a theatre during the first interval: the attraction to the stage is the field force, the rather less than dignified third-row spectator moving to the empty seat in the first row is the electron, and the empty seat is the hole. Soon the third-row seat gets filled by a spectator promoting himself from the tenth row, etc. 'Seat emptiness' moves in a direction contrary to that of the mobile spectators. Likewise, the apparent motion of holes is in a direction contrary to that of electrons. To sum up: in germanium doped with acceptor impurities, current is conducted predominantly by positive carriers, holes, moving within the valence band. (Diode and transistor action relies on both positive and negative carriers, so these devices are often referred to as *bipolars*, in opposition to field effect devices in which one type of carrier only is responsible for current transport.) Note, finally, that in germanium and silicon the mobility of holes is two to three times lower than that of electrons, making the reasons for hole sluggishness obvious. Having examined N-doped and P-doped materials makes us almost ready to proceed to the study of the P–N junction proper.

3.1.4 The Fermi level

The so called 'Fermi level' concept is so helpful in the understanding of junction physics that a few moments spent on it will be well rewarded. Figure 3.2 shows a dot-dash-dot line which, so far, has not been explained. The corresponding energy is specified on the drawings as E_f, the Fermi level. This is a kind of imaginary half-way demarcation line between levels of equal degrees of occupancy (above it) and of emptiness (below it). In other words, there are just as many filled levels above as there are empty ones below it. (For a full treatment of Fermi level, see Ref. 14, pp. 52, 99, 110, 123; Ref. 28, p.16; Ref. 29, pp. 39, 60, 69.)

In an undoped (intrinsic) semiconductor E_f lies just half way between the edges E_c and E_v of the conduction and valence bands, respectively. As this is in the middle of the forbidden gap, no available states lie there. If there were (like, for example, in a metal) their chance of occupancy would be equal to their chance of emptiness – a mathematical probability of ½. Doping a semiconductor strongly affects the position of the Fermi level. The addition of N impurities increases the concentration of free electrons in the conduction band and thus augments the chance of occupancy of its levels, without affecting the chances of emptiness of levels in the valence band. As a result, the Fermi level rises (see Figure 3.3(a)). Similarly, the addition of P impurities lowers the Fermi level, E_f (Figure 3.3(b)). In an electrically non-connected piece of semiconductor material in equilibrium, the Fermi level is uniform throughout.

3.1.5   The potential barrier and the depletion layer

In the course of manufacture of semiconductor devices, numerous P–N junctions are obtained through localised diffusions of impurities into the substrate, adjacent portions of which are made into the P- and N-types of material shown in Figure 3.3. Figure 3.4 shows what happens to the energy diagram of a portion of crystalline silicon with P and N regions brought so closely together. As the lattice is continuous, a self-adjusting process takes place resulting in an equilibrium condition with a uniform Fermi level throughout. Some free carriers have diffused through the junction: holes from the P into the N side, electrons from the N into the P side. As there are now fewer holes in the valence band of the P material than negative ions, near the P–N border, a highly localised *negative* charge develops there. Similarly, the desertion of the N side, near the border, by some of its

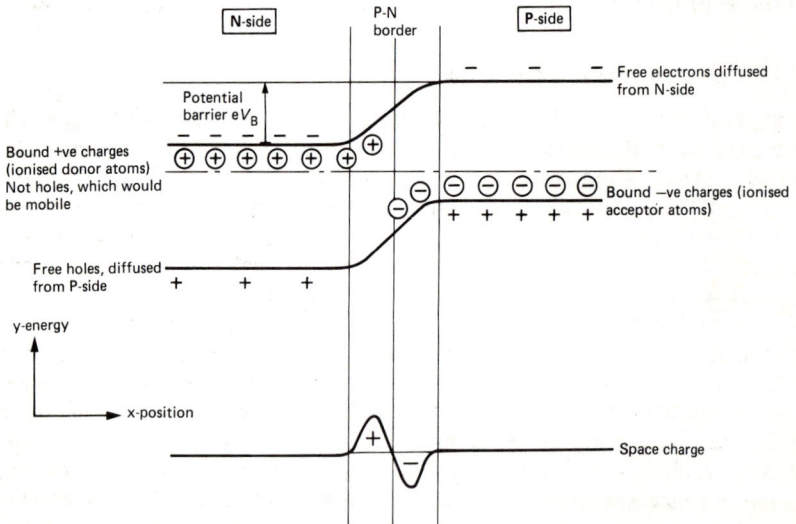

Figure 3.4 The P–N junction. Situation at room temperature for zero bias, zero external current (diffusion current = drift current), zero illumination

electrons has left behind uncovered positive ions, forming a highly localised *positive* charge. The junction has become self-polarised. We are faced with the situation of the N side being *positively* and the P side being *negatively* charged! The resulting electrical field at the junction is strong, owing to the smallness of the distances involved. Any free charges finding themselves there will be swiftly swept away to the one side or the other. A moment's reflection will show that the direction of this field is such as to oppose further electron travel from left to right and hole travel from right to left and to arrest them once the charge created by the uncovered donors and acceptors has reached a certain quota: a *potential barrier* has formed. It will also show that a thin central layer must be depleted of free charges. This justifies its name: the *depletion layer*. The overall equilibrium is a dynamic one (like all the previously described equilibrium conditions), with a statistical rather than absolute meaning, i.e. numerous charge movements and transitions take place all the time and the illustrations given pertain to averages. In the P–N junction two opposing 'pressing forces' exert their influence on free carriers all the time: diffusion and drift. Time and space localised 'micro-currents' caused by both exist in the crystal, but, with no connection to an outside e.m.f., the equilibrium of Figure 3.4 prevails.

3.2   The usefulness of P–N junctions to the electronics engineer and the electro-optics specialist

To the electronics engineer, the enormous usefulness of the P–N junction stems from the fact that its potential barrier can be lowered or raised, regulating the current flow through it. To the electro-optics specialist, the usefulness of the junction is greater still: the current flow can be controlled by means of light and the generation of light by means of current flow.

3.2.1   Reverse biasing the junction

By applying an external voltage to the junction to make the N side still more positive with respect to the P side than it is in its natural state, the potential barrier is raised. With, say, 2 or 3 V added to V_B, even the most energetic free electrons of the N side are now unable to climb it and cross the blocking junction. Free holes attempting to cross the junction in the opposite, P to N, direction suffer from an identical predicament. There are, however, always a few thermally generated free electrons in the P material and a few holes in the N material. These minority holes are unhindered and can roll down the potential hill (i.e. travel downfield) and contribute to the transport of current in the N to P (inside the crystal) direction. A little reflection shows that thermally generated electrons in the P material, which are minority carriers too, also contribute to this current. Small as it is at ordinary room temperature, this *reverse current*, I_R, grows quickly with heat increase, doubling for every 7°C in germanium and for every 11°C in silicon. This is the *leakage current* of a diode and the *dark current* of a photodiode. Its value at the absolute zero temperature, I_0, is called the *reverse saturation current*.

P-type material   N-type material

eV_B

E_f

(a) The physicist's view

PSU

Reverse
current I_R

I_R

N → P

(b) The materials scientist's view   (c) The electronics engineer's view

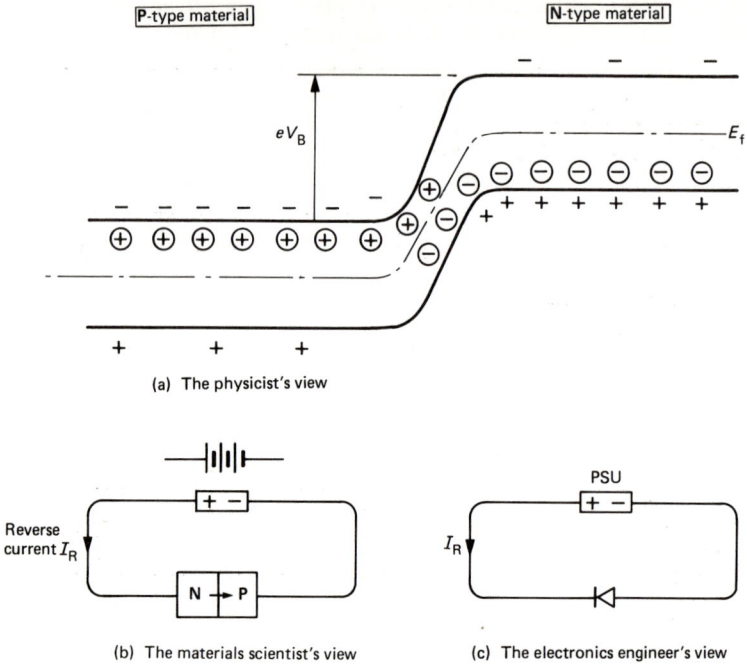

Figure 3.5 Three ways of representing a reverse-biased P–N junction

It is important to remember that the leakage current, I_R, is:

1. Transported by *minority* carriers.
2. Highly temperature dependent.
3. Almost entirely independent of the value of the applied reverse voltage, from a fraction of a volt to the reverse breakdown voltage, V_{BR}.

Finally, regarding the Fermi level, as the junction now transports electricity, E_f is no longer uniform across the device (see Figure 3.5).

3.2.2   Forward biasing the junction

The immediately noticeable consequence of the application of a forward bias is the *lowering* of the potential barrier, V_B (Figures 3.4 and 3.6). It is left to the reader to establish the mechanism by which this facilitates the diffusion of *majority* carriers and makes the P to N direction (inside the crystal) the passing sense of the diode. Equation 3.1:

$$I = I_0 \left[\exp \left(Ve/kT \right) - 1 \right] \tag{3.1}$$

summarises the diode behaviour in both directions. It uses the saturation current I_0, explained earlier, the electron charge e and k, the Boltzman constant. V is the controlling voltage, 'the bias'. The value of the fraction e/kT is 25 mV at room temperature. Equation 3.2 represents a useful

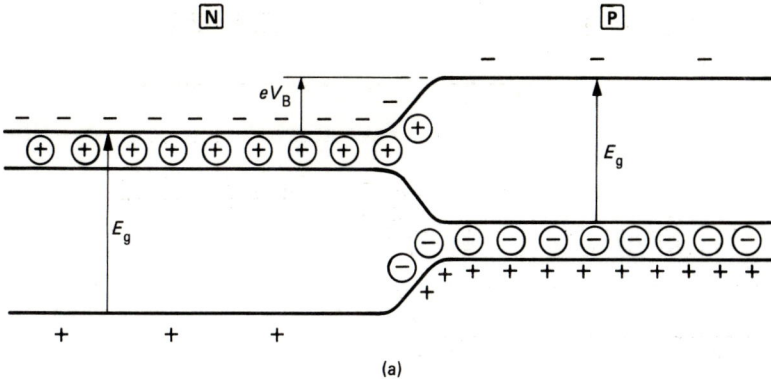

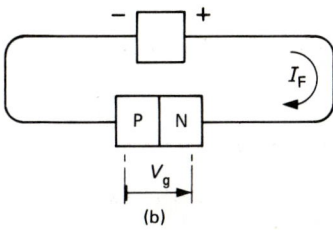

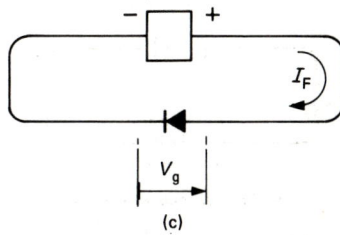

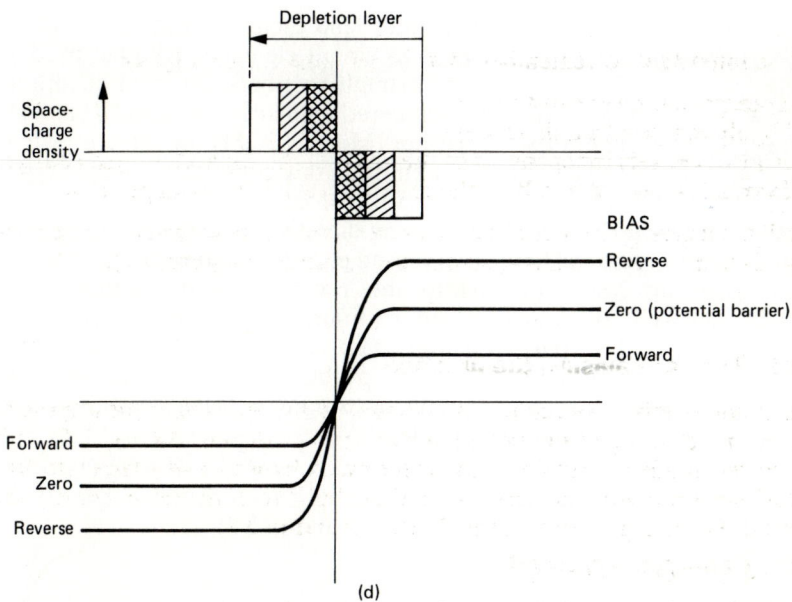

Figure 3.6 (a–c) The forward biased P–N junction. (d) The space charge and the potential at a P–N junction for three values of applied bias voltage. The effect of space charge layer widening with higher potentials is evident

approximation of Equation 3.1 for forward current only, and for voltages in excess of +0.1 V at room temperature:

$$I = I_0 \exp(V/0.025) \tag{3.2}$$

showing that I_F doubles for every +17.3 mV across the junction. It is useful to remember that:

1. A small residual internal potential barrier remains in the forward current condition (Figure 3.6), any excess power supply voltage being absorbed by the external circuit.
2. Electrons are injected into the P side and holes into the N side. Once on the other side of the junction the injected carriers are short-lived and short-travelled, as they quickly recombine with carriers of the opposite polarity. Thus, current transport is by majority carriers. Fresh supplies from the power supply sustain the current flow. The relevance of this remark will become apparent in the study of LEDs and semiconductor lasers.

All this, then, shows the physics of the rectifying action of a P–N junction as known to the electronics engineer in daily practice. (For a fuller treatment of the P–N junction see Ref. 29, chapters 1–8.) We shall now relate this action to electro-optics.

3.3   The three quadrants and the six operating modes: interaction with light

The familiar voltage/current coordinates lend themselves rather well to a global representation of P–N diodes as circuit elements, whether used as photoreceivers, as light emitters or as simple rectifiers. The four quadrants of the voltage/current surface are numbered anticlockwise: I, II, III and IV (Figure 3.7(a)). The First Quadrant Mode (FQM) of Figure 3.7(b) corresponds to forward biased light emitting devices of the injection type: LEDs and semiconductor lasers as well, of course, as the more trivial rectifiers and signal diodes in the passing direction. The Second Quadrant Mode (SQM) appertains to the non-biased photoreceiving junction and the Third Quadrant Mode (TQM) to the reverse biased photoreceiving junction (as well, of course, as to rectifiers and signal diodes in the 'blocking' sense). The fourth quadrant is not used (which is just as well, as abbreviating it to FQM would create problems!). The actual working conditions of bias, load and illuminations can be easily visualised and clearly interpreted on the diagram [48].

A gallium arsenide (GaAs) junction can be operated in any one of the three quadrants.

3.3.1   First quadrant mode

In FQM, the recombination of majority carriers is radiative in this material. When an injected electron fills a hole (or an injected hole captures an electron), a *photon* is *emitted*. This event represents the conversion of an individual electron's energy, eV, into E, the energy of an

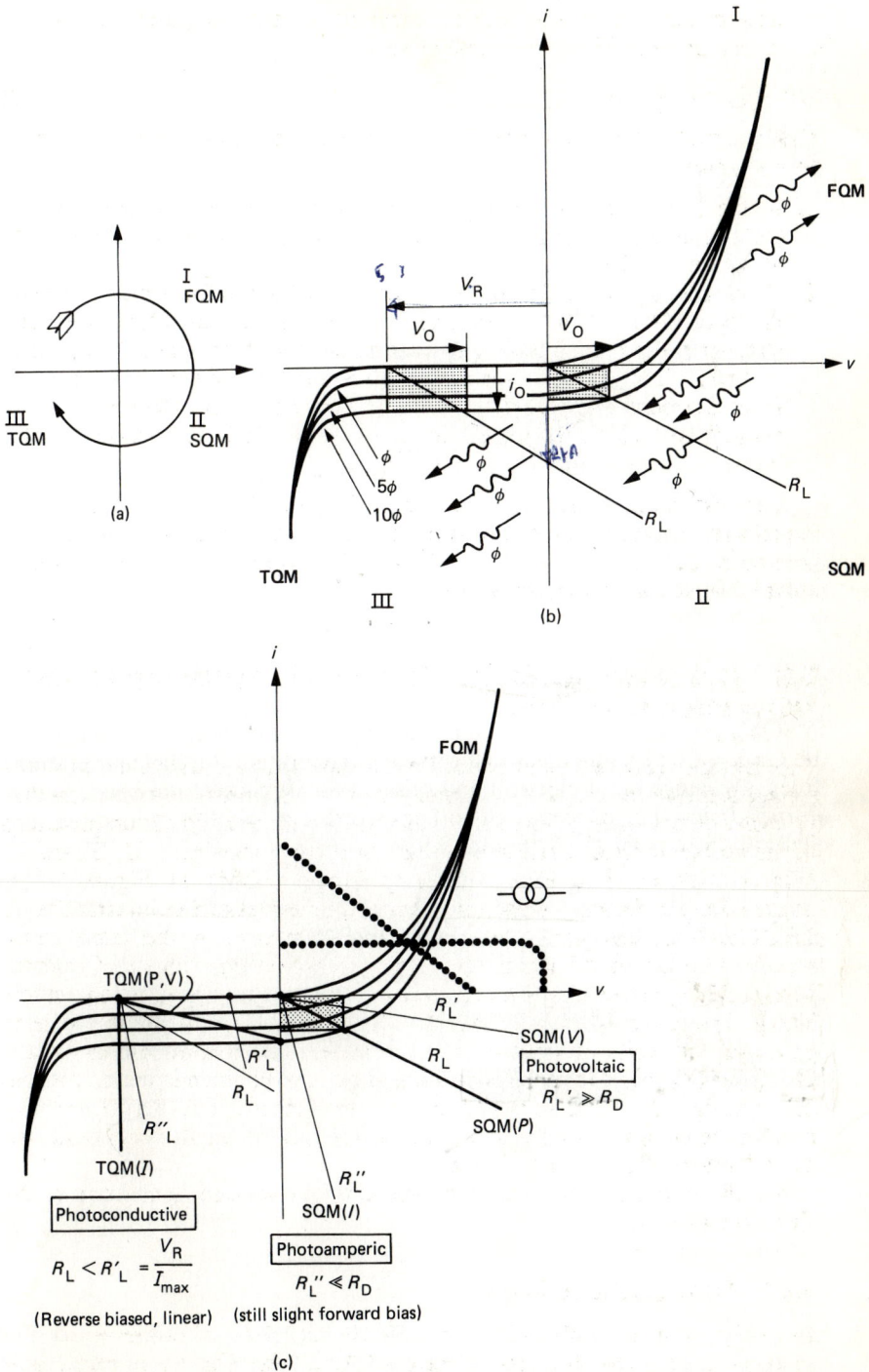

Figure 3.7 (a,b) The three quadrants and (c) six operating modes of an opto-electronic P–N junction

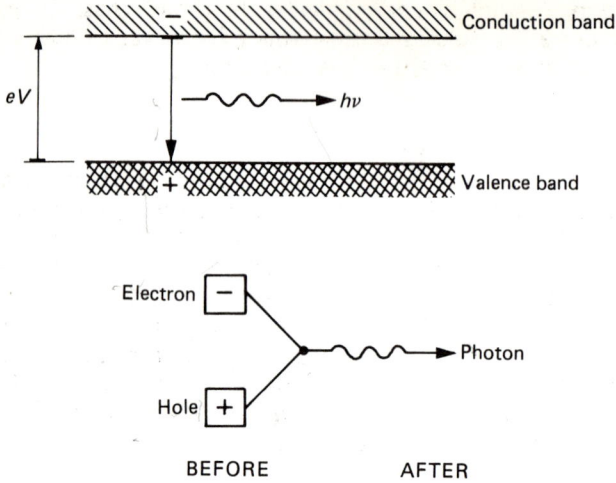

Figure 3.8 Radiative recombination of carriers in FQM

individual photon, the shower of which makes up the radiated light flux, φ. The mechanism is recalled in Figure 3.8.

3.3.2   Second quadrant mode

In SQM, conversion takes place in the opposite direction: light falls onto the GaAs junction and electrical current is generated. An incident photon 'frees' a valence bond electron and raises it to the conduction band, being *absorbed* in the process. In other words, a photon generates an electron–hole pair, as per Figure 3.9.

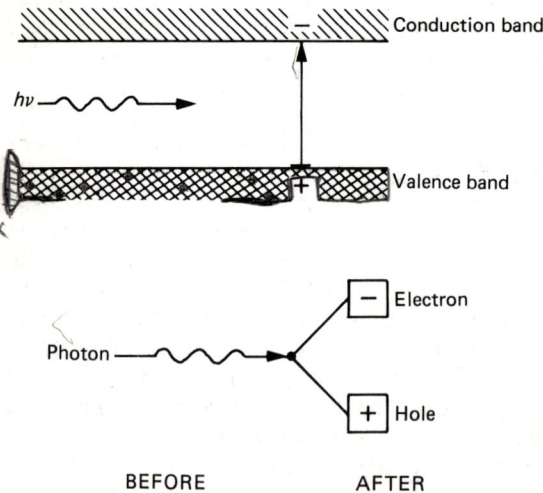

Figure 3.9 Carrier generation by photon absorption in SQM

The photon's energy $E = eV$ is communicated to the electron for its promotion. The two processes, photon *emission* and photon *absorption* are complementary and both obey the Einstein's Equation

$$E = h\nu \tag{3.3}$$

one of the cornerstones of modern physics.

This is an extraordinary equation: it relates quantitatively a *corpuscular* phenomenon (the energy drop or rise of an electron) to a *wave* phenomenon (the emission or absorption of an elementary quantum of light, the photon). In Equation 3.3 h is Planck's constant (6.625×10^{-34} J s) and ν is the frequency of the radiation. We know that wavelength and frequency are related by $\lambda = c/\nu$ and that $E = eV$. Thus, the underlying statement of Equation 3.3 is that *the wavelength of the emitted (or absorbed) light is numerically determined by a difference between two electronic energy levels.* Einstein's equation $E = h\nu$ links colour to voltage! It works out that $\lambda = hc/eV$ or, in an easy to remember approximation:

$$\lambda = \frac{1234}{V} \tag{3.4}$$

with V in volts and λ in nanometres. This colour/voltage relation links 3 V to blue, 2 V to red and 1.4 V to the infrared radiation of 900 nm, typical of GaA junctions operated in the FQM regime. More will be said about photons and Equation 3.3 in subsequent chapters. Let us return now to the operating modes of P–N junctions of Figure 3.7(b).

3.3.3   Third quadrant mode

In TQM, external reverse bias is applied to the junction. In comparison with the non-biased condition of SQM, reverse bias benefits us in two ways: it increases the output power and broadens the frequency bandwidth. The first effect will be readily understood by looking at Figure 3.7(b). Here, the shaded rectangles, representative of the instantaneous output power $p_o = v_o \times i_o$, are shown for both working modes, the load resistance R_L and light flux variation 10ϕ remaining unchanged. The TQM rectangle, already larger than the SQM one, could be made larger still by optimising R_L, while the SQM one has reached its maximum. The second effect results from a reduction of the dynamic junction capacitance C_j. This capacitance is caused by the depletion layer (Section 3.1.5), separating the P and N sides of the junction. It will be remembered that the depletion layer contains no free carriers. As such it behaves like an insulator. The P and N sides are conductive. Hence the condenser-like behaviour of the P–depletion layer–N sandwich. Figure 3.9 shows that the application of a reverse bias to the junction widens the depletion layer and thus reduces C_j. (The voltage modulation of C_j is the basis of voltage-sensitive condensers, sold under various proprietary names, used in voltage-turned circuits.) The physics of the TQM is the same as that of the SQM: light generates carrier pairs (Section 3.3.2). Free electrons and free holes roll down the potential barrier (Figure 3.4) and are swept through it to the opposite side of the junction. The generated current is thus transported by minority carriers in both the SQM and TQM cases. In both

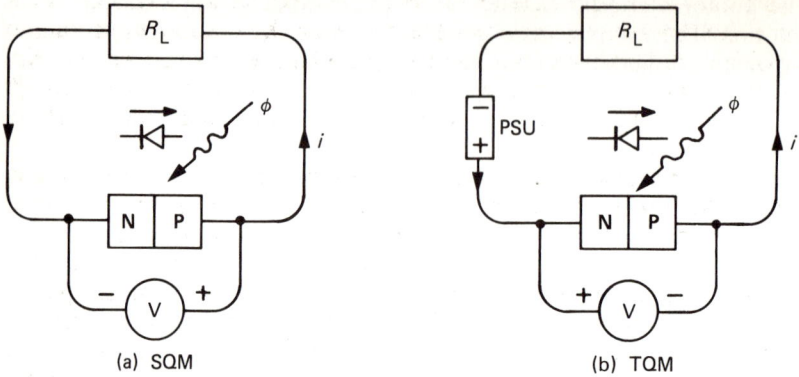

(a) SQM   (b) TQM

Figure 3.10 SQM and TQM. Across-the-junction voltages are of opposite polarities. Current flow direction is of the same polarity, as is $dv/d\phi$

cases it is *reverse* current. The big difference concerns the polarity of the externally measurable voltage across the junction: positive for SQM, negative for TQM. A careful examination of Figures 3.4 and 3.7(b) will confirm the polarities shown in Figure 3.10. However, $dv/d\phi$ is positive in both cases.

3.3.4 Operating modes

As the purpose of an electro-optically used P–N junction changes, so does the associated circuit resistance. This leads to the six basic operating modes shown in Figure 3.7(c).

In FQM, the LED driving circuit should have a high internal resistance, approaching, if possible, the ideal 'constant current' condition. (A justification of this is given in Section 5.1.) While there is only one truly useful operating mode in FQM, SQM has three, depending on the use to which the light receiving junction has been dedicated. When a *voltage* signal is required, the highest output will be obtained along the V axis, i.e. with a very high resistance R_L' of the circuit, approaching the ideal open circuit case as much as possible. We shall call this mode SQM(V). In this mode, the output is proportional to log ϕ, which makes it eminently suitable for lightmeters. However, if *current* is required, the junction should be loaded by a very small R_L'', approaching the ideal case of a short circuit. This is almost as easily done as said, by the use of the 'visual earth' principle of operational amplifiers with feedback. The operating mode will be termed SQM(I). Circuits of this kind are of immense value in optical communication and metrology, as the output is now proportional to the incident light flux, ϕ.

You must have noticed that neither SQM(V) nor SQM(I) delivers much power: when gradually pushed to their theoretical limits, $R_L = \infty$ and $R_L = 0$, respectively, they cover a vanishing surface area $p_o = v_o \times i_o$ of the cross-hatched rectangle of Figure 3.7(c). When output *power* is what we want, we must optimise R_L so that this area and p_o are maximum. The natural way of naming the corresponding operating mode is SQM(P).

Nothing could exemplify better the usefulness of SQM(P) than the solar battery. Energy conversion is not the only case, however, for wanting to use SQM(P). Information-carrying light, too, can use it to advantage with the help of impedance matching networks or transformers. Reference 48 describes a method for the determination of the optimum value of R_L in SQM(P).

The two remaining operating modes are concerned with the externally biased TQM receiver function of a junction. They are the current mode TQM(I) and the voltage/power mode TQM(VP), which no longer require an explanation. It will be instantly deduced from Figure 3.7(c) that TQM(VP) is simultaneously the mode in which the greatest output power and the highest output voltage (without polarity reversal) are obtained. Hence, the total number of basic operating modes of an electro-optically worked P–N junction is six. Circuit design problems related to these modes will be discussed, in some detail, in Chapter 6.

Chapter 4

Photometric and radiometric quantities

It is easy to be in a muddle over light units. Most electronics engineers are. An electro-optics specialist cannot afford to be. Hence this chapter. It represents a return to basic concepts underlying light measurements, and it deliberately cuts out the dull listing of units and tabulation of conversion factors, concentrating on four physical quantities: flux, intensity, luminance/radiance and illuminance/irradiance. The treatment emphasises this physical character of light units, to make them tangible to engineers. If you are not in a muddle over light units, go on to Chapter 5 now.

One of the units of photometry is called the nit – a word with unfortunate connotations in English. *Chambers's Twentieth Century Dictionary* defines it as:

> **nit**, *nit, n.* the egg of a louse or other vermin:
> a young louse: a term of contempt.

Another, more often encountered unit for light measurement is the candle. (Romantic, perhaps, but not very practical.) We also have noxes, stilbs and apostilbs, sea-mile candles, foot-lamberts, carcels, lumens, luxes, heffners and other talbots, without mentioning the radiometric unit of watts per steradian per metre square per nanometre used by CRT specialists. How then do we get out of this jungle? Simple. By going straight to the basic concepts of light measurements.

These concepts are but four, relating to four physical qualities: flux, illuminance/irradiance, intensity and luminance/radiance. Equipped with these you will be able to put into the right place every single one of the two dozen or so existing units. Articles dealing with stage illumination, with camera sensitivity, with the light performance of LEDs, CRTs, incandescent and other light sources, with photodiodes, phototransistors and other light receivers will become clear, catalogues will become intelligible, and comparisons of components from different sources possible.

4.1 Luminous flux

The first and truly fundamental concept is that of *luminous flux*; the remaining three derive from it. The idea of flux is closely associated with

Figure 4.1 A shaft of light in the Pantheon, Rome, illustrating the concept of flux (Reproduced by permission of the Conway Library, Courtauld Institute of Art)

that of flow: think of the flow and you 'feel' the flux. For example the flow of people in Oxford Street. How many per hour? Think of the water flow of a mountain stream. How many gallons per minute? Think of your Company's cash flow. Try to remember now the shaft of light you once saw pouring through a stained glass window, or imagine a torch shining on a pitch-dark night (Figure 4.1) – this is light flow – and you will have grasped the notion of light flux.

Light is a form of energy. The luminous flux is the rate of flow of this energy through a certain area or out of a certain solid angle. For instance, in the case of the shaft of light, this will be the 'energy' rate of the light beam traversing a particular fragment of the stained glass window or the whole of it; in the case of the torch, the total flux is the 'power' radiated into the light cone of the torch, out of its apex.

Photometric units are designed to convey a sense of strength of human responses to light and *not* to give an objective measure of the power carried by a beam of light. (This is why 'power' is in inverted commas above.) Being physiologically dependent, photometric units of flux are colour-related. Radiometric units are not. They alone represent *genuine* power. They alone have licence to use the watt as a unit of flux. The practical consequences of the unequal sensitivity of the human eye to various colours is that even though two fragments of stained glass, one green, the other red, may be transmitting equal amounts of true power (such as would be measured in absolute terms and hence expressed in watts) their

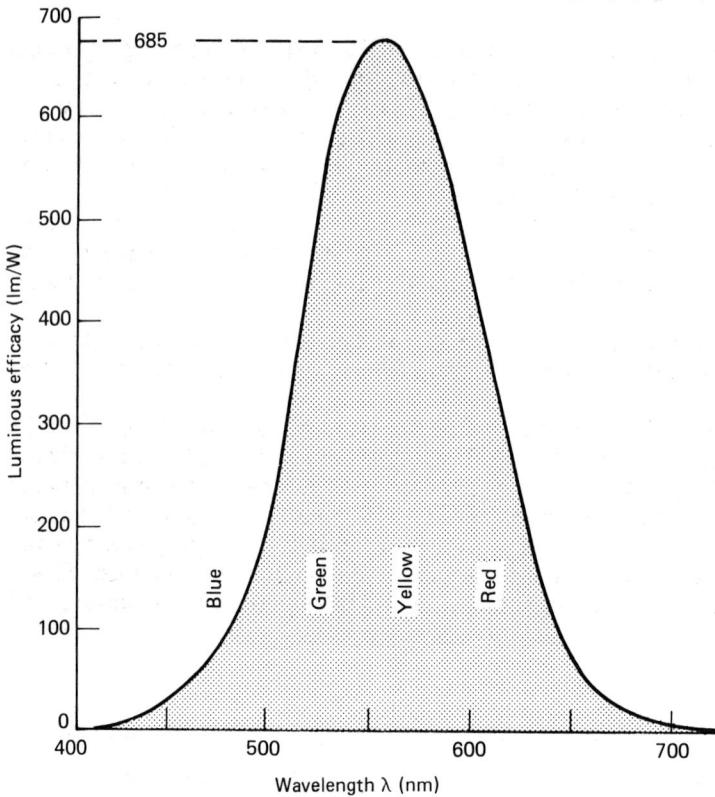

Figure 4.2 The internationally agreed relationship between lumens and watts as a function of wavelength

photometrically assessed fluxes will be different, the human eye being more sensitive to green than to red light. The photometric unit of luminous flux is the lumen. For pure colorimetric green light 1 lumen corresponds to 1.47 mW. For red light ten times more is required to produce the same physiological sensation and so, here, 1 lumen corresponds to 15 mW. Green and red colours as used above correspond to monochromatic radiation of 550 and 650 nm wavelength respectively. An internationally agreed lumen/watt relationship, called the visibility curve for the whole range of colours, was established many years ago based on an 'average eye', the result of numerous measurements made on a large sample of humans (Figure 4.2). This curve gives an immediate answer to a common question of the type: "My gallium arsenide diode emits 0.7 mW. How many lumens is that?" As GaAs LEDs emit at a wavelength of 900 nm, the answer is zero. This is how it should be, as infrared radiation produces no visual effects.

4.2   Illuminance–irradiance

The text you are reading is illuminated. So is the theatre stage (though sometimes dimly), the shop window display and the road. What they have in common is the fact that they all receive light shed onto them, unlike a television screen, for example, which is self-luminous. This distinction must be clearly received and firmly rooted in the mind for the remaining three of the basic four to be understood.

Illuminance is the area-density of light falling from an external source onto a surface. Hence it is represented by lumens per square metre. The unit used in photometry is the lux, with one lux representing an illuminance of one lumen per square metre: $1 \text{ lx} = 1 \text{ lm/m}^2$.

When light from more than one source falls onto an area, the individual fluxes are added. (Laser light requires a specialized treatment. See Chapter 10.)

The radiometric conceptual (not numerical) equivalent of the lux is the watt per square metre (W/m^2). Here, the area density of incident flux is called *irradiance*. You will have noticed the identity of the basic concept linking illuminance and irradiance. It is obvious from Figure 4.3 that the more the surface is tilted with regard to the incident rays, the larger the area lit by the same flux and the smaller the illuminance/irradiance. This is what is expressed by saying that the sun is hotter midday than morning and evening.

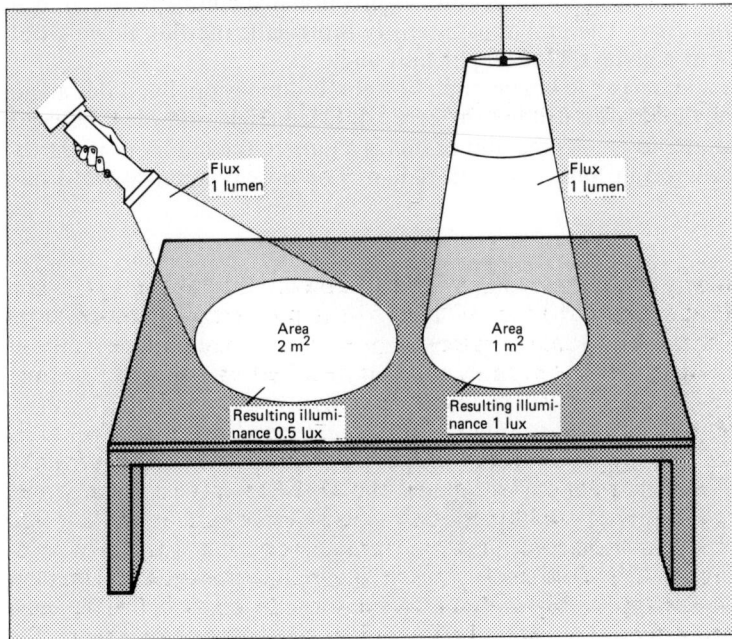

Figure 4.3 The area-density of light falling onto a surface is represented by illuminance, i.e. lumens per square metre, for both divergent light and parallel light

Before going on to the next item of the basic four, it is of utmost importance to emphasize that neither illuminance (lux) nor irradiance (W/m^2) gives the slightest idea of how bright an area appears to us. Consider the example of Figure 4.3. The illuminance of a matt black table top will be exactly the same whether or not it is covered with a snow-white table cloth. This fits the definition of illuminance which, like irradiance, is concerned with the area density of the oncoming and not the outgoing radiation.

Just how strong a lux is and what practical magnitude a W/m^2 is can be judged from these few examples:

1. Moonlit landscapes receive 0.01 lx.
2. A comfortably lit desk is illuminated by 300 lx.
3. St Tropez sunbathers receive 1.5×10^5 lx.
4. A 2 mW helium–neon laser (red) produces an illuminance of a few thousand lux, or an irradiance of 200 W/m^2.

4.3 Intensity

Few real light sources radiate with the same vigour in all directions. Some, such as a torch, are directional by design. Some, meant to be omnidirectional, fail in this respect through unavoidable manufacturing or exploitational constraints. Such is the case of a spherical light bulb (Figure 4.4), in which the unavoidable contact-bearing base impedes the light penetration into a part of the surrounding space. Clearly, to characterise the strength of the radiation in a certain direction, a directional quantity is required – *luminous intensity*. The luminous intensity represents the flux flowing out of a source in a given direction per unit angle.

Because a light source beams radiation three-dimensionally, a flat angle unit such as the degree will not do here. A space angle unit must be used instead: the steradian. As the unit of flux is a lumen, the luminous intensity will be measured in lumens per steradian. For brevity a single word has

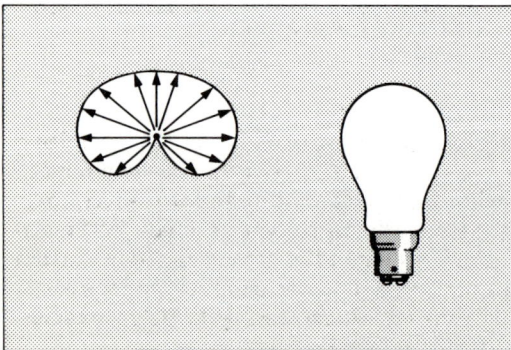

Figure 4.4 As few real light sources radiate equally in all directions, a directional quantity is needed to characterise strength of radiation in a particular direction. Candelas are lumens per unit solid angle

been internationally agreed, the candela, to stand for one lumen per steradian.

The choice of a steradian for a unit of spatial angle is unfortunate: a steradian is a very large chunk of space and as such it does not impart well the sense of directionality. Steradians are seldom used in other fields and it will certainly help to describe an easy way of visualising their size. To form a steradian, take an orange or an apple and cut it into six as if sharing it equitably between six people. Then make a fourth, horizontal cut through the middle (Figures 4.5(a) and (b)). You have 12 equal portions. Each one of them contains at its apex a space angle of one steradian (within a 4% error). A corner of a room contains approximately 1.5 steradians.

Within the context of light intensity measurements, it might be more helpful to visualize the spatial angle not as the hollow of a three-sided structure, but as the interior of the tip of a cone. A hypothetical cornet with a rounded-off 'filler' surface having an area just equal to r^2 would make exactly one steradian at its tip (Figure 4.5(c)).

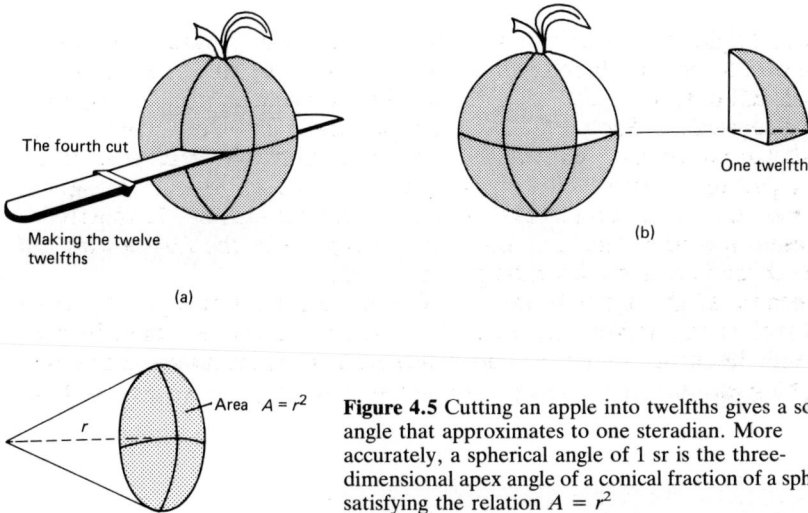

Figure 4.5 Cutting an apple into twelfths gives a solid angle that approximates to one steradian. More accurately, a spherical angle of 1 sr is the three-dimensional apex angle of a conical fraction of a sphere satisfying the relation $A = r^2$

In radiometry, the third basic concept corresponds to the power radiated into a unit solid angle. This is named radiant *intensity* and is measured in watts per steradian. The intensity concept is valid only for sources small with regard to the surrounding space, aptly called point sources. As long as the linear dimension of the radiating element is some ten times smaller than the distances of interest around them, one can call them point sources and use the intensity concept. This is mostly the case with bulbs, candles, LEDs or CRT spots, but not with large panels.

Finally, the values of both luminous intensity and radiant intensity in a given direction are independent of the distance from the source at which it is measured, as seen from Figure 4.6.

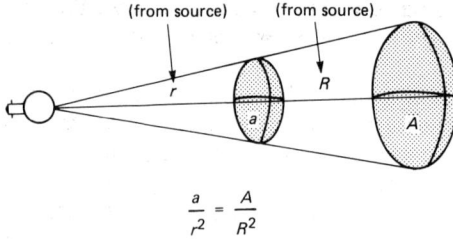

$$\frac{a}{r^2} = \frac{A}{R^2}$$

Figure 4.6 Values of both radiant and luminous intensity are independent of source distance

4.4 Luminance–radiance

The last of the basic four concepts of photometry is that of *luminance*. Imagine you are viewing a tiny, compact filament shining through its bulb of clear glass. The bulb, in fact the filament, is so bright that it hurts your eyes. Then imagine that the glass is opalescent. The device emits now very nearly the same amount of light as before, but the eye perceives it unhurt. The total flux is constant to a first approximation, but the opal glass envelope spreads the radiation over a much larger surface which re-diffuses it. Luminance expresses the brightness of the source in a given direction.

  The surface area of the source has a large part to play. Imagine that the milky spherical bulb containing the filament broke and was replaced by another, twice its diameter (Figure 4.7). The new bulb will appear four times less bright, despite the constancy of its wattage and its total flux. To convey these effects of source brightness, the luminance expresses luminous intensity per unit surface area of the source. This is, of course, the same as the luminous flux per steradian per unit area.

  Thus we have a unit of luminance:

candela/metre^2 or   lumen/steradian $\times$ metre^2

It is a unit that characterises outgoing radation, to be used with objects that emit or re-emit light; a filament, a bulb, an illuminated lamp shade, a

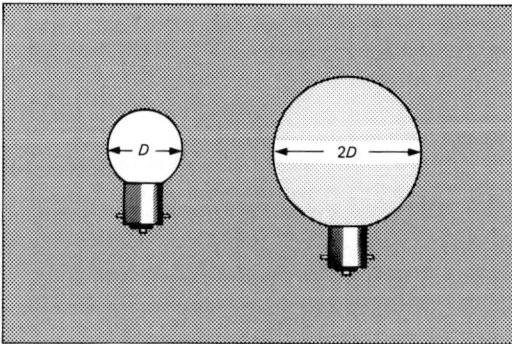

Figure 4.7 Luminance expresses brightness of source. A large bulb appears four times less bright than a smaller bulb for the same power and flux. Luminance is luminous intensity per unit surface area (the same as flux per steradian per unit area)

working screen or an illuminated table top. As an idea of its size, the UK standard for screen luminance in film viewing rooms is 37.5 cd/m^2 at full illumination.

Luminance is a directional quantity, as is intensity, one of its two constituents. The surface area, the second constituent, must be taken as the projection of the physical radiation area on the plan perpendicular to the direction in case. With certain emitting or re-emitting devices the intensity versus viewing angle variation is such that luminance remains constant. This is so because as the observer looks more obliquely at such a source, the projected unit reduces in the same proportion as the intensity does. Such sources, called lambertian, are exemplified by the moon, flashed opal glass, chalk and good Bristol board. However, this directional independence must not be taken for granted, as most devices and materials are not lambertian. Their luminance varies with direction.

Finally, the radiometric sister of luminance is *radiance* and I think that nobody will show puzzlement any longer at the fact that it is usually measured in

W/sr $\times$ m^2

and sometimes (I am sure you will know where and why) in

W/sr $\times$ m^2 $\times$ nm

And yet 'watts per steradian per metre square per nanometre' must have sounded puzzling when first met in the opening paragraph of this chapter.

One final word of guidance. When you come across an unknown exotic unit try to establish, first of all, to which of the basic four denominations it belongs, and whether it is photometric or radiometric. The subsequent working out of numerical conversion factors should come easily.

Chapter 5

The LED: heart of the light transmitter

The LED (Light-Emitting Diode) is basically an FQM-operated P–N junction. The principal semiconductor used for LEDs is gallium arsenide (GaAs). Phosphorus (P) and aluminium (Al) are also being used, the first either in a GaP compound or as an adjunct to GaAs for wavelength shifting, the second as an additive in the so-called heterojunction devices. Silicon is not suitable for electro-optic FQM, as its electron transitions are of the 'indirect gap' kind which do not produce radiation. (The excess energy goes into lattice vibrations instead.)

An electro-optic light transmitter (TX) is usually made up of three components: the driver, the electro-optical (EO) converter and the beam-shaping optics. In most TXs it is an LED that produces the EO energy conversion and imprints to the light beam whatever electrically conveyed intelligence reaches it via the driver. Clearly, then, the LED is the heart of the transmitter.

5.1 Wavelength varieties : IREDs, VLEDs and LWLEDs

The bulk of EO linkages use LEDs emitting in the 780–930 nm range of wavelength, i.e. in the near infrared region of the spectrum. We shall call them IREDs for InfraRed Emitting Diodes. Display and panel signalling LEDs, green, red or amber, will be termed Visible Light Emitting Diodes or VLEDs. Finally, the name LWLED will be given to relatively recent newcomers to the optical communications scene: diodes operating at wavelengths in excess of 1 μm (typically at 1.3 or 1.55 μm). When no valid reason exists for making a distinction between IREDs, VLEDs and LWLEDs, the all-embracing name LED will be used, despite the fact that to a linguistic purist the use of the epithet 'light-emitting' may seem improper in connection with infrared emitters, as there is never any light whatsoever to be *seen* coming out of these devices. It has become customary, however, to extend the usage of the word 'light' to infrared wavelengths up to some 2–3 μm, which is justified by the fact that in this region most glasses and many plastics behave very nearly in the same way as in the visible part of the spectrum.

Let us now examine *practical* LEDs from the mechanical, electrical and optical points of view.

5.2 Packaging the wafer

Seen from the outside, most LEDs look like older-type transistors with a glass window at the top (Figures 5.1(a) and (b)). It is a pity that transistor-like encapsulation should have been adopted for light-emitting devices, as this is not the best type of package for them. Firstly, TO-5 and TO-18 derived cans are not truly cylindrical but slightly conical (owing to their press-tool taper), and hence cannot easily slide inside a bore for collimation or fine focusing. They are also too short for this. Secondly, chip (wafer) positioning on the header is often haphazard, spoiling the desirable on-axis features of the emerging light flux. Thirdly, the glass-to-metal seal

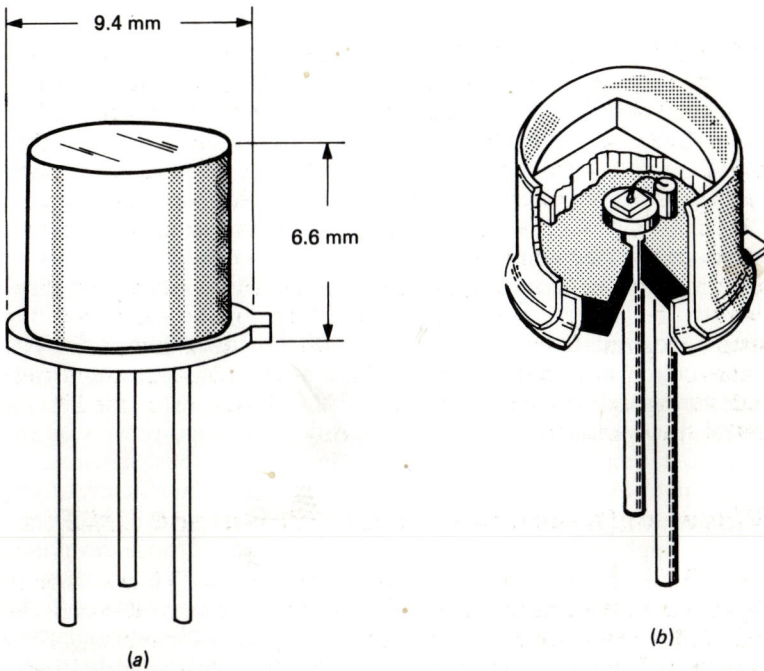

Figure 5.1 Packaged LEDs: (a) modified TO-5 can with flat window; (b) Modified TO-46 with domed window

process, used for securing the window (often called the 'lens' – to make any engineer with some optical sense shudder) to the can, distorts the glass because of the high temperature involved. Fourthly, heat sinking, so important for the efficiency of the electricity-to-light conversion (see Section 5.1.5), suffers from this type of encapsulation. If the can is gripped by the rim of the header, all chances of sliding the LED inside the retention bore are lost altogether; if it is not, cooling is poor. The reason (or the excuse) for using TO-46, TO-18 or TO-92 packages, all derived from transistor technology, is historic: jigs, tools, machinery and skills were readily available at the time LEDs came to the fore, ready for production.

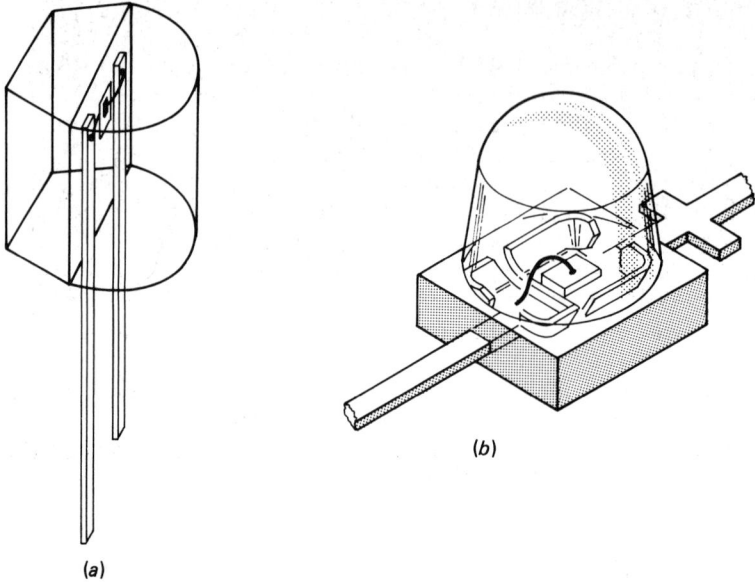

(a)

(b)

Figure 5.2 Packaged LEDs – 'all plastic' encapsulated types. In (a) the transistor legacy is again noticeable

'All plastic' encapsulation (Figure 5.2) is used for lower power devices, with some cooling obtained through the leads. Here, too, the transistor legacy can be sensed.

The term 'low-power devices' refers mostly to average powers as the instantaneous luminous power (flux) can, in some of these devices, be surprisingly high. Hence, a careful examination of the data sheets of any such LED should replace the prejudice against 'plastics'. At the other end of the scale, high-technology devices are put into purpose-developed packages. This is especially true of the so-called pigtailed emitters, in which a small length of optical fibre, secured to the wafer, comes with the device (Figure 5.3), but it is also true of VLEDs and IREDs characterised for producing a closely defined small, circular irradiance zone ($d < 400$

Figure 5.3 Pigtailed LEDs (Courtesy of Optilas Ltd)

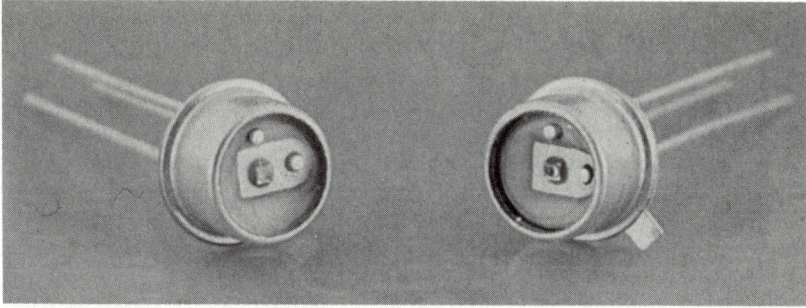

Figure 5.4 Honeywell's 'Sweet Spot' LED (Courtesy of Honeywell Control Systems Ltd)

μm), e.g. Honeywell devices sold under the nearly ambiguous name of 'Sweet Spot' LEDs (Figure 5.4) and a few other 'specials'.

All this does not change the fact that an overwhelming majority of LEDs come in TO-18, TO-46 and TO-92 packages.

5.3   Wafer geometries

We have just looked at LED housings. What dwells inside? – obviously a light-emitting chip and two lead attachments. The chip (wafer) rests on a heat-sinking header, the base of the housing, which often also forms one of the electrical connections. A thermobonded gold whisker linking the chip to a glass–metal feedthrough provides the other. Let us now examine the principal geometries of light emitting wafers. There are surface emitters and edge emitters and either category can use a homojunction (HJ), a single heterojunction (SHJ) or a double heterojunction (DHJ).

5.3.1   Surface emitters

Figure 5.5 shows the cross-section of a standard red VLED. The layers above the GaAs substrate are epitaxially grown. (For a description of this

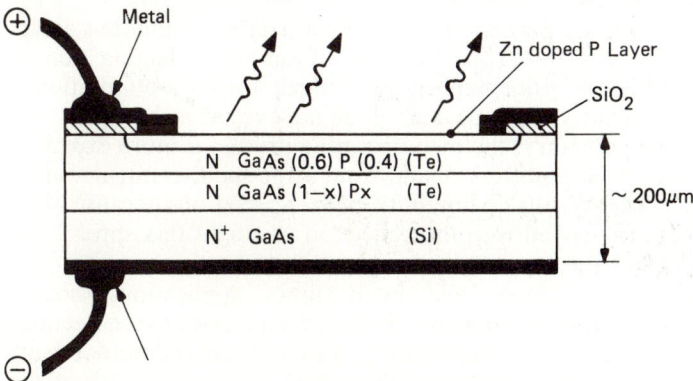

Figure 5.5 A standard surface-emitting VLED. Elements in brackets are dopants

process, see for example Ref. 8, p. 113.) The P layer is formed by diffusion doping with zinc (Zn). The passivation layer prevents junction edge exposure.

Light transmission through a flat crystal/air interface is rather inefficient. Indeed, total internal reflection (TIR) (Sections 2.2.2 and 7.2.1) prevents a large proportion of obliquely incident light rays from ever leaving the crystal, especially in the case of a material like GaAs, which has a very high refractive index ($n \simeq 3.6$ giving a critical angle of $\simeq 16°$). External add-on resin domes, so-called windows, can be helpful in this respect (critical angle reduces to $\simeq 25°$) but introduce temperature T_{max} problems. Shaping the outer surface of the crystal into a hemisphere seems to be an interesting answer to this problem (Figure 5.6). Here, most of the rays

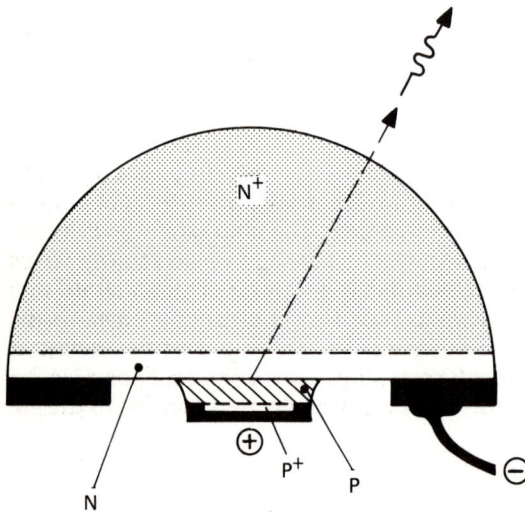

Figure 5.6 Domed GaAs Si-doped emitter

striking the interface are perpendicular to it (or nearly so) and thus escape TIR trapping [8, pp.209–212]. The use of an antireflection coating represents the ultimate refinement in the struggle for the optimisation of the conversion efficiency of the device. There are two snags, however, with the hemispherical geometry: the manufacturing process is more expensive than in the case of flat emitters, made from large multi-element wafers, and the practical user results do not quite meet expectations because of the optical problems associated with the collection of a light flux spread over such a wide angle.

In optical communications and many other applications, source *radiance* is a more important parameter than *radiant power* (see Sections 4.2 and 4.4). Wafer cross-section (Figure 5.7) exemplifies geometries using *current confinement* to achieve high radiance. Only about a tenth of the relatively large upper junction area (in the centre) conducts current and

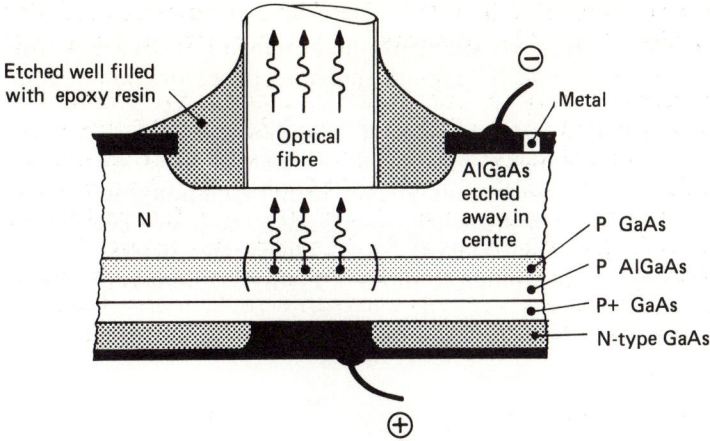

Figure 5.7 Burrus current confinement and etched well high-radiance structure

luminescences. High current density calls for effective cooling. The remainder of this junction area and the whole of the lower one help to evacuate heat from the carrier-congested centre. Note that the lower P–N junction does nothing else, as it cannot conduct current in its reverse biased condition. In older geometries, the lower N layer was a stratum of insulating material (SiO_2), which is a poorer heat conductor than GaAs. The well in the top N layer is obtained by etching (Burrus, 1969). It facilitates the accurate bonding of fibres, thus increasing the light coupling efficiency. The centre portion of this layer is made very thin ($\simeq 10 \, \mu m$) to minimise light absorption.

5.3.2 Heterojunctions

Further improvements can be obtained by the inclusion of *heterojunctions* into the device architecture (see Section 9.3.2 and Ref. 28, p. 112). A heterojunction is a junction between two regions of a semiconductor usually (though not always) of the same polarity (either N or P), but with different energy gaps. The E_g variation is obtained by varying the proportions to two components of a ternary compound, for example, by altering the value of x in the $GaAs_{(1-x)}P_{(x)}$ or the $Ga_{(1-x)}AsAl_{(x)}$ recipe. The result is threefold: a very low absorption by the wide-gap layer of the radiation produced by the narrow-gap material, a difference of carrier concentration, and a difference of optical refractivity between both materials. Heterojunctions enable designers to produce wafer architectures with *both* current and light confinement. How both types of confinement have been put to good use by wafer designers will be shown in the description of an *edge emitter*, to follow. It should be noted, however, that heterojunctions can also be used advantageously in surface emitters (Section 5.3.1). An example is the device in Figure 5.5, in which an SHJ has been shown without an explanation [49, p.514; 50, p.89].

5.3.3 Edge emitters

In the geometry of Figure 5.8 the upper heterojunction, formed by the interface between the N and N^+ regions, prevents light from being propagated upwards.

Similarly, the second heterojunction P/P^+ inhibits downward propagation. This *light confinement* results from the sudden variation of the refractive index at the heteroboundary. The overall radiation propagation effect is similar to that of a microstrip waveguide. The active conversion layer is extremely thin, some 0.05–0.2 μm. Transverse current confinement is obtained by means similar to those of Figure 5.7, and longitudinal

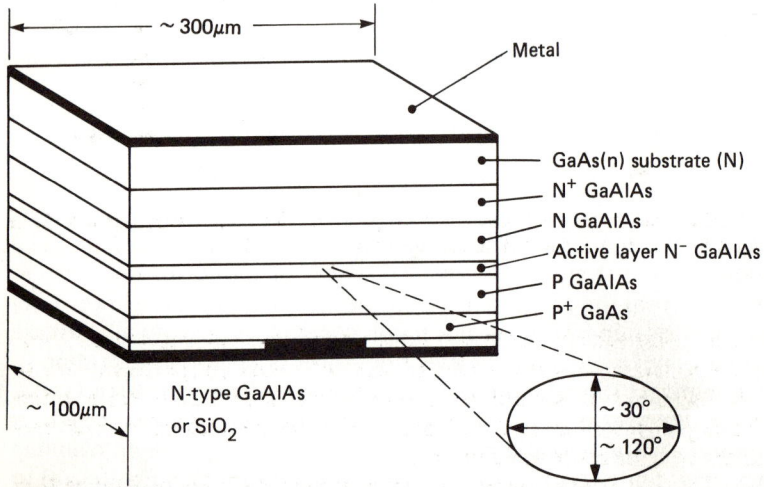

Figure 5.8 Example of double heterojuntion (DHJ) edge emitter. The light output is lower than for types shown in Figures 5.5–5.7, but flux directionality often more than makes up for this when it comes to fibre optics coupling

carrier confinement by means of E_g variation. (Higher carrier concentration in the narrow-gap extra-thin active layer is shown in the N and P layers [49, sections 16.2 and 16.3].) In some structures the layer P^+ is omitted, leaving the light-guiding task to metallisation and insulation layers. This results in better cooling. The above geometry is capable of yielding enormous radiance figures, e.g. some 1500 $W/cm^2 sr$, a value several times higher than that of the 'brightest' surface emitters. The smallness of the emitting area, a strip of merely 3 μm × 10 μm, typically, can be extremely valuable in high-resolution position sensing and, of course, in fibre optic (FO) communications (light coupling into thin fibres).

5.4   The LED as a circuit element

To an electronics engineer, the LED means a *diode*. A circuit designer has to distinguish, however, between VLEDs and IREDs/LWLEDs and,

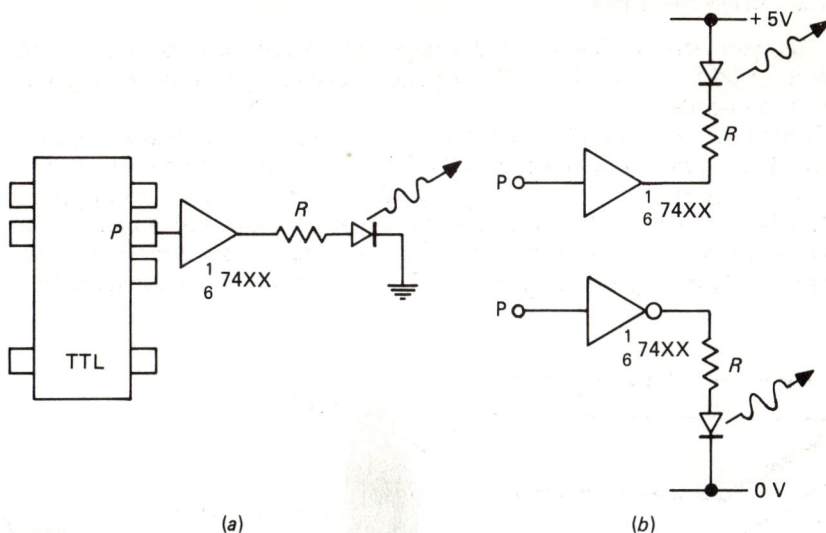

Figure 5.9 LEDs used for state indication on logic circuits: (a) a way of signalling a 'high' at point P of a TTL IC; (b) two ways of indicating a 'low' at point P

though to a lesser extent, between homojunctions and heterojunctions. Only very occasionally is use made of the rectifying properties of these diodes. In all cases it should be remembered that LEDs have steep current/voltage characteristics and hence should be treated as current-driven and not voltage-driven devices.

The VLED is disarmingly simple – just a current-defining resistance $R = (V - V_f)/I_f$ in series and the VLED will do its job of signalling the presence of the voltage V, e.g. the 'on' condition of a power supply unit. In logic circuits, where VLEDs are often used for state indication, it is useful to reduce the drain on the monitored point. Figure 5.9 illustrates this for TTL boards. 270 Ω is a popular value for R in these situations.

IREDs and LWLEDs are used for communications, metrology and positioning, all applications in which light modulation becomes a necessity. The beauty of LEDs in such applications lies in their capability of being modulated *directly* by the driving current. In serious professional driver design work, the equivalent circuit of an LED will often be called upon. One such circuit is shown in Figure 5.16. Although digital modulation covers the majority of cases, the analogue method has its own attractions (e.g. in short-haul transmission), and this will also be discussed. In fact, it is there that we shall begin. A linear section of the Φ versus I_F characteristic (Figure 5.10) will be chosen, for signal integrity preservation (e.g. voice). This will minimise the generation of harmonics and thereby the risks of cross-modulation in carrier multiplexing situations.

The first of two reasons for adopting a *current* rather than a *voltage* drive is the fact that the light output Φ is far more likely to be proportional to the current I_F (think of the number of carriers per second crossing the junction) than to the voltage V_F across the device. The second reason is the

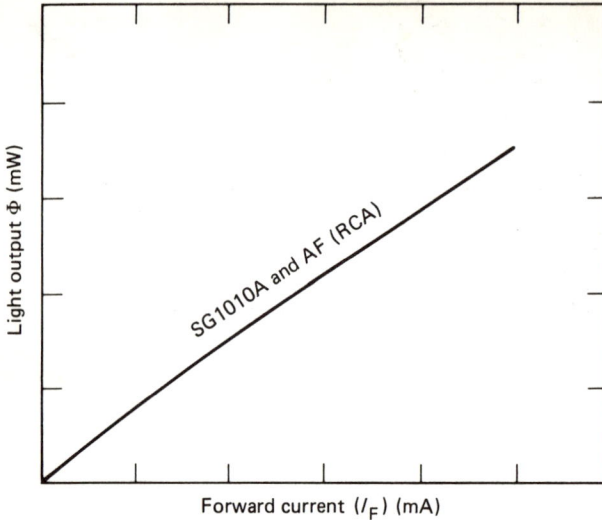

Figure 5.10 Typical light output versus current of a commercial IRED shows the linear dependence of ϕ upon I_F (except for very low values of I_F)

thermal stability. If we examine closely the I_F versus V_F family of curves for three junction temperatures T_j (Figure 5.11(a)), we notice that a voltage imposed working point is intrinsically unstable. Indeed, in a voltage-driven circuit, shortly after switching the device on, the working point P_0 is certain to move up, towards P_1, owing to self-heating. This increases I_F and the $I_F V_F$ area, which unfortunately is representative of the power turned into heat. Not only do we now get a lower conversion efficiency (owing to the rise of T_j), but we also face a real risk of thermal runaway of the device, because the newly generated extra heat will shift the working point further up, onto the curve corresponding to a yet higher junction temperture T_j, and so on. This *positive* thermal feedback could lead to a catastrophic failure of the device. Figure 5.11(b) shows, however, that a constant current drive is inherently stable, being self-regulating through the now *negative* thermal feedback.

5.5   Driver circuits

The circuit of Figure 5.12 represents a possible, very simple way of separating the control of the d.c. working point from that of the baseband modulation. This circuit is eminently suitable for the first time electro-optics experimenter. It will work, for example, as a TX driver for an unassuming speech link and will provide its constructor with the kind of 'feel in opto' that hams get, in radio, from their first transmitting set.

Figure 5.13 shows a circuit for a.m. carrier frequency communications. For reasons familiar to electronics engineers, this technique is suitable for greater distances than the direct a.m. system of Figure 5.12. A further improvement can be obtained from the use of pulse modulation, especially

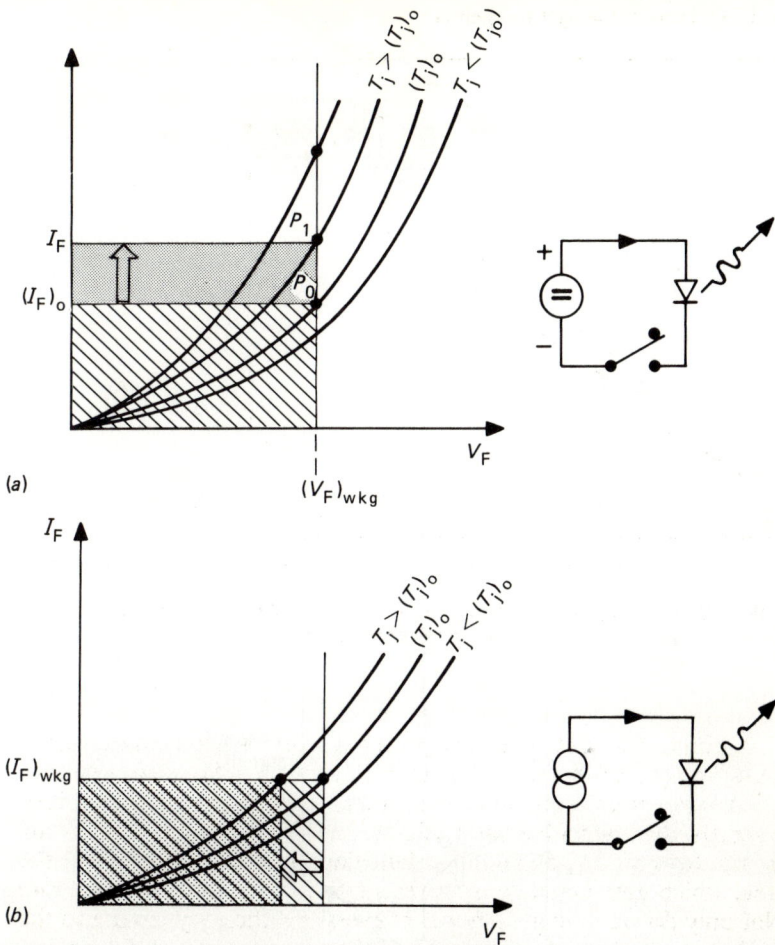

Figure 5.11 (a) A voltage-imposed drive point is inherently unstable. (b) A current-imposed drive point (constant current drive) is inherently stable

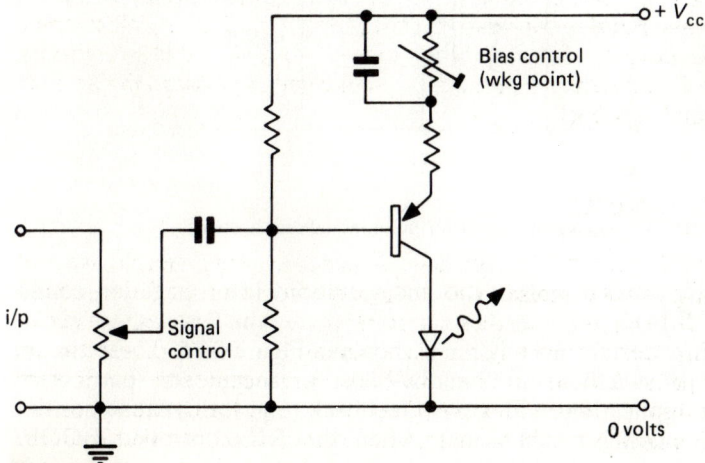

Figure 5.12 Basic circuit for a 'constant current' driver with separate bias and signal controls. Suitable for simple audio links

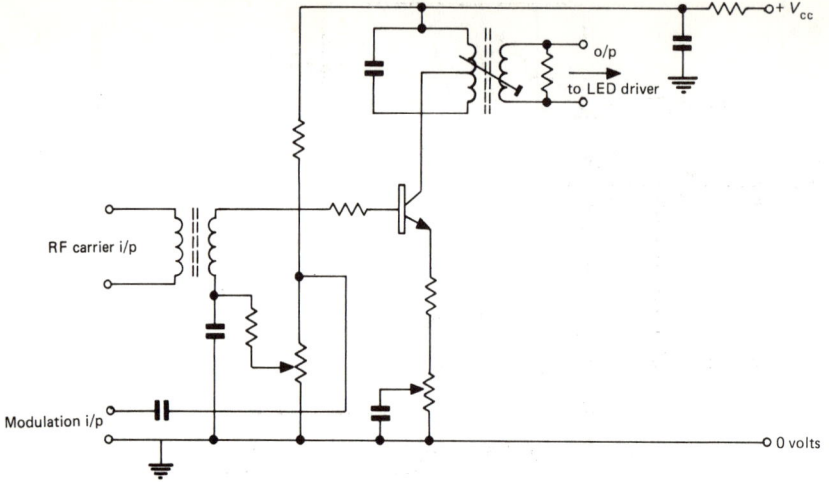

Figure 5.13 Carrier frequency a.m. modulator. RF carrier will be usually derived from a crystal oscillator

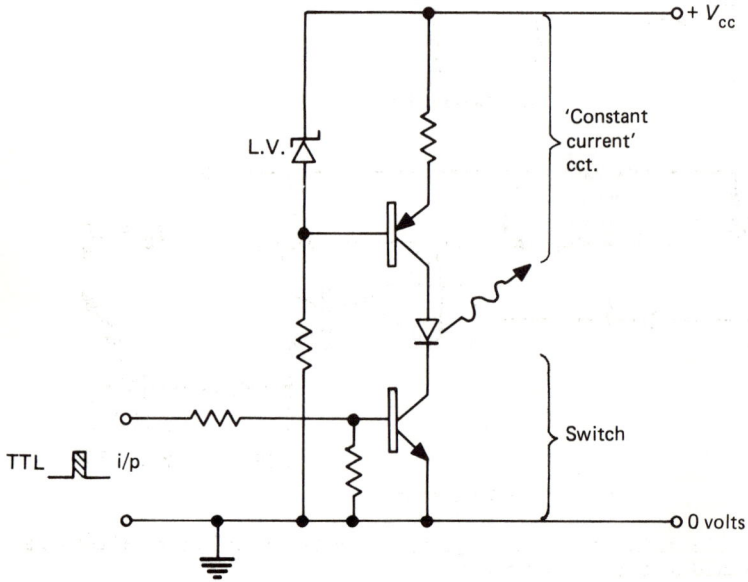

Figure 5.14 On/off TTL compatible driver for pulse modulation systems

with regard to signal-to-noise ratio and variations in propagation conditions (Figure 5.14).

A pulse-on-pedestal transmission is shown in Figure 5.15. These circuits are not on/off switches but standby-boost arrangements, particularly attractive for applications with *system* feedback (e.g. for dynamic position detection in moving mechanisms), when the RX controlled BOOST command can generate brief, but enormously intense spurts of the light output. With some LEDs, the presence of a standby current speeds up

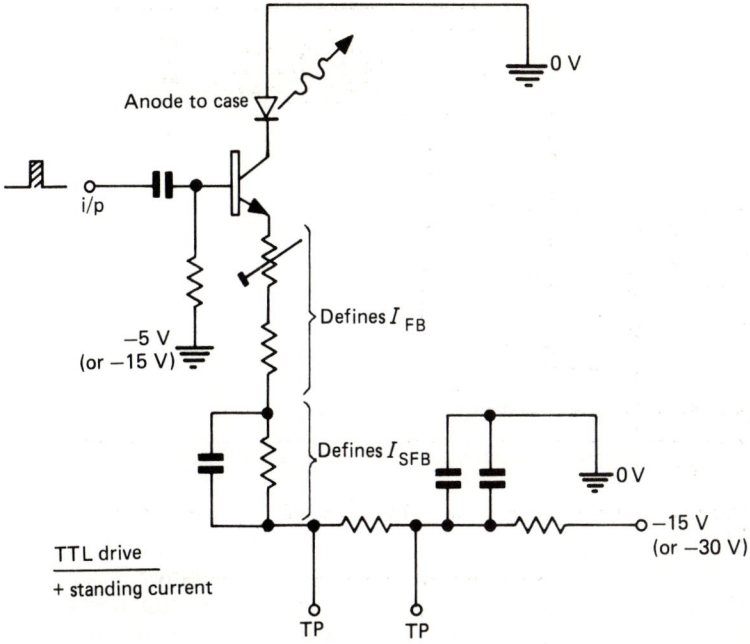

Figure 5.15 Pulse modulation with stand-by I_F: (a) anode-earthed LED, positive boost input; (b) cathode-earthed, negative boost input

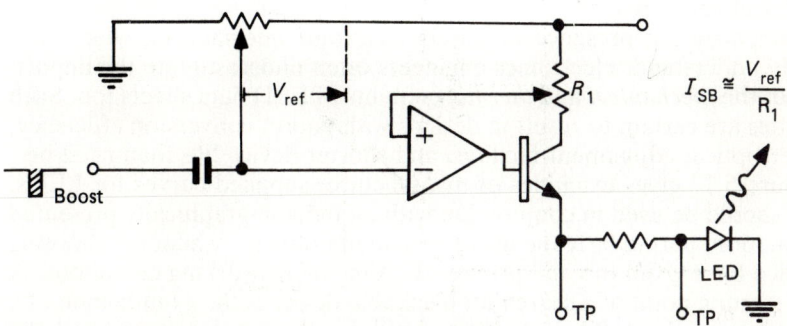

switching and reduces edge 'ringing' which can be of value in work with very short ($t < 1$ μs) pulses. Finally, Figure 5.16 shows an equivalent circuit of an LED, the understanding of which is useful in the process of designing high-frequency drivers.

5.6 Making full use of data sheets

Perhaps because of their unfamiliarity with light units and a feeling of discomfort with regard to optical concepts, most electronics engineers pay

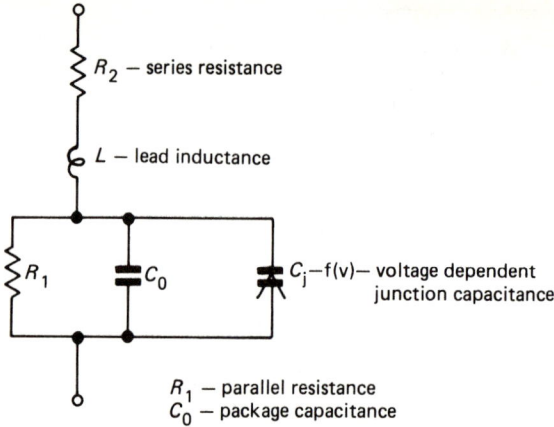

Figure 5.16 A possible equivalent circuit of an LED, useful in high-frequency applications

no more than lip service to the optical part of LED characterisation sheets. And yet, treating the LED more like a diode than a radiator is certainly wrong when it is radiation they are after to fulfil the EO function of the design. It is hoped that Chapters 2 and 4 will have given the reader a reasonable surefootedness in the marshlands of millicandles, lumens, luxes and mW/sr/cm^2 and removed all apprehension in front of f-numbers or numerical apertures.

Blinded by the presence of purely electronic information, which they readily understand, electronics engineers often underestimate the import-ance of the *mechanical* and *thermal* contents of LED data sheets too. Such attitudes are certain to result in designs with poorer conversion efficiency, poorer optical adjustment facilities and shorter device life than need be.

Figure 5.17 gives examples of manufacturer-supplied curves for LEDs. They should be used in conjunction with all the non-graphically presented information, particularly the list of 'maximum ratings', whenever choosing a device for a given job and prior to deciding on its working conditions. A good starting point of a search for a suitable device is the examination of a pack of data sheets likely to fit the bill for the top frequency and the collectable luminous power. Shortlisting the pack for frequency or rise/fall time presents no problems. Just how much of the total radiant flux ϕ_T given by the data sheet can be collected by an optics with a given $f_\#$ or NA can only be found after the relationship between ϕ_{coll}, the flux contained within a cone having an apex half-angle $\alpha/2$ and the angle itself

$$\phi_{coll} = f(\alpha/2) \tag{5.1}$$

has been established. A curve of this kind is seldom provided by manufac-turers. What always is provided, however, is the intensity versus direction curve (Figures 5.17(a, b)):

$$I = f(\alpha/2) \tag{5.2}$$

sometimes in the form of a polar diagram (Figure 5.17(a)). Appendix 3 gives the procedure of converting Equation 5.2 into Equation 5.1, which

Hewlett Packard 4950 series

(a)

(b)

HAFO/ASEA 1A 119F

I_F forward current (A)

10.0

1.0

0.1

0   2   4   6   Volts

V_F

(c)

MOTOROLA MLED 93–94–95

P_0 – normalised power output

3.0
2.0
1.0
0.7
0.5
0.3

−75 −50 −25 0   25   50   75 100 150

T_j junction temperatures (°C)

(d)

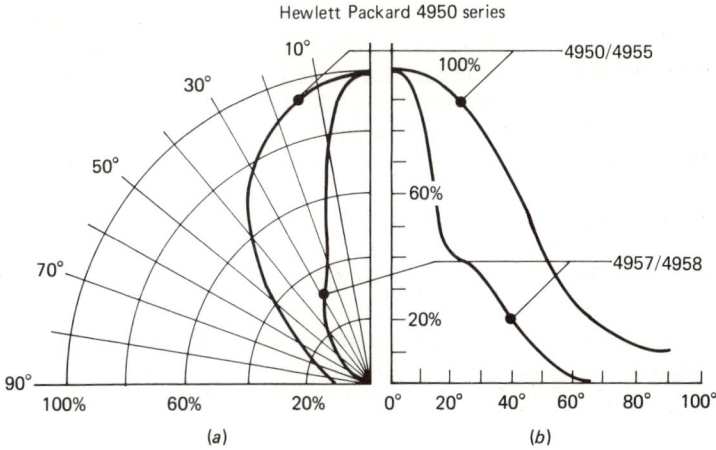

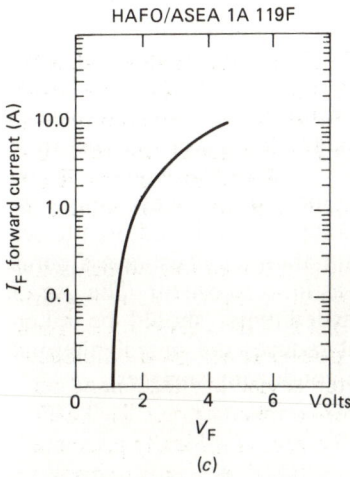

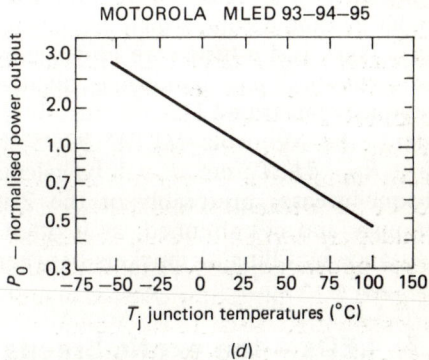

Figure 5.17 Examples of manufacturers' curves for LEDs: (a, b) relative radiant intensity versus direction in polar and cartesian coordinates; (c) forward current versus forward voltage (pulse condition); (d) output power versus junction temperature

includes the conversion of flat angles $\alpha/2$ into space angles θ. In many applications it is the irradiance at the RX end that is more important than the transmitted flux which, by the time it reaches the RX, could be spreading over several square metres and thus become uncollectable. Free Space Optical Communications (FSOCs) are an example. In this type of application the designer will be looking for devices with high axial radiance rather than power. The transmitter optics obviously plays its part in the overall performance of the linkage and two or three design trials might be necessary before the balance is struck between expensive IREDs and cheap TX optics and vice versa, while giving due consideration to the size and weight constraints of the apparatus. To establish the RX irradiance, the size of the image might have to be calculated (unless it is certain that the image is smaller than the TX sensor). The *overall* optical magnification of the TX/RX system will then have to be calculated first (which in the case of identical TX and RX optics equals unity), and allowances for the blur factor made. Should the dimensions of the TX wafer be absent from the data sheet, the manufacturer will provide them, upon request. (Figures of this kind supplied by an agent or distributor over the telephone should be treated with great caution!)

Mechanical information is relevant for two reasons: the distance of the light-emitting portion of the wafer and its axiality with regard to the TX optics are optically important, and accurate dimensional package information is important in the design of efficient junction cooling. For 'fixed focus' TXs the dimensional tolerances ought to be examined for unit-to-unit LED repeatability. The wafer-to-window distance and its tolerance have sometimes to be asked for in writing. For adjustable designs it is the ease of sliding the device or that of seizing it in a slidable barrel that needs looking at. The importance of designing for efficient cooling cannot be overemphasised. The curves in Figures 5.17(a) and (b) show how much the conversion efficiency η drops with rising junction temperature T_j. Statistical data are held by some manufacturers to prove that LED time degradation is far more pronounced at higher T_j values. Cooling of LEDs is normally obtained by convection, usually natural, via a heat sink (except for VLED indicators, which are left to manage themselves the best they can). Thus, just a little care in designing a good thermal contact with the heat sink will pay handsome dividends. Good data sheets (e.g. the previously illustrated Hafo professional LED type 1A119F and, to a lesser extent, the Motorola MLED 93/95) contain thermal resistance figures R_{thjc}, R_{tha}, dI/dT_j, etc. T_j will be calculated as for a transistor. The use of silicone grease, preferably of the ZnO loaded type, should be taken seriously and not shunned, as it often is; it reduces the overall thermal resistance by filling air pockets between the conducting surfaces.

5.7   LEDs – the exotic breeds

When we see for the first time a red VLED panel indicator turn green, we marvel and wonder how this can happen; how can the energy gap of a wafer suddenly change from some 2.0 eV to about 2.4 eV in order to satisfy numeric Equation 3.4?

$$\lambda = \frac{1234}{V} \tag{3.4}$$

The answer is: it can't and it doesn't. We are being deceived – this LED is a cheat. Within it, two tiny wafers are bonded, in close proximity, onto the same header and electrically connected in parallel/opposition (Figure 5.18). As the polarity of the drive reverses, D_1 switches off and D_2 gets turned on or vice versa. A bulbous diffusing plastic top helps to produce the 'one lamp' delusion. It is the electronics engineer and not the physicist who has designed the bi-colour LED.

The practice of enclosing the indispensable current-limiting series resistor (or even the current-regulating IC) inside the package is spreading

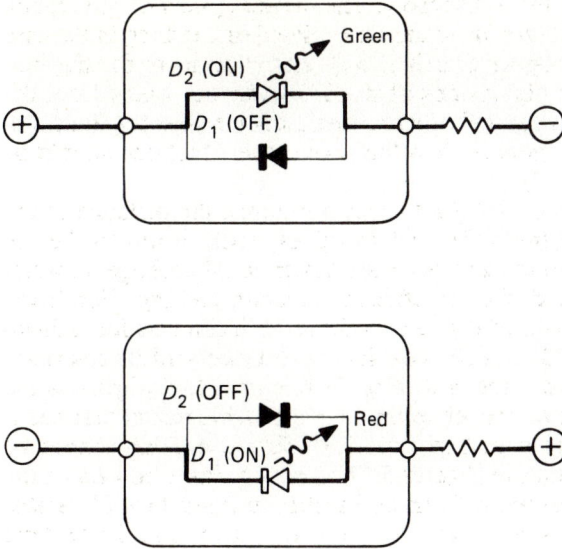

Figure 5.18 Bi-colour VLED

rapidly among manufacturers of VLEDs, so that this particular variety will probably soon cease to be 'exotic'.

Previous chapters have praised the LED for the smallness of its light-emitting area. In display and signalling devices, however, highly localised emission is not an advantage. Thus, most VLEDs are capped with diffusing tops, usually of tinted plastic. Shapes other than circular begin to appear: opal squares, rectangles and triangles of all three VLED colours are already here. Arrows, miniature man-like cutouts and a whole range of other ideographic signs, symbols and mimic displays are most likely to follow suit. At the time of writing, an American was reported contemplating the manufacture of hand-held (Space-Invaders) games with LED aeroplanes and spaceships, etc.

Some signalling devices include their own function-orientated circuitry. Such is the case of a 'bleeper' IC (RS Components types 587080 and 588011). Optics-enriched devices are available too: the Hewlett-Packard HP HEDS 1000 TX/RX combination contains a twin plastic lens, the TX part of which projects a tiny ($d \leqslant 0.4$ mm) deep red spot of light on to a plane 4.7 mm away from its front window.

The most impressive of all the 'specials' is, to me, the so-called 'pigtailed' LED (Figure 5.3). Destined almost exclusively for fibre optics communications, this high-technology 'exotic' shifts the extremely difficult task of making the wafer–fibre bond from the user to the much better equipped LED manufacturer. The user is left with the fibre-to-fibre connecting job, which he or she needs to master anyway. The wafer–fibre bond must be not only optically efficient but also mechanically strong to be secure. Another way of simplifying the user's job is to include the LEDs in a standard fibre optic connector (Figure 5.19). In the former, a short-fibre

Figure 5.19 Various LED/photodiode housings (Courtesy of Optronics Ltd)

pigtail is protected by a stainless steel tube. Alignment is of the 'jewel-bearing' type (see Chapter 8). The pigtailing technique proved so success-ful that it soon spread from LEDs to light-receiving devices.

5.8   The attractiveness of the LED as a light source

Within 15 years the LED has become a firmly established source of infrared and visible radiation, advantageously replacing neon and small incandescent bulbs in a myriad of applications. Straightforward lighting and projection alone have escaped the invasion. Precious little remains unconquered in the fields of photoelectric control, automation, intruder alarms, component and materials inspection, smoke detection, punched tape reading, panel indicators, instrument signalling, etc. Thanks to LEDs, older jobs are done better, and new ones, hitherto impossible, have become possible. Optical communications (both by fibre optics and in free space), fine ranging and positioning, as well as optical signal coupling in electronic circuitry, are here to witness the new openings. What makes the LED such an attractive light source?

Uppermost comes the *modulatability* of all LEDs: VLEDs, IREDs and LWLEDs. It is not only possible, but easy, to modulate the emitted light flux, both proportionately to the controlling signal strength and at frequen-cies up to the gigahertz region. Thanks to modulatability, electro-optics is to photoelectrics what cinematography is to still photography. In both cases we have the addition of a new dimension to an older technique: time. The next attraction of LEDs is their *geometrical confinement*. We have seen that radiating areas of LED wafers are not only small: they are also flat. Light emerging from a two-dimensional source can be focused or collimated far more effectively than that originating from a three-dimensional filament of an incandescent bulb. Both operations rely on the use of the focal plane of a lens. There is, however, one and only one focal plane on each side of a lens and it cannot take a three-dimensional source in its entirety.

Collimation is treated in Section 8.3.1(a). See particularly Figures 8.3 and 8.4(a) to appreciate that (a) the smallness of an emitting area assists collimation, and (b) the in-depth third dimension of a source renders collimation impossible as light from points outside the focal plane either diverges (in front of the focal plane) or converges at first, only to diverge further afield (behind the focal plane).

The smallness of the emitting area makes it into a quasi-point source which, by reducing divergence, assists further with collimation problems. The outgoing radiation can be thus more easily shaped into a quasi-parallel long-ranging beam. (It also makes it easier to diaphragm shape the beam cross-section into squares, rectangles, slots, etc.) Next to dimensional compactness comes *spectral compactness*: the radiation of an LED is quasi-monochromatic. That of an incandescent source has a broad spectrum. When current modulated, an LED holds its spectrum well, while an incandescent source changes its colour dramatically. Monochromaticity facilitates signal from 'noise' (e.g. ambient light) separation in an electro-optic link, by optical filtering. IRED-produced radiation is invisible to most animals and insects and to all humans. This invisibility, an invaluable asset by itself, presents an enormous attraction to a whole class of applications, such as biological research, security and secret communications, when coupled to modulatability. The fifth and certainly not the least attraction of LEDs is their *solid-state, transistor-like character*: they are small, light, running nearly cool (in comparison with incandescents) and very long lived (say $100\,000$–$1\,000\,000$ h).

With such advantages, it is not surprising to see the use of LEDs spreading into so many applications that to cover them all would require another book. Thus, only a broad categorisation of LED applications will be attempted in this volume, illustrated by a small number of examples. Because in most of their applications LEDs are used in conjunction with photoreceiving P–N junctions, the chapters on applications will be preceded by the description of the work such junctions do.

Chapter 6

The photodiode: core of the light receiver

6.1 The non-amplifying light-receiving junction

A photodiode is essentially a P–N junction operated in the second or third quadrant mode (SQM or TQM). Junction physics was covered by Sections 3.1 and 3.3.2 and mode distinctions were illustrated in Figure 3.7(a). More particularly, the generation of free carriers by incident light, a result of electrons being liberated by photons, is represented by Figure 3.9, while the polarity of the e.m.f. and the direction of current flow in basic SQM and TQM circuits with regard to the junction are shown in Figure 3.10. Equations 3.3 and 3.4 hold inasmuch as the incident radiation must have a wavelength equal to or smaller than λ of Equation 3.4, which simply means that the impinging photons must have sufficient energy to make electrons jump from the valence to the conductor band, or

$$hv \geqslant E_g \tag{6.1}$$

(where E_g is the energy gap of the semiconductor used).

For the sake of convenience, the gist of the information pertaining to the basic photodiode mechanism, culled from Chapter 3, has been repeated in Figure 6.1.

Useful photodiode action can be obtained with both gallium arsenide and silicon, as it can with germanium and many other, mostly low-gap materials. It is worthwhile to note in passing that because of this photodiode capability GaAs emerges as the electro-optic semiconductor *par excellence*, as it lends itself well to the manufacture of EO junctions capable of operating in all six modes: FQM for LEDs; SQM(I), SQM(V) and SQM(P) for high-efficiency solar cells; and TQM(I) and TQM(V,P) for photoreceivers.

The electro-optic light receiver RX is, like the TX, usually made up of three components: the beam-shaping optics, the EO converter and the electronic circuit. In most RXs it is the photodiode that produces the EO (strictly speaking the OE) energy conversion and imprints to the electrical output the beam-carried intelligence. Clearly, then, the photodiode is the core of the light receiver.

62

E

Conduction band
Electron

E_g

$h\nu$

Valence band
Hole

BEFORE and AFTER
the impact of a photon

E

Electron

Electron

E_g

Hole

N Hole P

Electron

N P

Cathode — Anode

i

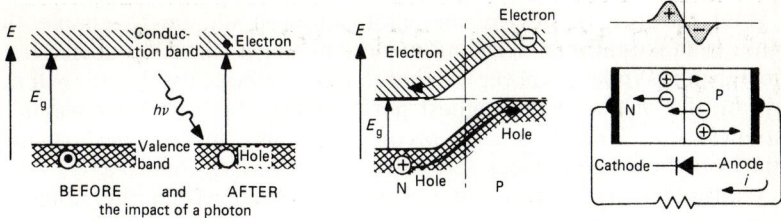

The Physicist's (left) and the Engineer's (right) aspects

The intrinsic photoelectric effect

The incidence of a photon creates a pair of free carriers when $h\nu > E_g$, junctions notwithstanding. Valid for either P or N material.

Junction photoelectrics

The vicinity of the P–N junction causes the pair to split. 'Ex-spouse' migrates and crosses the border. (Two such pairs are shown.) The resulting conventional current flows from cathode (N-side) to anode (P-side) inside the photodiode and from anode (P-side) to cathode (N-side) in the external circuit. Only one carrier per created pair is transported across the junction.

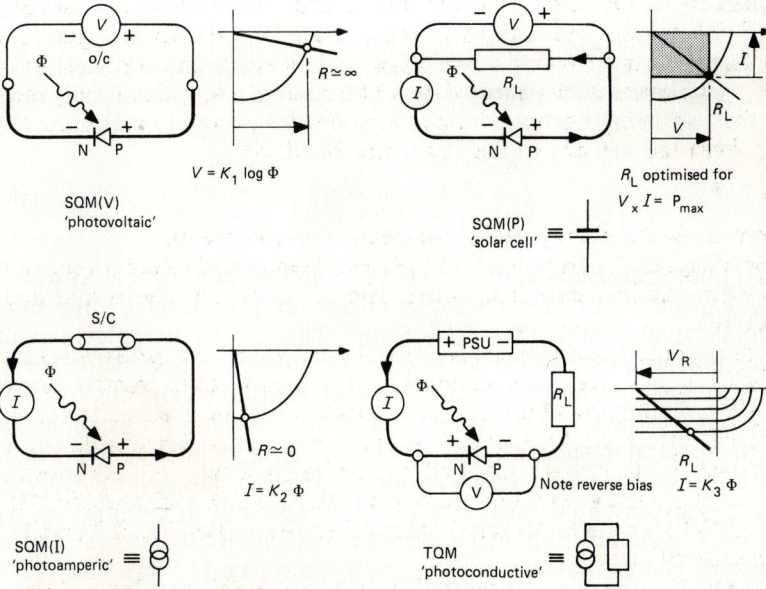

V o/c Φ

$- +$

N P

$R \simeq \infty$

$V = K_1 \log \Phi$

SQM(V)
'photovoltaic'

V Φ R_L

I

$- +$

N P

R_L

R_L optimised for
$V \times I = P_{max}$

SQM(P)
'solar cell'

S/C

Φ

I

$- +$

N P

$R \simeq 0$

$I = K_2 \Phi$

SQM(I)
'photoamperic'

$+$ PSU $-$

Φ R_L

I

$+ -$

N P

V

Note reverse bias

V_R

R_L

$I = K_3 \Phi$

TQM
'photoconductive'

Principal operating modes with voltage polarities and current senses shown
Note the polarity reversal across the device in TQM.

Figure 6.1 Photodiode action at a glance

6.1.1   Device characterisation

The chief purpose of device characterisation should be to enable the designer to make rational decisions when selecting a particular device for a particular job. While it is accepted that the 'best' device for the job will not always be the one with the highest performance (perhaps for reasons of economics, size, weight or even second sourcing), making performance comparisons between devices should, nevertheless, be made easy. Alas, with photoreceivers, this is not always so – even within the range of photodiodes alone. Because the electrical response to the optical stimulation of a photodiode can be gauged in several ways, a straightforward comparison of this response is sometimes impossible. Table 6.1 exemplifies such a case.

Table 6.1 The lack of uniformity in ways of expressing the electrical response of a photodiode to optical stimulation

Device	Manufacturer	Units used to characterise response	Terminology	Notes
MRD510	Motorola	μA/mW/cm^2 at 5 mW/cm^2 tungsten colour temp. 2870 K or at 0.5 mW/cm^2 at $0.8\,\mu$m	Radiation sensitivity	Surface area not explicitly stated
SGD100A	EEG	A/W at 0.9 μm	Spectral sensitivity	Surface area given but window effects not specified
BPW34	Mullard (also made by Siemens, AEG and others)	nA/lx at E_v = 1000 lx, T_c = 2856 K unfiltered tungsten, V_R = 5 V and μA at E_v = 100 lx, T_c = 2856 K unfiltered tungsten, V_R = 0 V	Light sensitivity / Light current	Surface area calculable

Had phototransistors and field effect devices been looked at as well – in many a project they are at least as eligible as photodiodes – the jungle would be thicker still. (See Ref. 5, pp. 290, 294.) Parameter and unit conversion attempts aimed at finding a common denominator for the devices under scrutiny may prove not only discouragingly lengthy (if not altogether frustrated by the absence of some vital data, for example, chip size or spectral response), but sometimes also pointless, because the electrical response of the device to optical signals is only half the story: the device's self-generated noise has to be looked into in conjunction with sensitivity, especially when the received optical signal is weak. Indeed, what is the good of having a highly sensitive receiver if its own noise covers up the signal? In such situations, no valid comparison can be made between devices for which no noise figures are given. This disconcerting disparity of characterisation parameters used by various manufacturers stems, probably, from the main field of applications for which their wares were originally destined. (In a cross-field application an experimental functional comparison of the contenders for the job may well be the only solution. Not a very sound one, at that, unless the position of the tested samples within the production spread of their category is known, e.g. 'min', 'max' or 'typical', a rather rare occurrence.) Let us examine these parameters in turn.

(a)   Light (reverse) current, $I_{R(L)}$

This is a very simple *static* parameter, usually expressed in microamps and given for a specified illuminance, usually expressed in lux. It is valid only for the specified colour temperature (always remembering that a figure in lux gives only an idea of how bright a surface appears to the human eye). For devices with an integral infrared filter, the current is given for a single wavelength, e.g. 930 nm, and a specified irradiance, e.g. 1 mW/cm². Variations of $I_{R(L)}$ with T_{amb} are often shown by a graph.

(b)   Luminous sensitivity, N

This parameter, also called light sensitivity, is a very simple *incremental* parameter, usually expressed in nanoamps per lux. It is valid only for the specified colour temperature, often 2700 K or 2856 K. When working with weak signals, care is to be taken to find out for what static conditions N has been given (illuminance, reverse bias if in TQM, dark current) as these affect internal noise. Regarding devices with integral filters, the same remarks as for light (reverse) current above apply.

(c)   Quantum efficiency, η_Q

Simply stated, this is a number telling us what proportion of incident photons are being successful in generating electrons. It is also called quantum yield. As not all of the generated electrons reach the external circuit (some recombine within the semiconductor) there are two types of η_Q, namely, the internal quantum efficiency, η_{Qi}, and the external quantum efficiency, η_{Qe}. η_{Qi} is of little interest to engineers. η_{Qe} is to be used with caution, as it does not illustrate the *energy conversion* efficiency of the device. The reason for this is that the energy of a photon depends on its wavelength (Equations 3.3 and 3.4), thus direct comparisons are valid only between devices characterised under identical spectral conditions, e.g. for the same wavelength of monochromatic radiation. For non-amplifying junctions, η_Q is obviously always less than unity, although some authors claim exceptions, when photon energy exceeds twice the energy gap of the material (see Ref. 10, p. 4.2).

(d)   Responsivity, $\mathcal{R}$

Also called flux responsitivity or radiant sensitivity, this is one of the most useful performance parameters in electro-optical engineering. Taking into account the wavelength dependence of photon energy it tells us directly, in simple terms, how many microamps of current to expect from a photodiode for every microwatt of radiant power falling upon it. Older literature uses the 'amp per watt' figure (A/W) which is, of course, numerically equal to today's µA/µW. $\mathcal{R}$ is wavelength dependent. *The spectral response curve* of a device defines this dependence. At their peak spectral response ($\approx$ 900 nm), modern silicon photodiodes exhibit $\mathcal{R}$ values of 0.5–0.6 µA/µW.

(e) Incidence response $\mathscr{R}_E$

This parameter gives the ratio of the photocurrent to the so-called 'incidence'. It is expressed in A/mW/cm². Instead of incidence we can use here the already familiar irradiance (see Section 4.2), bearing in mind, however, that devices are $\mathscr{R}_E$ calibrated at perpendicular incidence. (As some devices may have rimmed cases and/or lenses, cosine corrections will not always work.) This parameter is useful in cases in which a small area photoreceiver is exposed to a uniform field of known flux density, overfilling it. In such cases, photocurrent calculations based on the knowledge of $\mathscr{R}_E$ and A may give incorrect results, owing to the edge effects of the device (see Ref. 10, pp. 4.3, 7.2 and 7.3).

(f) Noise limits of usefulness

Noise equivalent power (NEP) This is a most useful characterisation parameter for OE linkages with weak RX signals, a category into which the majority of optical communication applications fall. NEP specifies the strength of the signal for which the signal-to-noise ratio (SNR) is unity. ('Noise' means the RMS value of the unwanted self-generated electrical fluctuations of the device and 'signal' stands for the power of a sinusoidally modulated light flux.) NEP tells us what is the weakest possible optical signal the photoreceiver will handle, whilst preserving a chosen, designed-in SNR. Should you choose an SNR of say 3, 10 or 20 dB, NEP will serve you in your calculations as the 'zero dB' reference level. In digital transmissions (data or coded speech or video) the NEP value helps to predict the statistical Bit Error Rate (BER). (Section 8.4.4). The noise voltage (or current) is proportional to the square root of the bandwidth, $\triangle F$, of the receiver terminal, giving NEP the slightly strange dimensions of $W\,Hz^{-\frac{1}{2}}$. This is a direct consequence of the noise *power* being proportional to $\triangle F$. This 'per root hertz' aspect must be always remembered, as some manufacturers quote NEP in watts, omitting, for simplicity's sake, the $Hz^{-\frac{1}{2}}$ dimension. NEP is wavelength dependent. In addition, the real device noise is not a 'white' noise and hence the modulation frequency for which NEP is quoted should be specified. As a result NEP can be sometimes found written:

$$\text{NEP} (0.9, 1000, 1) = 1.2 \times 10^{-14} \, W\,Hz^{-\frac{1}{2}}$$

the figures in brackets denoting, respectively, $\lambda = 0.9\,\mu m$, $f_{mod} = 1000\,Hz$ and $\triangle f = 1\,Hz$. The wavelength chosen corresponds usually to the peak responsivity of the device. The 10^{-14} factor gives a realistic order of magnitude of NEP values of practical devices. The ambient temperature and reverse bias voltage (TQM) should be quoted as well, as they have a bearing on leakage current which, in turn, influences the thermal noise strength.

   Detectivity, D There is an awkwardness about NEP which the parameter D removes. The awkwardness resides in the fact that the *better* the device for the detection of weak signals, the *smaller* the numerical value of NEP. This is unnatural to designers. The remedy is straightforward: taking the reciprocal of NEP gives a factor which grows with the device goodness. This is D.

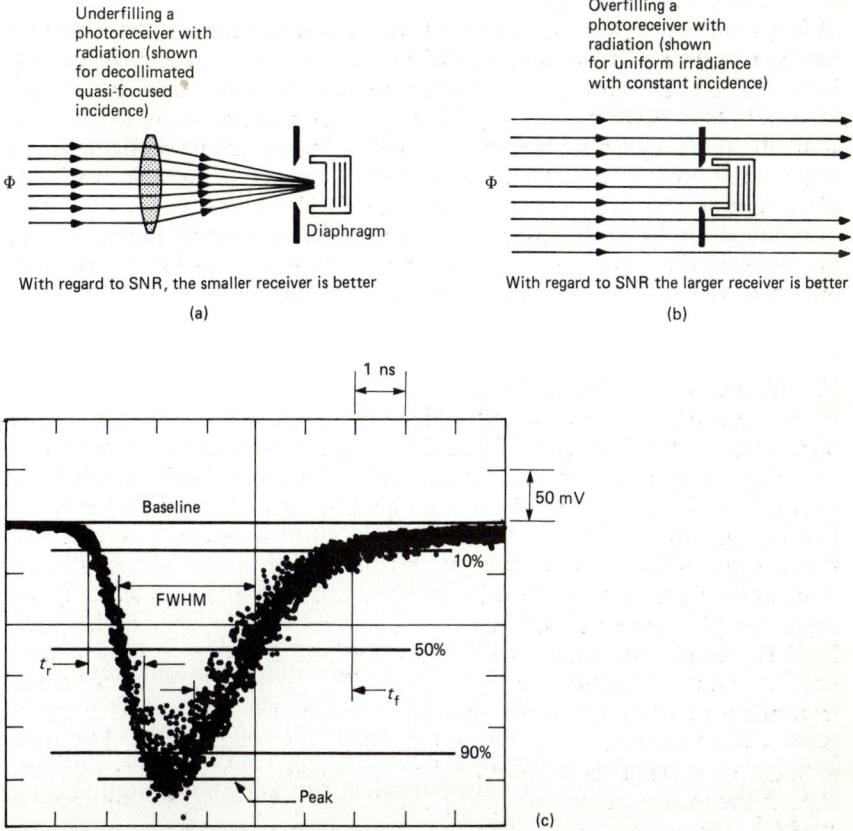

Figure 6.2 (a, b) Small and large area receivers. (c) Output current pulse showing rise time, t_r, and fall time, t_f, characteristics and full width at half maximum points

$$D = \frac{1}{\text{NEP}} \qquad (6.2)$$

The units are obviously $H^{1/2}W^{-1}$. D is a figure of merit of the kind engineers like.

Specific detectivity, D^* ('dee star')   If we compare two photoreceiving junctions produced by the same process from the same semiconductor material, one small, the other k times larger (respective areas A and kA), in a uniform overfilling radiation field (Figure 6.2), the larger device will give a reception purer by a factor $\sqrt{k}$, the ratio of improvement of the SNR factor. SNR has improved $\sqrt{k}$ times, because the signal is directly proportional to the collected flux, hence k times greater for the larger device, while the noise, based on current or voltage and not on power, is proportional to $\sqrt{k}$ only. NEP, too, is proportional to $\sqrt{k}$, making D proportional to $1/\sqrt{k}$. To remove this surface area dependence, D^* was introduced by Clark Jones in 1959 [126, p. 51]:

$$D^* = D\sqrt{A} \qquad (6.3)$$

D^* is yet another figure of merit, particularly useful for comparing devices of different shapes and sizes in situations of uniform, overfilling radiation fields. It is measured in $cm\,W\,Hz^{1/2}$.

(g)   Note on noise in real projects

It is my experience that the noise limits as low as those implied in the parameters NEP, D and D^* are seldom reached in 'on-bench' situations. Great care ought to be taken there to minimise amplifier noise and noises of other origin which often mask the unavoidable self-generated noise of the photodiode–load combination (the so-called 'shot' and 'Johnson' components). Thoughtful *optical* diaphragming and screening will minim*electrical* screening and grounding will pay off by reducing interferences and hum. The Johnson component of the self-generated noise will itself be reduced if the 'd.c.' component of irradiance (optical bias) is made small. Any further reduction can be obtained only by cooling the photodiode and, ultimately, the 'front' portion of the RX electronics.

(h)   Speed of response

Fast as they are, photoreceivers, like all physical devices, are not without inertia. The speed of response to optical stimulation can be expressed either directly, as a time delay – usually in microseconds, nanoseconds or even picoseconds – or indirectly, in terms of frequency response – usually in kilohertz or megahertz – of the limiting value of the modulation frequency. The rise time t_r and fall time t_f of the output current in response to a positive or a negative step function-like light signal are defined in the usual 'from baseline to 90%' and 'from peak to 10%' manner so familiar to electronics engineers. Sometimes, however, the FWHM (Full Width at Half Maximum) is given for pulsed incidence of known duration (Figure 6.2(c)).

The high frequency limit of a device f_0 is that frequency of a sinusoidally modulated optical incidence for which the output current drops 3 dB below that observed under very low (or zero) frequency conditions. Photodiode parameters t_r, t_f, FWHM and f_0 are usually given for the TQM(I) situation, i.e. a nearly short circuit condition (e.g. for $R_L = 50\ \Omega$ closing a 50 Ω coaxial cable) and a specified reverse bias, V_R. Finally, passing from the time domain to the frequency domain parameters can be easily accomplished by the use of the relation:

$$t_r = \frac{0.35}{f_0} \tag{6.4}$$

In cases in which a high f_0 is at a premium, the junction capacitance C_j of a photodiode (see Section 3.3.3) of a given category can be minimised in two ways, used separately or jointly: (a) by applying a higher reverse bias and (b) by selecting a smaller area device of the range while preserving its radiation intercepting capability through the use of immersion optics, which increases its apparent size [2, pp. 232–234].

(i) Spectral response $\mathscr{R}_\lambda$

This is a straightforward matter. Given usually in the form of a graph, the spectral response of a device gives us the variation of its responsivity $\mathscr{R}_\lambda$ in

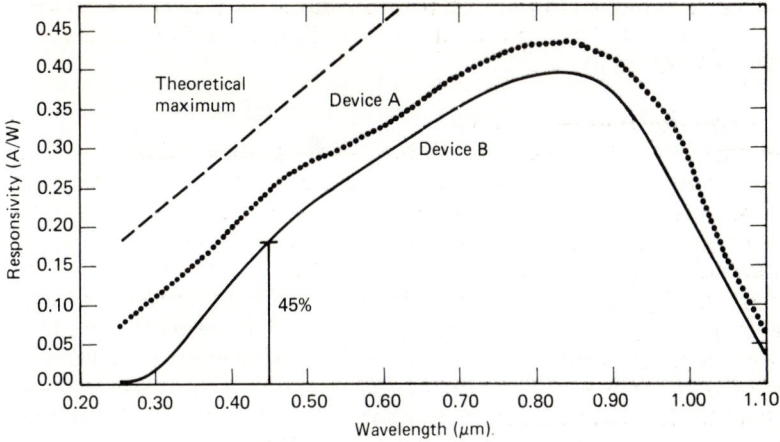

(a)

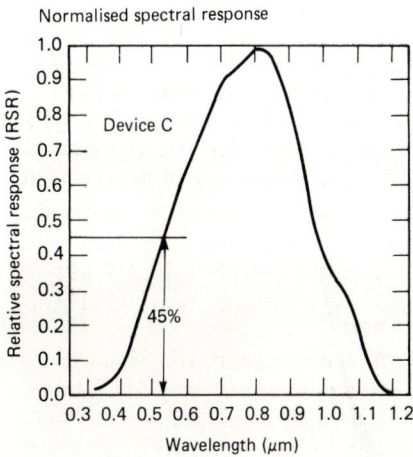

(b)

Figure 6.3 Absolute (a) and relative (b) spectral response of typical devices

$$\left(\frac{\mu A}{\mu W}\right)_{\lambda}$$

with the wavelength λ of the incident radiation. In some catalogues, instead of the *absolute* spectral response we find a graph of the *relative* spectral response referred to the peak response of the device, supported by a figure of $\mathcal{R}_{pk}$ (see Figure 6.3).

Some relative response curves, characteristic of devices of a certain kind, have been standardised and assigned 'S' number designations, such as S-1, S-11 or S-20. Germanium photodiodes, for example, have an S-14 and silicon an S-37 relative spectral response. The explanation of the physics of the shape assumed by the spectral response of junction photodiodes makes good reading. Unobtrusive texts on this subject can be found, for example, in Ref. 8, p.130 and Ref. 10, p. 4-2.

6.1.2   The non-amplifying junction photodiode as a circuit element

In most signal-handling applications the photodiode is worked in TQM (see Figure 6.4). In this mode its equivalent circuit can be represented as in Figure 6.5. R_L is very small (ohms or tens of ohms rather than kilo-ohms) in TQM(I) and rather large, perhaps as large as its limiting value:

$$(R_L)_{max} = \frac{V_R}{I_{max}} \qquad\qquad (6.5)$$

for TQM(V).

In the whole of the third quadrant the photodiode behaves like a current generator. The signal current I_S is proportional to the radiant flux. The leakage current I_L ('dark current') follows the temperature dependence law of all P–N junctions reverse current, which is to say that, for silicon, it doubles for every temperature rise of $\simeq 10$ °C. At room temperature, I_L of modern devices is very small indeed – picoamps or nanoamps rather than microamps. The noise current I_N hs two components (neglecting the usually filtered out extra-low frequency ($f < 25$ Hz) 'flicker noise'): the 'shot noise' (also called Schottky noise) and the 'thermal noise' (also called Johnson noise), both of which are temperature and bandwith dependent. The shot noise is a function of current and thus of both, I_S and I_L. Its temperature dependence is indirect: it stems from the growth of I_L with T. The Johnson noise depends on the value of both resistive elements of Figure 6.5, R_D and R_L. The matter is rather complex and as it deserves full

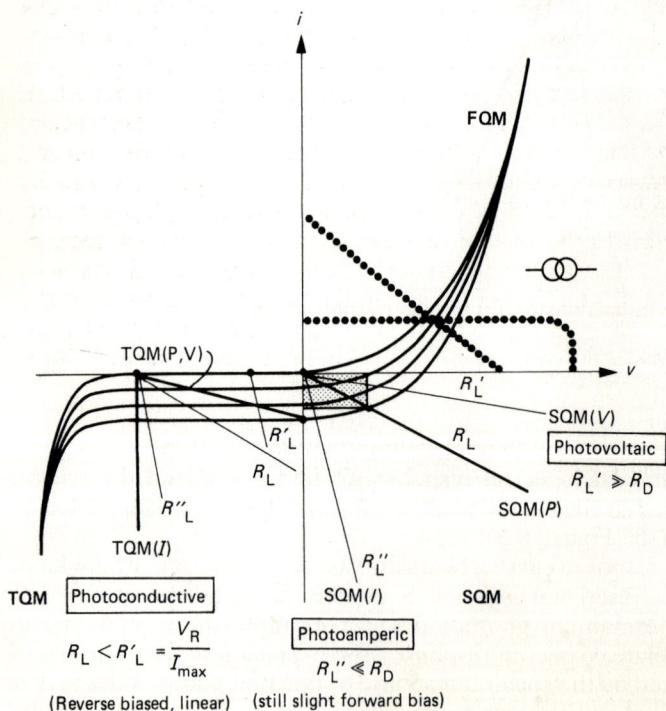

Figure 6.4 The three quadrants and six operating modes of an opto-electronic P–N junction

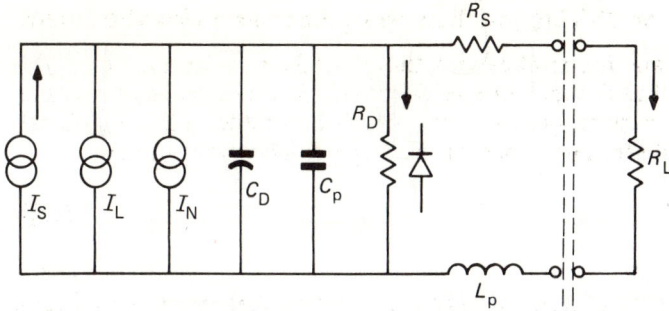

Figure 6.5 Equivalent circuit of photodiode in TQM: I_S, signal current; I_L, leakage (dark) current; I_N, noise current; C_D, depletion capacitance; C_P, package capacitance; R_D, dynamic diode resistance; R_S, series resistance; L_P, package inductance; R_L, external load resistance

attention in rare occasions only, its treatment has been relegated to Appendix 5. (For supplementary treatment, see Refs 276, 274 and, for more specific topics, Refs 27, 275, 192, 218 and 278.) C_d is the capacitance of the condenser-like depletion layer of the junction, explained in Section 3.3.3. Its value decreases with an increasing reverse bias V_R, reducing to approximately 70% of its value for every doubling of V_R. (This rule does not apply for $-1\ V < V_R < 0\ V$, a condition unlikely to occur in TQM operation.) This explains the interest of large V_R values in high-speed circuits. C_p is the fixed value of the package capacitance. The effective capacitance of the diode is $C_d + C_p$.

The horizontally drawn portions of the V/I characteristics in TQM (Figure 6.4(b)) do, in fact, have a very slight slope. R_D, the dynamic resistance of the diode, is represented by the reciprocal of this slope, $R_D = 1/\tan \alpha$. In modern devices, R_D has a very high value, tens or hundreds of MΩ rather than MΩ. The series resistance R_S is the combined resistance of the bulk semiconductor material, the contacts, the metal vaporisation and the spurious resistance of the package and its leads. The value of R_S usually lies in the vicinity of 1 Ω or less. L_S, the package inductance, has probably always a sub-microhenry value and becomes relevant in very high frequency work only. (It is worthwhile noting, nevertheless, that the value of ωL_1 for 1 μH is around 30 Ω at 5 MHz and that of a mere nanohenry around 3 Ω at 500 MHz.) In the majority of the SQM region the photodiode behaves like a voltage generator and its equivalent circuit is quite simple (Figure 6.6).

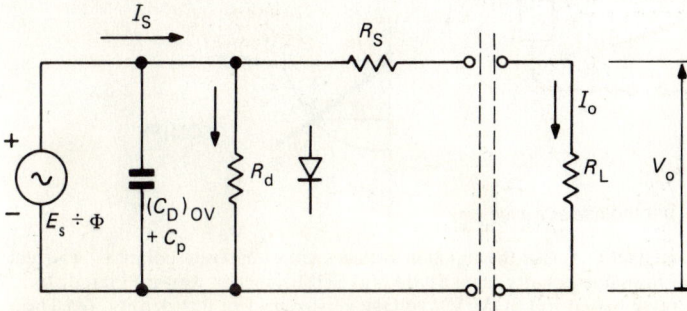

Figure 6.6 Equivalent circuit of photodiode in SQM

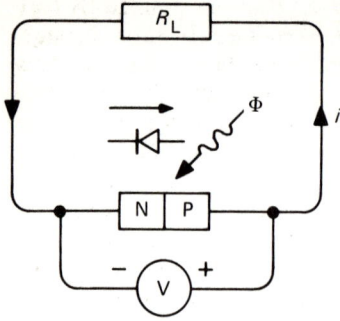

(a) SQM. Junction forward biased

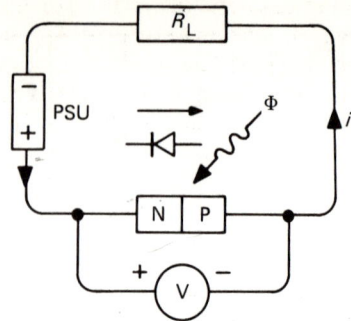

(b) TQM. Junction reverse biased

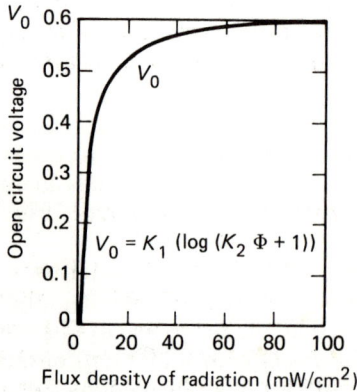

(c) SQM (V) Voltage vs incident flux density

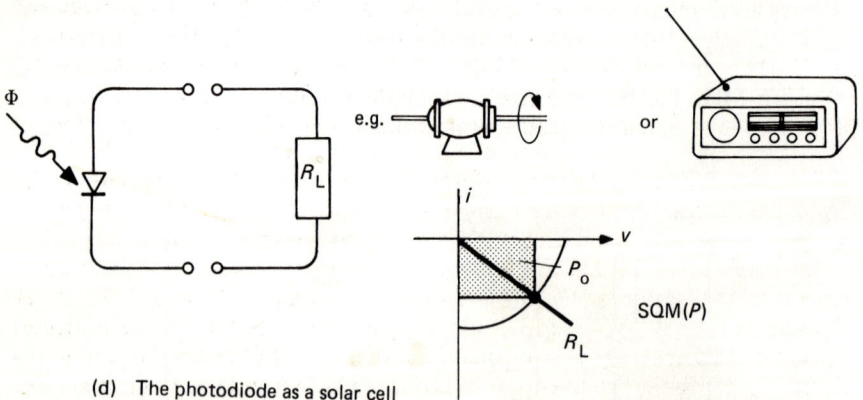

(d) The photodiode as a solar cell

Figure 6.7 SQM and TQM. Across-the-junction voltages are of opposite polarities. Current flow direction is of the same polarity, as is $dv/d\phi$. (a) SQM, junction forward biased.(b) TQM, junction reverse biased. (c) SQM(V), voltage versus incident flux density. (d) The photodiode as a solar cell

V_R being zero and R_D quite small, the noise reduces significantly (see Appendix 5). The influence of R_D, however, is being felt more strongly here than in TQM because of the diode action of the device. Indeed, now forward biased (Figure 6.7(a)) the diode 'robs' the photovoltaic generation of some of its current. In SQM(V), with $R_L = \infty$, the totality of the photoelectrically generated current is 'eaten up' by this diode, a fact reflected in the V versus I curve of the device (Figure 6.7(c)). Note the asymptotic drive of V_0 towards V_B, the barrier potential. Because of the diode effect, 'under-filling' (Figure 6.1) becomes highly inadvisable, as the non-irradiated parts 'rob' the photodiode without contributing to the photocurrent. If we were to progressively reduce R_L we would end up by having the photodiode working in SQM(I), for $R_L \simeq 0$. Thanks to the 'virtual earth' connection of an op.amp, this is not an impractical case. Figure 6.4(c) shows at a glance that the photodiode once again becomes a current generator, with $V_S \simeq 0$ and $I_S = K\phi$. This comes as no surprise, as the SQM(I) regime is a boundary case, almost equivalent to TQM(I) with $V_R = 0$. The adverb 'almost' is justified by the difference of the slope in the V/I characteristics in these two modes, when non-degenerate, to the left and to the right of the borderline case $V = 0$.

6.1.3 The photodiode and its load

Except when used for direct energy conversion, the photodiode has to be followed by an amplifier. The *direct energy conversion* case is the simplest.

The photodiode, here called a 'solar cell', is connected directly to its resistive load, R_L (Figure 6.7(d)). Most often, however, R_L will represent the d.c. resistance of an electrical energy storing buffer feeding an appliance, rather than that of the appliance itself. The working mode will be SQM(P), optimised for P_o. This optimisation process, i.e. the determination of the value of R_L giving the highest conversion efficiency of the arrangement, will be found in Ref. 48, section 4a, especially figure 7.

In communications and instrumentation alike, photodiodes handle *signals* rather than *power*. These signals are seldom strong enough to operate output loads directly and thus amplification is nearly always required. Moreover, a preamplifier located in the immediate vicinity of the sensing photodiode is nearly always used, in order to reduce load capacitance and pick-up noise. The preamplifying circuits fall into classes corresponding to the operating modes of the photodiode, namely, SQM(V), SQM(I), TQM(I) or TQM(P). We shall deal with the two fundamental and, at the same time, the most frequently encountered configurations: SQM(V) giving a logarithmic output and TQM(I) giving a linear output.

In *logarithmic operation* R_L is made as high as possible. With modern operational amplifiers (FET input stage) R_L values as high as 1×10^{11} Ω can be achieved, so that the ideal case of the SQM(V) open circuit operation (Figure 6.7(c)) is approached. Figure 6.8(a) shows the configuration. We recognise at once the standard non-inverting voltage amplifier with a very high input impedance and a gain $G \simeq 1 + (R_f/R)$. The special case of $R_f = 0$, $R = \infty$ (Figure 6.8(b)) is of some interest. This circuit is sometimes called the 'plus-oner' because its gain equals unity (more exactly, ~ 0.9999), giving the experimenter direct high-impedance access to

$$G = \frac{R + R_f}{R} = 1 + \frac{R_f}{R}$$

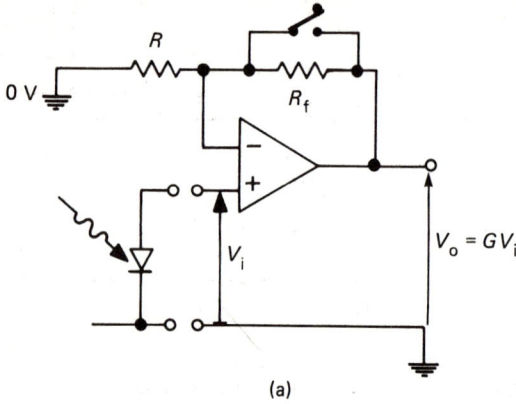

(a)

$$\text{For} \left\{ \begin{array}{l} R_f = 0 \\ R \longrightarrow \infty \end{array} \right\} G = 1$$

The circuit becomes a non-inverting 'plus oner'

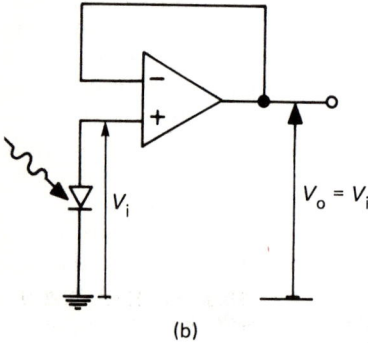

(b)

Figure 6.8 SQM(V) configuration giving $V_o \propto \log \phi$

the now open circuit photodiode. The output voltage of the more general circuit in Figure 6.8(a) is strictly proportional to the photodiode output which, in turn, is a logarithmic function of the signal ϕ (see Figure 6.7(c)),

$$V_o = GV_s = GK \log \phi \qquad (6.6)$$

provided the d.c. offset output voltage is balanced out.

Attractive as this configuration may appear, it has three serious weaknesses:

1. It is very slow. Indeed, if we assume an amplifier with $R_i = 1 \times 10^{11}$ Ω and the total input capacitance $C_i = C_d + C_{strays} + C_{opamp} = 5$ pF, the time constant $R_i C_i$ equals 0.5 s!
2. It is as temperature sensitive as the photodiode itself in this mode, which is considerably.
3. It is very noisy, despite its narrow bandwidth (flicker noise).

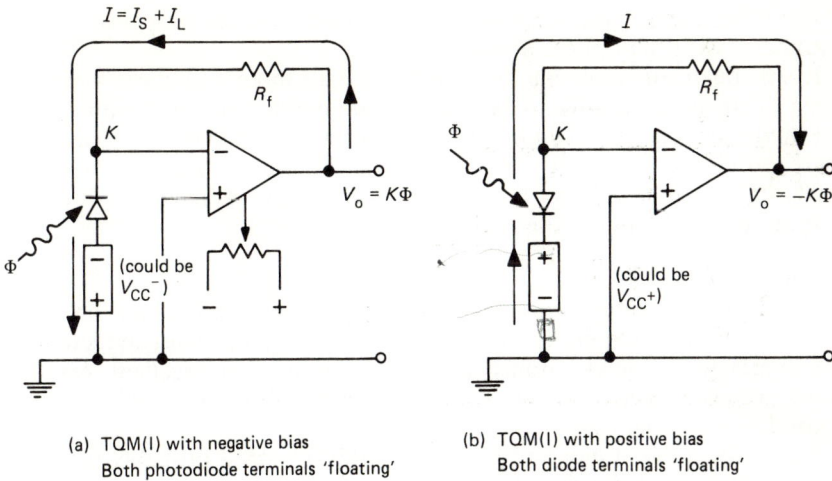

(a) TQM(I) with negative bias
   Both photodiode terminals 'floating'

(b) TQM(I) with positive bias
   Both diode terminals 'floating'

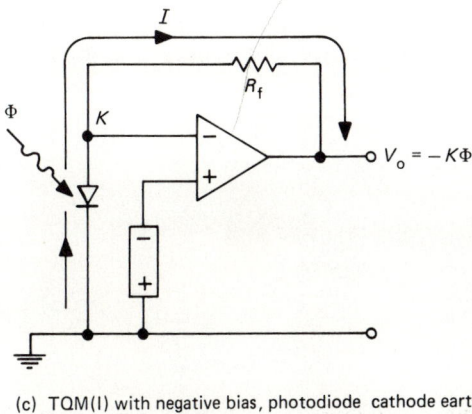

(c) TQM(I) with negative bias, photodiode cathode earthed

Figure 6.9 Three TQM(I) circuits

The second disadvantage weights so heavily against this circuit that it looses ground even in its prime application as an exposure meter or a photometer amplifier. (More stable operation of these instruments is being obtained either by the use of an external, high-quality diode replacing R_f in the feedback loop of a current amplifier, the photodiode operating in TQM(I) if not in SQM(I), or by having a linear TQM(I) amplifier, followed by a specialised log function chip.)

Superbly *linear operation* over several decades of amplitude can be obtained through the use of TQM(I) circuits (Figure 6.9). Configurations (a) and (b) differ only by the polarity with regard to earth of the reverse bias V_R. Configuration (c) is a variation of (b) offering the advantage of an earthed photodiode terminal, here the cathode. Devices manufactured with an anode-to-case connection can, of course, be accommodated in a similar way by reversing the polarities of both the photodiode and the source of V_R. The biasing differences between (a) and (b) result in a 180°

phase difference of output voltages, as shown, namely 'in-phase' or 'in antiphase' with the optical input.

It will be recognised that the point K of all three circuits is a virtual earth. This makes the R_L of the photodiode nearly zero, in keeping with the mode conventions of Figure 6.4. Such a small R_L assists with speed and noise problems. With good op.amps (e.g. FET input stage) practically 100% of the current flows through the feedback resistance, R_f, and the output voltage V_o obeys the relation:

$$V_o = IR_f \tag{6.7}$$

This operation corresponds to a current-to-voltage converter or to a *transconductance* amplifier, with I_i/V_o representing the transconductance. Commercially available transconductance chips give excellent results. Some are even integrated with the photodiode itself (being of silicon) and housed in the same case.

All in all, stable, quiet and fast, the TQM(I) is most probably the best preamplifier circuit for the great majority of photodiode applications.

6.1.4 Photodiode varieties

The biggest distinction amongst light-sensitive junction receivers is that between *generators* and *detectors*. The major differences are those of

Figure 6.10 Interesting application of larger cells – a 60-module system providing 445 W continuous power for a rural telecommunications microwave network in Kenya (Courtesy of Solapak Ltd)

emphasis on size, speed of response, leakage current, noise and quantum efficiency. Silicon and gallium arsenide are the basic semiconductors used for both groups.

Without dwelling on OE power generators (solar cells), recall that the emphasis here is on large sensitive areas (square centimetres rather than millimetres), high conversion efficiencies, ruggedness and longevity. Figure 6.10 shows a typical solar cell array. (For further reading, see Refs 14, 48, 125.)

Turning now to signal detectors we find that, here, the emphasis is on low noise, high speed, small leakage current and responsivity, rather than conversion efficiency. A small surface area, A, more often than not helps the first three characteristics (see Section 6.3.1 on underfilling and overfilling), hence A measures usually less than 1 mm^2.

(a)   PIN photodiodes

Small as these devices usually (though by no means always) are, they do not owe their name to their size but to their geometry: the photoelectrically active part of the wafer has a sandwich-like structure, with P (positive) I (intrinsic) and N (negative) silicon layers (Figure 6.11(b)).

The role of the *naturally* present *depletion layer* in a P–N junction has been explained in Section 3.3.3. Remember that the application of the reverse bias, V_R, increases its thickness. To increase the chances of electron–hole pair creation near or on a high field region, this depletion layer must be: (a) made thicker than its natural, V_R expanded state, and (b) brought nearer the irradiated, outer surface of the device. The artificial introduction into the wafer structure of a dopant-free I layer (I for intrinsic) near the now very thin, light-collecting P layer does just that (Figure 6.11(b)). The presence of the I layer is beneficial in more than one way: it increases the speed of the device by shortening the collection time of the photon created carriers and reducing the junction capacitance, C_d; it reduces leakage current and noise; and it improves device linearity. Because of their advantages, PIN photodiodes dominate today's photoreceiver field.

(b)   Surface barrier (Schottky) photodiodes

Replace the P$^+$ silicon layer of the conventional planar diffused P–N device of Figure 6.11(a) by a thin layer of gold, and you get Schottky photodiode (Figure 6.11(c)). The evaporation deposited gold layer is so thin ($\simeq 0.012$ μm) that much light goes through it. The proximity of the depletion layer (swollen by the application of V_R) to the plane of incidence results in high-energy (short-wave) photons being absorbed in it. The generated free carriers can thus be swiftly swept away by the high field. This gives the device a good blue and violet sensitivity on the one hand and a high speed of response on the other. Responsivity at larger wavelengths (red and near infrared), however, is reduced by the high surface reflectance of the gold layer in this part of the spectrum – unless, of course, the device is anti-reflection coated. Amplitude linearly is excellent. Schottky photodiodes are not suitable for high-temperature or high-irradiance operation.

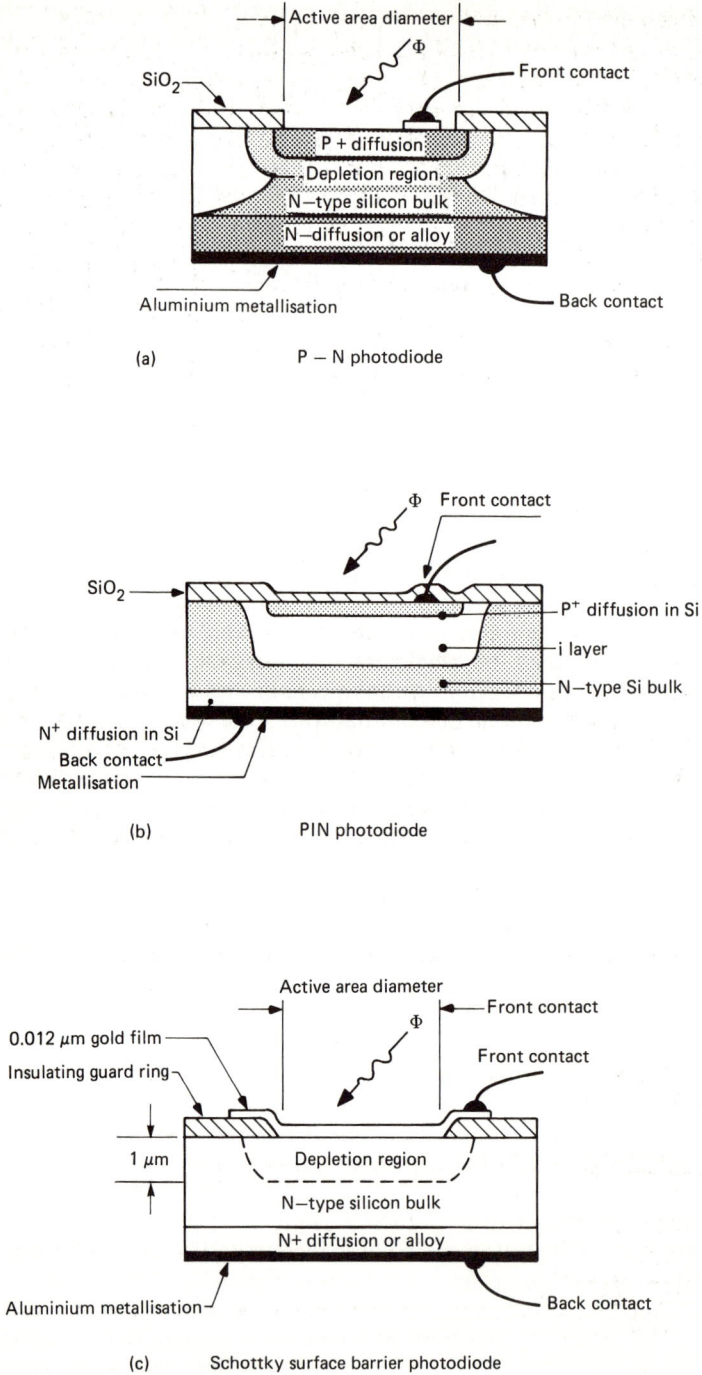

(a)   P − N photodiode

(b)   PIN photodiode

(c)   Schottky surface barrier photodiode

Figure 6.11 Photodiode wafers

(c) Filter fitted photodiodes

The expansion towards the blue of the Schottky diode's spectral response was accomplished by a structural change of its wafer architecture. Quite apart from various other wafer technology means, the simple technique of adding a colour filter to the wafer is sometimes employed for modifying the spectral response of a device. The technique can, of course, be only restrictive. One reason for wishing for such a restriction is the need for a cheap silicon device giving some degree of match to the CIE visibility curve (Section 4.1). The Texas Instrument TIL77 does just this by fitting a green filter inside the package of a planar diffused silicon photodiode. Another likely reason is to exclude visible light from an infrared opto-electronic linkage. Mullards have achieved this aim with their BPW50 PIN photo-diode by fitting it out with an internally worn infrared (low-pass) filter (Figure 6.12).

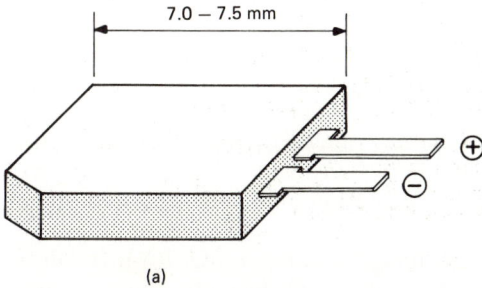

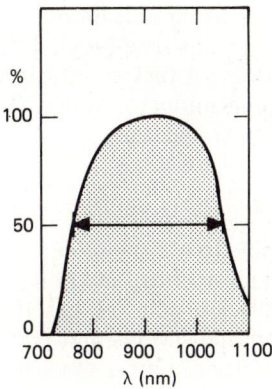

Figure 6.12 Filter fitted photodiode (BPW50 by Mullard) and its relative spectral response curve

(d) Long-wave photodiodes

For a number of years optical communications relied on infrared radiation of a wavelength in the 850–950 nm region. This is precisely the region in which silicon photoreceivers are peaking. The match was perfect until longer wavelength radiation was found to be much more attractive. Silicon then ceased to be the eminently suitable receiver semiconductor and the search began for materials with high spectral responsivities, first at 1300 nm and later at 1600 nm. The search may go on for some time to come but, at the time of writing, indium gallium arsenide is the strongest contender. As an example, RCA's InGaAs C 30979E photodiode [56] has a responsivity of 0.8 μA/μW at 1600 nm.

Modern optical communications could not do without high speed of response. Here, too, InGaAs is very good – the t_r and t_f of the above-mentioned device is less than 1 ns. The term 'long wave' will be used in the remainder of this book in connection with equipment working on a wavelength in excess of 1000 nm.

(e) 'High-voltage' photovoltaic chip

We know from Section 6.1.2 that the photoelectrically generated voltage in silicon cannot exceed 0.6 V (Figure 6.7(c)), and yet a tiny (0.9 mm × 1.0 mm) silicon device is on the market [128] – the D1 16V8 – producing a full 8 V when illuminated by no more than one LED! Here there are in fact several (16) tiny silicon elements in series on the same chip. (If you feel that calling 8 V 'high voltage' is an exaggeration, remember that everything is relative.)

The device delivers a few microamps, practically in SQM(V). It can be useful in controlling FET circuits, solid-state switches, etc.

Multi-element photodiodes with a common cathode (arrays), extremely useful in control engineering, do not come under this heading ($(V_o)_{max}$ being $\simeq$ 0.6 V) and will hence be treated in Section 12.2.2(b).

(f) 'Pigtailed' photodiodes

Figure 5.3 shows a pigtailed LED for optical communications. The pigtailed photodiode is the RX counterpart of this pre-bonded TX component. Here, also, the user is spared the difficult task of efficiently and reliably securing the optical fibre to the semiconductor wafer. The pigtailed device is sometimes built into a standard connector.

6.1.5   Two-way EO junctions: transreceivers

As early as 1969, I reported that it would be feasible to use the same GaAs P–N junction for both transmitting and receiving functions, and demonstrated its practicability with a two-way semi-duplex speech link based on such junctions [48]. The range of the supporting developmental equipment was a mere 100 m or so, in free space – certainly most modest by today's standards. Since then, however, workers in the field have scored greater feats of performance [285] proving that P–N junctions capable of operating in all three quadrants of Figure 6.4, namely in FQM, SQM and TQM regimes, were possible. The great attraction of using such transceivers (TRXs) for free space optical communications is that of requiring only *one* set of lenses at each terminal, instead of two. This reduces not only the size, weight and equipment cost but also alignment problems. Similarly, in FO communications with TRXs, one lightguide in lieu of two will do, without the use of beam splitters, optical mixers, etc. If the transceiver remains today something more of a laboratory curiosity rather than a commercial reality, and its applications lag a great deal behind those of single-purpose TXs and RXs, it is because the design and engineering criteria are quite different and often contradictory (e.g. the size of the working surface area) for the transmitting and receiving functions of the active element.

6.2 The amplifying light-receiving wafer

This category includes avalanche photodiodes, phototransistors, photo-darlingtons, photofets and photodiode + blind active combinations.

6.2.1 Avalanche photodiodes

In the depletion layer of a PIN photodiode, the strong electric field sweeps away the photon-generated free carriers. If the field is very strong and the path of a free electron sufficiently long, the continually accelerating particle can gather sufficient momentum to knock out, like a projectile, another electron from an atom on its way. Should this happen, we get two electrons for one, and, if conditions are favourable, the multiplication may continue, giving rise to a geometrical growth of the number of free carriers. Hence the device's name.

Figure 6.13 shows the layer structure of a popular type of avalanche photodiode (APD), the so-called Reach through Avalanche Photodiode (RAPD). The electric field profile to the right shows the two distinctly different regions: (a) the moderate field *conversion/drift* zone, and (b) the extra high field *multiplication* zone. (The device owes its name to the fact that under the strictly controlled bias conditions that are essential for

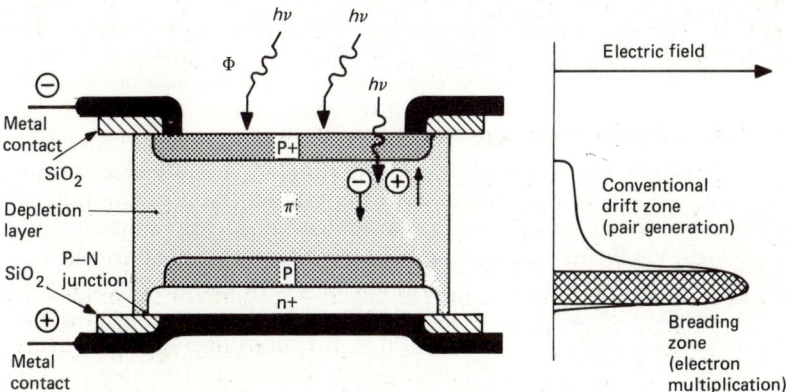

Figure 6.13 Architecture of a reach through avalanche photodiode

correct operation, zone (b) extends 'upwards' and *reaches through* to the low-field, much less depleted zone (a).) Created in zone (a), the 'parent' carriers drift into the breeding grounds of zone (b), where even more generations of electron–hole pairs are born. The field-driven electrons cross the P–N junction (downwards, in Figure 6.13), and either contribute directly to the photocurrent or knock out more electrons, from layer n^+. These knocked out electrons and many others that they then knock out create holes when leaving the valence bond. Under the influence of the field, these holes will cross the junction 'upwards' and drift towards p^+, thus contributing profusely to the photocurrent. This is how avalanche photodiodes work. Practical gain figures range from 50 to some 300. The principal interest of APDs lies not in the extra gain itself but in the fact that the noise increase is only a fraction of the signal gain, thus raising the SNR well above that of a non-amplifying photodiode.

What are the penalties in return for the advantages? Firstly, a reverse bias V_R of as much as 300–500 V is necessary to create a field intensity that will ensure avalanching. This is not always easily accommodated, especially in portable equipment. Secondly, the exact operating value of V_R is not only critical within 0.1 V, but also temperature dependent. These two constraints call for accurate stabilisational tracking of V_R, causing complication and expense. Thirdly, the gain is also temperature dependent so that gain compensation or stabilisation circuits are often required – yet another servitude. Despite these drawbacks, the APD is one extremely useful often irreplaceable photoreceiver.

6.2.2 Phototransistor and photodarlington

Almost every low-power transistor is a phototransistor. So much so, that early units which were glass encapsulated had to be painted black to exclude light. The working of the phototransistor is quite simple: the flow of photon-generated carriers takes the place of the conventional base current and, like it, is amplified by the transistor section of the device.

Figure 6.14(b) shows one of the existing phototransistor architectures. The similarity to that of a common planar transistor in Figure 6.14(a) is striking.

Figure 6.15 shows the transition from a photodiode + transistor circuit to a single two-terminal phototransistor. The collector–base junction of Figure 6.15(c), which is reverse biased (see Figure 6.15(d)), acts as a TQM-operated photodiode. The base–emitter junction is forward biased (FQM). The transistor conducts. The internal photocurrent i_{ph} (equivalent to base current) is amplified β times in the usual, transistor-like way, i.e. the reduction of the barrier potential of the emitter–base junction allows more free carriers to flow between emitter and collector (electrons 'up', holes 'down'), via the extremely thin base. As there is no need for the base lead to be brought out (Figures 6.14(a) and 6.15(c)), the phototransistor is usually a two-terminal device. (Occasionally, though, a designer may want to access the base electrically. This could be for choosing a specific working point, for feedback or for switching/clocking purposes. To cater for such applications, some manufacturers produce three-terminal devices.) Transistor gains running into hundreds, the phototransistor has a very high

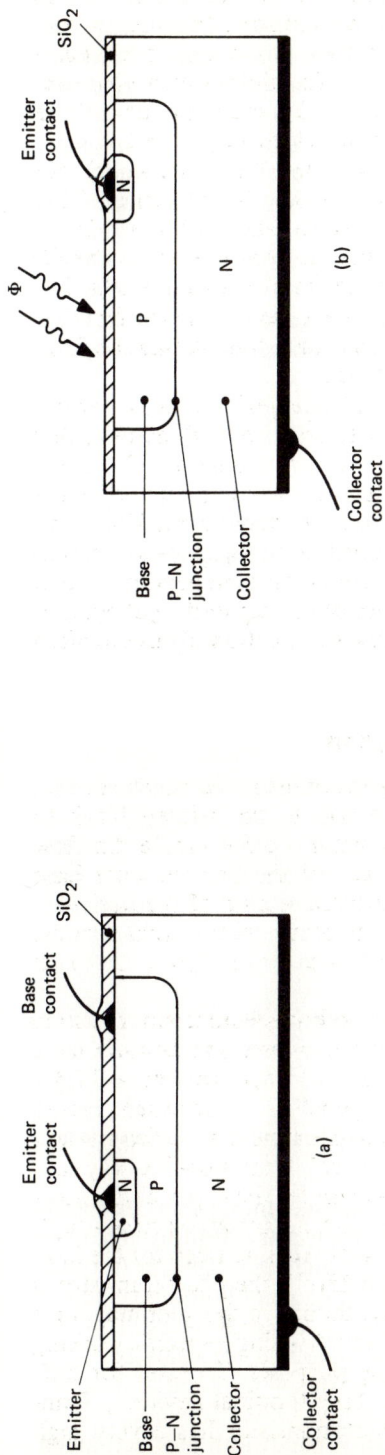

Figure 6.14 (a) Planar transistor and (b) phototransistor cross-sections. The resemblance is striking. Note the absence of base connection in (b)

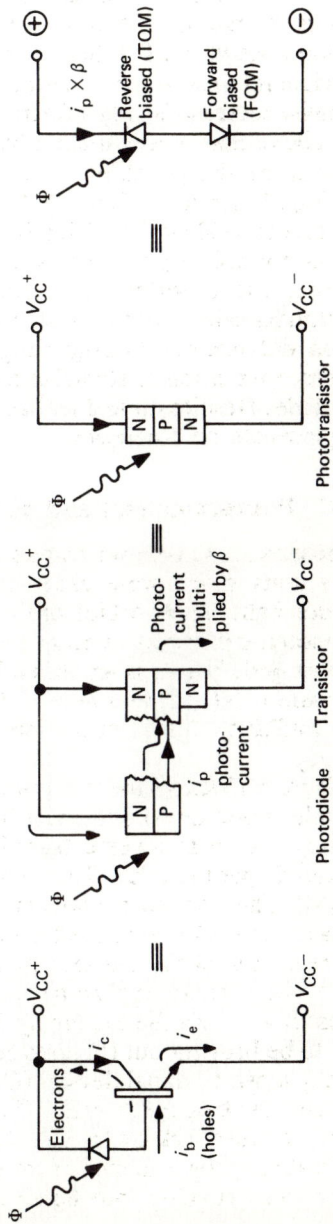

Figure 6.15 Transition from a photodiode and transistor combination to a phototransistor

responsivity. This attraction, together with that of cheapness – a result of the colossal world capacity for transistor production – has made the phototransistor an extremely popular device. There are snags, however: phototransistors suffer from poor linearity, are temperature sensitive and, above all, are intrinsically slow.

The photodarlington is, as the name suggests, a Darlington circuit made photosensitive by means of exposing to light a part of its structure. The photodarlington is, in fact, a small integrated circuit. (The Darlington connection consists of two cascaded emitter followers. For further reading, see Ref. 53, p. 274.) The darlington gain being enormous (up to 30 000), the photodarlington responsivity is enormous too. Leakage current, temperature sensitivity, saturation voltage and frequency response are its problems. For these reasons, the photodarlington is rather less popular than the phototransistor.

6.2.3   Photoelectric FET (photofet)

The photofet is to the FET (Field Effect Transistor) what the phototransistor is to the transistor. Light is allowed to fall into the gate–channel junction of the device. The photoelectrically generated gate current (TQM) increases the normally present thermally generated gate leakage current, and influences the gate–source voltage. The rest of the story hardly needs telling (see Figure 6.16). Readers unfamiliar with FETs will find an excellent description of these devices in Ref. 54 (chapter 14).

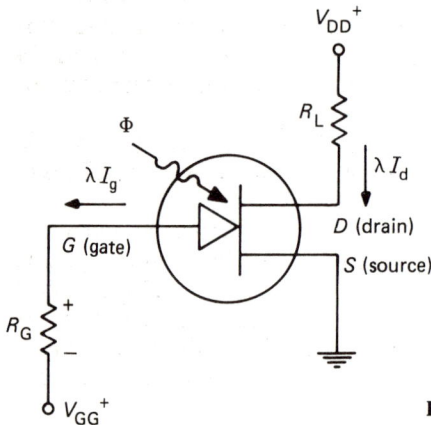

Figure 6.16 The photofet in a typical circuit

The most salient feature of the photofet, unique among semiconductor photoreceivers, is that its *luminous sensitivity* is electrically adjustable over enormous ranges and without loss of linearity. One manufacturer (Crystalonics Inc, whose photofets bear the proprietary name of Fotofets) claims a range of 10^6 : 1. This sensitivity control comes about as a result of the voltage dependence of the channel resistance. Double the value of R_G in Figure 6.16 and you have halved the amount of flux Φ required to produce the same V_{GS}. You have doubled the sensitivity. Other advantages over ordinary phototransistors include greater gain $\times$ bandwidth product, much greater sensitivity at low speeds, excellent switching properties and, in many situations, lower noise. The need of a bias voltage V_{GG} and, perhaps, a certain electrical fragility, seem to be the device's only disadvantages. Having so much in its favour and so little against it, it is difficult to understand why the photofet does not enjoy a greater popularity.

6.2.4   LASCR, optoschmitt and other integrated or hybrid devices

Many more than the above-mentioned four single-wafer devices exist in which the photoelectric effect of a P–N junctions is amplified. Some are photoelectric versions of their 'blind' (standard) counterparts; some others are standard integrated circuits enriched by a front-end photodiode. Worth mentioning in the first category is the LASCR (Light-Activated Silicon-Controlled Rectifier). Like the SCR, the LASCR is a P–N–P–N latching switch: once switched on by a light beam, it stays on, no matter what the light variation, until the *supply* voltage falls below the 'holding' level. With their periodic zero crossings, the a.c. mains are an ideal power source for LASCR controlled equipment. For such a supply an N–P–N–P version is also available, for complementary half-cycles operation. In both versions a gate connection is usually brought out, for greater design flexibility.

In relation to their size, LASCRs can handle enormous powers. For example, the L9B type (The General Electric Co, USA), packaged in a TO-5 style case, can control 1.6 A RMS, under a peak repetitive reverse voltage of 200 V. LASCRs for power engineering, handling kilowatts and tens of kilowatts also exist.

Like an ordinary Schmitt, the optoschmitt is a threshold device with hysteresis: it comes on for a certain irradiance and holds until this irradiance has fallen below a certain, preset level, when it switches off. In this device, voltages, currents and output powers are much lower than for the LASCR, usually in keeping with TTL techniques.

Containing at least two transistors and a few passive components, in addition to the P–N photodiode, the optoschmitt, unlike the LASCR, is in fact a small integrated circuit.

This brings us to photosensitive ICs in general. Forming a photodetector on a chip of silicon intended for an op.amp presents little difficulty. The advantages of such an integration are great and many, especially in optical communications. It is not surprising therefore that several manufacturers should have brought out new varieties of photosensitive ICs and that many others keep doing so. This chapter can do no more than mention them.

6.3   The photoreceiver – core of the RX antenna

Like its counterpart the TX, the RX is made up of three components: the beam-shaping optics, the OE converter and the electronic circuit. The sensing head – the RX antenna – contains the first two and at least a part of the third one – the circuit. Within the RX antenna it is the photoreceiver which produces the OE conversion and extracts from the light beam the intelligence imprinted to it by the TX. Clearly then, the photoreceiver is the core of the RX antenna.

6.3.1   Energy management

Efficient energy management is all-important. This involves as much the careful preservation of *signal* carrying energy as the exclusion of *unwanted* radiation. The manufacturer-supplied angular response curves or polar diagrams, much the converse of the radiation diagrams of LEDs shown in Chapter 5, ought to be checked against the numerical aperture of the RX optics and against the divergence of the incident beam $\alpha/2$ to make sure that: a very substantial proportion – or, better still, the totality – of the signal flux ϕ_E, reaching the entrance pupil is 'seen' by the photoreceiver (for further reading see Ref. 2, pp. 124–131), and that very little other than ϕ_E does so. Even non-varying spurious light (often, though quite incorrectly, called 'd.c. light') should be excluded as its ingress generates additional *shot* noise in the system and thus reduces its signal-to-noise ratio. (See Appendix 5 on noise in detector photodiodes.) Optical stops and baffles can help to restrict the angle of admittance and to minimise glare (intra-antenna reflections) (Figure 6.17).

Furthermore, the size of the imaged TX source must be checked to be filling the photosensitive area of the RX without underfilling it severely

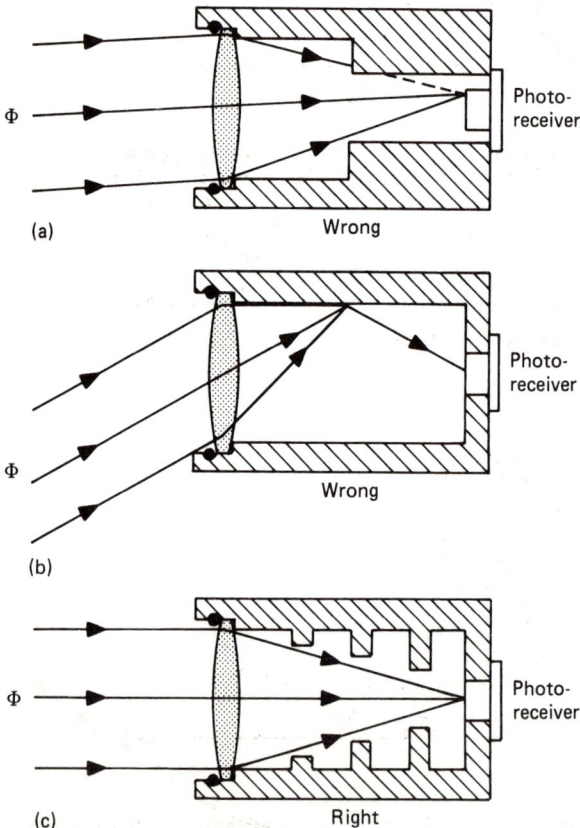

Figure 6.17 Action of an antiglare baffle. (a) Part of the useful radiation is obstructed; (b) part of the spurious radiation reaches the photoreceiver; (c) baffles are correctly located

(Figure 6.1). (Allow for blurr. When calculating skill is in doubt, make measurements, by using progressively reducing masking stops in the focal plane.) The mechanical contents of data sheets ought to be looked at with regard to secure receiver fixing and provision of adequate alignment facilities. While much of what was said in Section 5.6 applies here as well, with obvious transpositions, the *thermal* considerations do not. As power levels arc usually low, cooling is normally unnecessary. Only very exceptionally is the received signal so weak (e.g. in astrophysics) that forced receiver cooling is used to improve the SNR.

6.3.2   Spectral flux management

The flux management mentioned in the previous section was *spatial*. *Spectral* energy management is important too. This consists of:

1.   Ensuring that the relative spectral response (RSR) of the photoreceiver is such that the TX peak wavelength lies within its 0.8–1.0–0.8 ordinates (Figure 6.18).

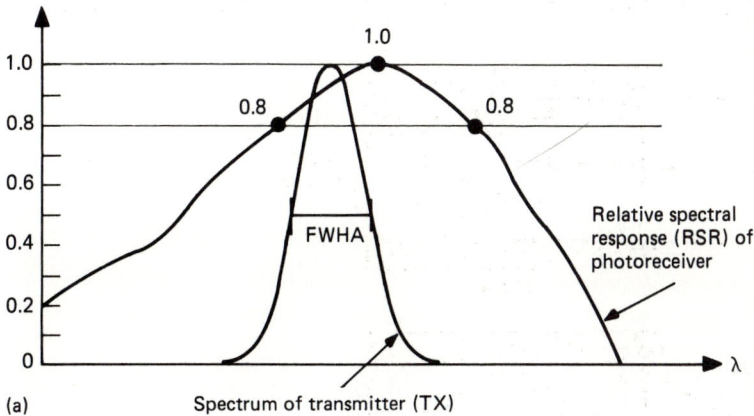

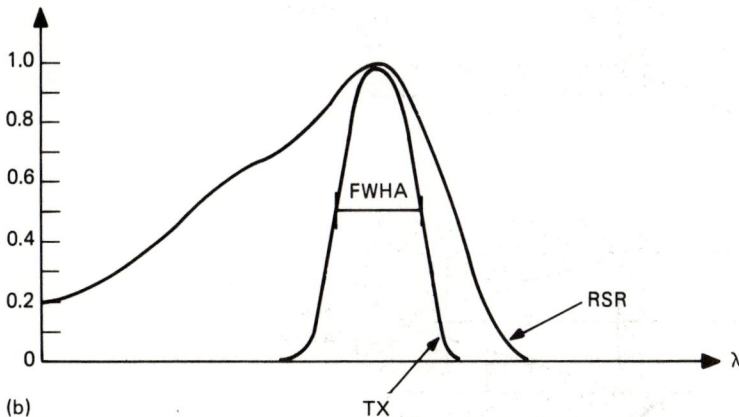

Figure 6.18 Spectral energy management. (a) The peak wavelength of the TX lies within the 0.8–1.0–0.8 ordinates of the RSE. (b) The FWHA, while nearer the long-wave limit of the RSR than in (a), remains comfortably within its limits

2. Ensuring that the FWHM lies comfortably within the RSR (Figure 6.18).
3. Minimising the intrusion of radiation from outside the FWHA region (background).

This will be achieved by the use of optical filters. Wratten gelatine filters (e.g. Kodak 87 or 87A for GaAs links) are thin, easily cut to size and inexpensive. They should not be used, however, in truly professional equipment because they are prone to damage by heat, humidity and plain ageing (wrinkles, peeling, etc.). Colour glass filters (e.g. Schott RG.850 pass long for GaAs links) are good. Extra narrow pass-bands (e.g. 13 nm at 940 nm, Ealing 354696) are characteristic of interference filters. The drawback here are high attenuation at the passing wavelength (e.g. 45% for the above-mentioned type), thickness (6 m/m here), little choice of size and shape, wavelength sensitivity to angle of incidence, and high cost.

The spectral match of GaAs-based TXs with silicon RXs presents no problems. Long-wave TXs match well with InGaAs receivers [56]. In more complex cases, e.g. when conditions 1 and 2 cannot be easily satisfied, the TX/RX spectral matching factor may have to be calculated [48] and the value obtained assessed in the context of the project as a whole. With regard to immunity to ambient light, the engineer has a few more weapons in his armoury, in addition to those already mentioned. Peruse more specialised work (e.g. Refs 110 (p. 20) and 127).

Chapter 7

The captive ray: fibre optics communications

7.1 Insulators compete successfully with conductors

Before long electricity will stop carrying our trunk calls. In fact, intercity telephone connections will be made of an *insulating* material! Hair-thin glass threads will carry the voice data or, to be more precise, will guide the light waves upon which these data are imprinted. Zigzaging or meandering on its long journey, trapped within glass fibre, the *captive ray* will be working for us.

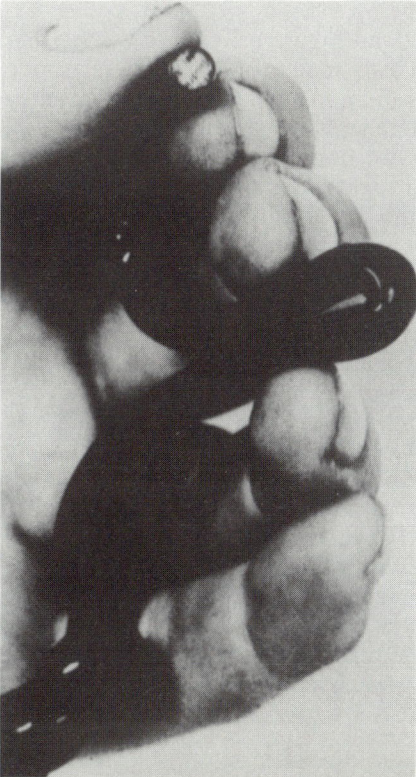

Figure 7.1 The knotted image-carrying FO cable (Courtesy of Circon ACMI)

The replacement of copper by glass is already under way and proceeding fast. The World's Telephone Administrations are not the only people putting the captive ray to good use. Quite simple bundles of fibre are capable of illuminating *inaccessible* places (e.g. inside the stomach), while slightly more elaborate ones can carry images through most tortuous paths belying the 'straight line' theory we learned at school. Chapter 12 will show how both features help the doctor, the surgeon and the drain inspector, to mention but a few (Figure 7.1).

Fibre optics can convert circles into squares and straight lines into circles, reduce or enlarge images, flatten or curve optical fields, etc. Thus, not all that is fibre optics is optical communications. (We shall see in Chapter 8 that, conversely, not all that is optical communications is fibre optics.)

7.2   The fundamentals: fibre optic physics

What is it that makes the ray *captive*, and what makes the captive survive journeys as long as those of intercity trains?

7.2.1   The captive ray

Return for a moment to Section 2.1.2 and Figure 2.3. There we see that upon hitting an interface separating two media of differing refractive indices, wavefronts change their orientation, in a hinge-like way. This they do to accommodate the velocity differences in these media. Figure 7.2 grew from Figure 2.3. Looking at it, we notice what happens when we progressively increase (from left to right) the angle β the incident wavefronts make with the interface: the wavefronts in the *faster* medium lean progressively less and less to accommodate the wavelength λ_0 characteristic of that medium, until they become perpendicular to the interface. If β is increased beyond this critical value, β_c, there is no more propagation in medium F – the wave just doesn't penetrate it. What happens to the electromagnetic energy light carrier? The waves are *totally* reflected back into the medium whence they came (Figure 7.2(e)). In ray parlance, we say that at the critical angle of incidence, α_c (notice that $\alpha = \beta$), the refracted ray grazes the interface. For angles greater than α_c, we witness *total internal reflection* (TIR). In optical fibres it is TIR that makes the ray captive.

Figure 7.2 explains TIR intuitively. For those who prefer a more formal argument, a glance at Equation 2.1 will show that, for $n_1 > n_2$, α_2 must be greater than α_1, and as α_1 grows it reaches a value α_c for which $\alpha_2 = 90°$ and beyond which $\sin \alpha_2$ can grow no more. This confirms that for α_c – the critical angle of incidence – the refracted ray is a grazing one and that for $\alpha_1 > \alpha_2$ we have TIR. The value of α_c is obviously:

$$\alpha_c = \arcsin \frac{n_2}{n_1}$$

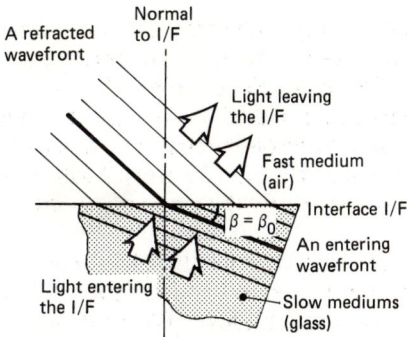

STAGE ZERO. This is the lower part of Fig. 2.3 rotated anti-clockwise to make the exit interface horizontal. Illumination from bottom left

(a)

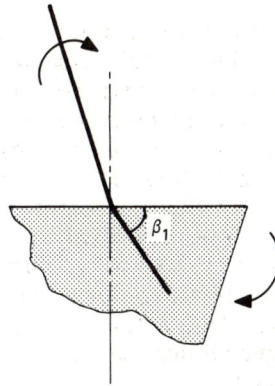

STAGE ONE Angle β is increased to give $\beta_0 < \beta_1 < \beta_c$

(b)

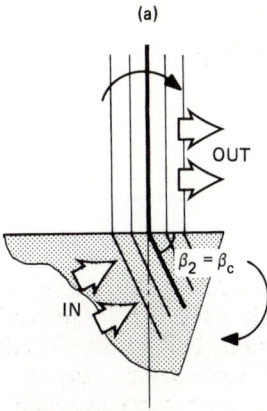

STAGE TWO Angle β is further increased to β_c value. Outward propagation in air stops

(c)

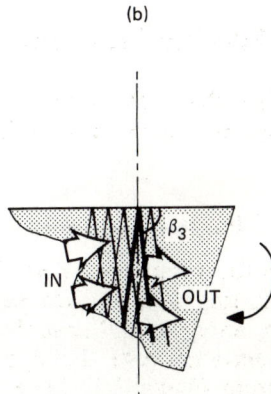

STAGE THREE Angle β_3 increased beyond β_c ($\beta_3 > \beta_c$). Total internal reflection takes place

(d)

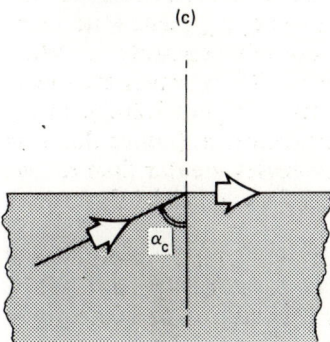

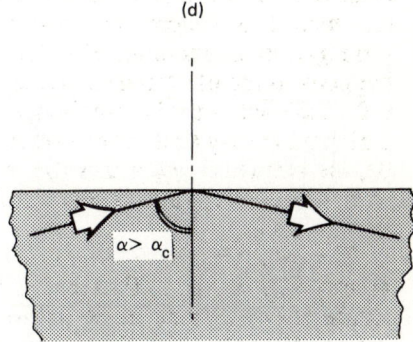

(e)

Figure 7.2 (a–d) Wavefront treatment of TIR. Sketches from left to right show what happens to refracted wavefronts when the angle β the incident wavefronts make with the interface is progressively increased. (For clarity, reflected wavefronts are omitted in stages 0–2.) (e) Ray treatment. The waves are totally reflected back into the medium from which they came

or, with the notation of Figure 7.1:

$$\alpha_c = \arcsin \frac{n_F}{n_s} \tag{7.1}$$

Should we now provide a second interface, in the proximity of and parallel to the first one, we would get a captive ray situation (Figure 7.3). How does light enter the medium S in the first instance? In the case of an optical fibre, we always have a butt-end coupling. Be the source a pigtailed laser or a lensed TX head, the radiation enters the fibre – and leaves it – in the butt-end fashion (Figure 7.3).

For a meridional incident ray, the explanation of Figure 7.3 holds, as the corresponding imaginary meridional slice of the fibre is optically equivalent to the flat slab or ribbon assumed there (see Figure 7.2(c)). Let us

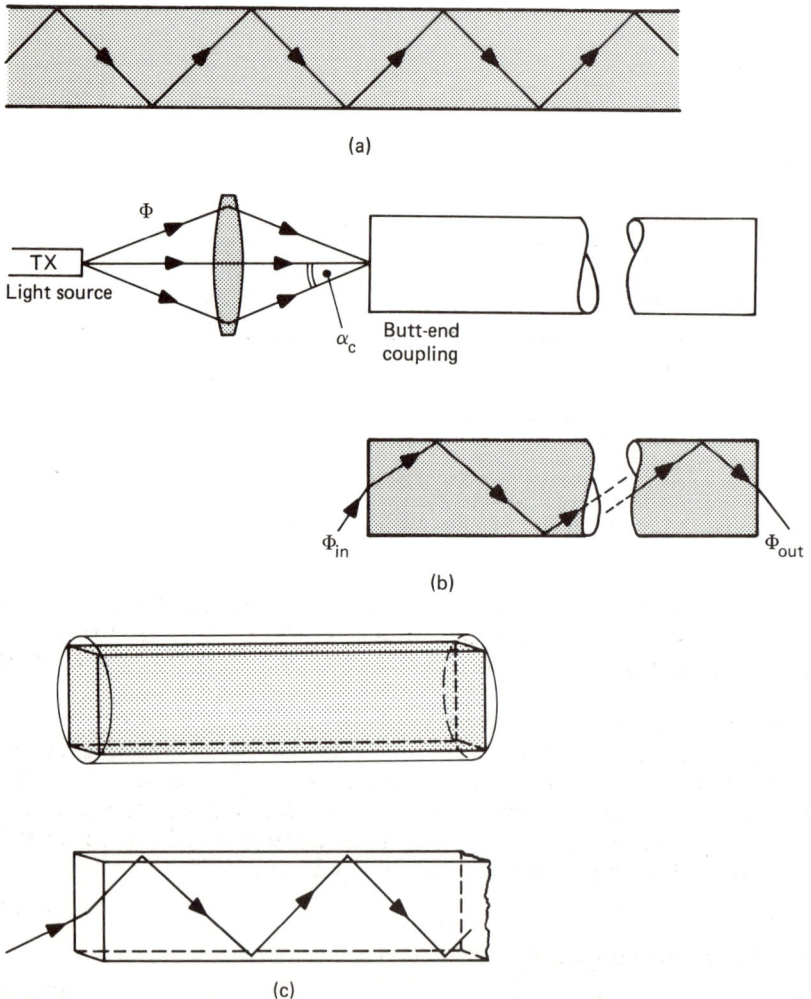

(a)

(b)

(c)

Figure 7.3 The captive ray in a fibre

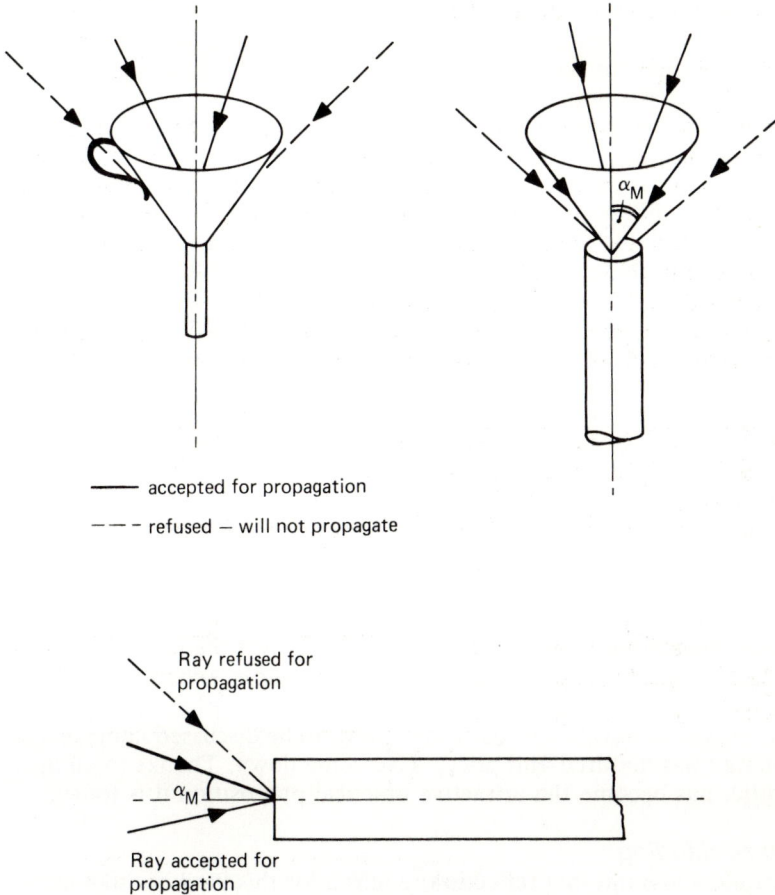

─── accepted for propagation

─ ─ ─ refused — will not propagate

Ray refused for propagation

α_M

Ray accepted for propagation

Figure 7.4 The funnel concept for maximum angle of acceptance

assume that the butt-end coupling is an air–glass one and that the fibre is made of homogeneous non-cladded glass. It will be easily seen from the previous drawings that, for meridional rays, the fibre has a *funnel of acceptance* (Figure 7.4(a)).

For rays entering the fibre at more oblique angles (from outside the funnel) the *internal* glass–air angle of incidence, α, is smaller than the critical angle, α_c, and there is no TIR: these are escaped would-be captives. The half summit angle of the funnel (or cone) of acceptance is called the *maximum angle of acceptance*, α_M, and is usually characterised by its sine, called the *numerical aperture* (NA) of the fibre:

$$NA = \sin \alpha_M \tag{7.2a}$$

It can be calculated (see Appendix 6) that

$$NA = \sqrt{(n_s^2 - 1.0)} \tag{7.2b}$$

with n_s being the refractive index of the slow medium (Figure 7.4(b)).

In the case of a non-meridional (skewed) ray, the situation is a little more complex, the path becoming a kind of broken line helix. The numerical aperture for skew rays is slightly larger than NA. Cladded fibres are dealt with in Section 7.2.2(a).

7.2.2   Attentuation

How does the captive ray survive mile-long journeys? It has many reasons not to and, until the enormous technological progress of the 1970s, it didn't. At every single reflection there are, inevitably, some losses and the zigzaging captive ray experiences several thousand of them in a single metre of fibre. A very simple calculation shows that at 7000 reflections per metre, a figure typical of a fibre $50\,\mu$m thick, a 0.1% loss per reflection would leave no more than $1\,\mu$W of an original input power of 1 mW after just one such 1 m journey (an attenuation of 30 dB/m). The situation improves rapidly, though, as reflection losses reduce; at 0.01% loss per reflection the attenuation falls to 3 dB/m (with 50% of signal preserved), and at 0.001% to 0.3 dB/m (93% remaining). The second threat to survival is the between-reflection absorption. The causes are scatter (by glass non-homogeneity and impurities) and straightforward residual glass opacity. The engineer's and physicist's answer to reflection losses is *fibre cladding*, and their rejoinder to attentuation is high *glass purity*. Together, physicists, chemists and engineers have produced over the last few years a spectacular progress in glass technology. For example, overall signal loss figures have been slashed to a mere 0.0006 B/m. Other transmission characteristics too have been vastly improved (to be discussed later), cable manufacture has matured and prices have come down. Thanks to all this, fibre optics has become the attractive practical proposition it is today.

(a)   Fibre cladding

The extra low loss internal reflection required for this kind of attenuation will only take place if the outer surface of the fibre is kept scratch free and impeccably clean. Any foreign matter in contact, e.g. moisture, oil, fingerprints, will frustrate them, causing losses. So, the claddings fulfil a *protective* function.

The second, quite obvious, role the cladding plays, is that of *optical insulation*, important in image transmission by fibre bundles; without it, rays would penetrate from one fibre into the neighbouring one (crosstalk) causing a loss of definition. Perhaps the most important in long-range communications is the capability cladding has of *controlling the NA* of a fibre. Equation 7.2 holds for an unclad fibre in air, a special case of the general situation in which three media – none of them necessarily air – are involved, obeying the general relation A6.4 derived in Appendix 6:

$$NA = \frac{1}{N_o}\ \sqrt{(n_{core}^2 - n_{clad}^2)} \qquad (A6.4)$$

In practice, the cladding of a glass core is most often of glass too, with just a slightly lower – by a mere few percent – refractive index. Equation A6.4 becomes:

$$NA = \text{const.}\ \sqrt{\Delta n} \qquad (A6.6)$$

which shows that the small difference $\triangle n = n_{core} - n_{clad}$ has a far stronger controlling action upon NA than n_{core} of an unclad fibre (Equation A6.6).

The reasons for which the designer may wish to control NA will be explained in Section 7.2.3.

(b) Glass purity

Shortly after optical fibres were first suggested as a propagation medium for long-range lightwave communications, critical opinions were raised stating that unless attenuation figures dropped below 20 dB/km the idea had no hope. Within a few years, optical cables with some 5 dB/km were developed and at the time of writing the 0.6 dB/km figure is a reality (at the preferred wavelength). The glasses developed for this purpose have such a level of purity and homogeneity that the depths of oceans could be seen through them, were they to replace water! Some experts believe that there is a 'fundamental' theoretical limit to glass fibre transparency, imposing an absolute limit to attenuation, namely 0.18 dB/km. However, laboratory samples exhibiting 0.2 dB/km exist already.

7.2.3   Modal dispersion

The captive ray of Figure 7.3 is not alone. At the TX head, an LED and lens combination radiates in a fan-out; the light rays enter the butt interface at a variety of angles of incidence. Ray 1 (Figure 7.5) enters the fibre at an angle α_1 and propagates within the core with an obliquity (β_1). Ray 2, having entered the core at an angle α_2, has a lower obliquity (β_2). Ray 2 proceeds towards the RX port of the fibre more slowly than Ray 1. The various possible paths in a fibre are called *modes*. Because light is a form of electromagnetic, wave-like energy, the fibre – a waveguide of sorts – will not sustain the propagation of all obliquities. The number of

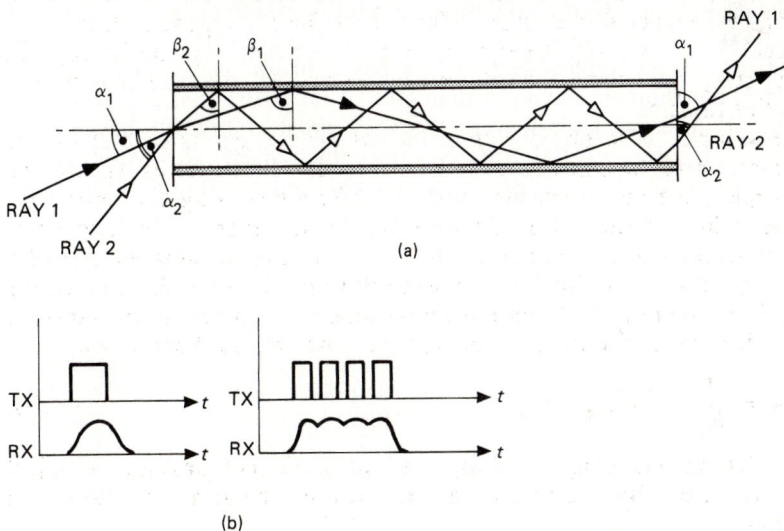

Figure 7.5 The captive ray is not alone: (a) modal dispersion and (b) its effect

allowable modes is very great, however – often several hundreds in this kind of fibre. Hence its name: *multimode*. If a nice, sharp-edged (top hat) pulse of light is launched into the fibre by the TX head, the *modal dispersion* will cause a sluggishly sprawled out pulse to emerge at the RX end (Figure 7.5(b)). Indeed, some photons will arrive there a little later than others. The high-mode travellers are slower, the low-mode ones faster. The statistical distribution of path lengths, together with attenuation differences (high-mode light travels longer paths and undergoes more numerous reflections), result in the type of amplitude distribution at the RX port shown.

It follows from what has been said that the transmission of high pulse rates requires small modal dispersion – otherwise, consecutive pulses become fused together (Figure 7.5(b)) and the information content is lost. Hence the need for a limitation of the numerical aperture NA in high pulse rate links, with the coupling efficiency being traded off for bandwidth. On the other hand, when pulse edges are gently sloped and repetition rates are low, large NAs will be preferred as they will give more efficient but-end couplings. All this shows the usefulness of an NA controlling agent. Equations A6.4–A6.6 shows that fibre cladding can be such an agent.

7.3   Fibre optic hardware. Phase velocity

The multimode fibre described in Section 7.2.2 is one of the three basic types used today. The remaining two are *monomode* fibre and *graded index* (GRIN) fibre. Figure 7.6 shows schematically all three types, together with their refractive index profiles. From it, it becomes obvious why the multimode and the monomode fibres (a) and (b) should be called *stepped index* types. In fact, the main structural difference between (a) and (b) is their core thickness. While the core diameter of fibre (a) lies in the 50–600 µm range (typical human hair has a 50 µm diameter), fibre (b) core is much thinner, having a diameter of no more than 2–8 µm. The functional differences, however, are very great. Fibre (c) has a core diameter of some 50–100 µm. Not too thick itself, this variety has, however, a fat and tubby close cousin outside the field of FOCs: the 2 mm or so thick *graded index lens* (see Chapter 14).

7.3.1   The multimode stepped index fibre

Here, a highly homogeneous, relatively thick glass core is covered by a 5–50 µm cladding of a material (glass or plastic) with a refractive index n_{clad} a few percent lower than n_{core} (Figure 7.6(a)). The NA range is 0.2–0.4. This, together with the relatively large core diameter, makes light coupling into and out of the fibre relatively easy. The same goes for splicing and connecting. The fibre has reasonably good mechanical strength. The penalty for this advantage lies in the already explained bandwidth limitations.

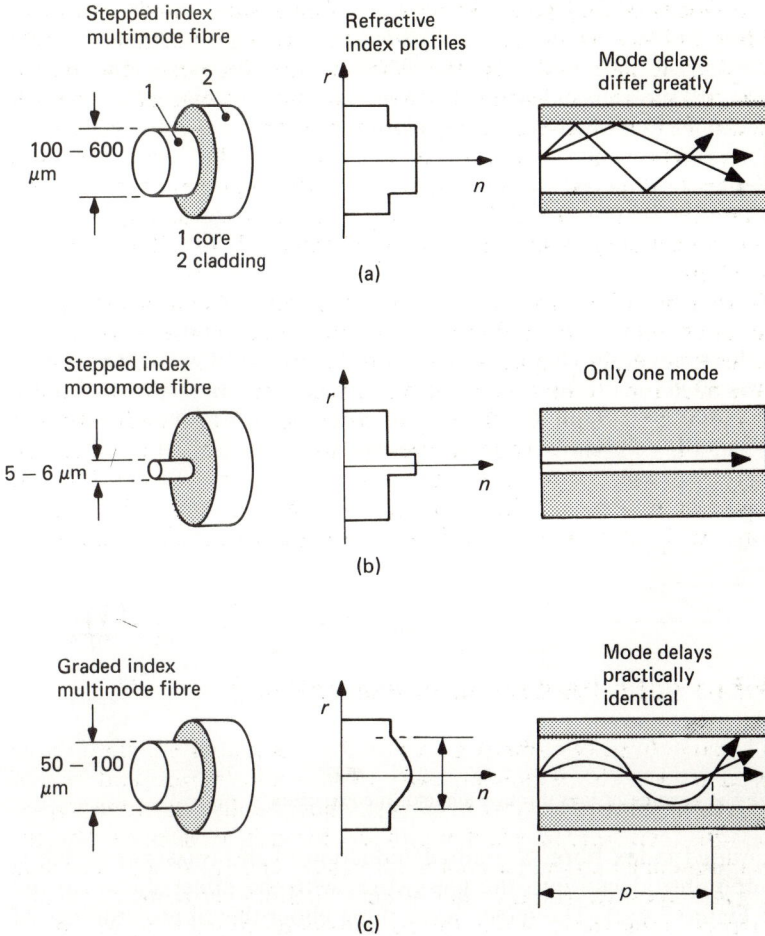

Figure 7.6 Three basic types of communication fibre (not to scale)

7.3.2 The monomode stepped index fibre

The physics of this fibre (Figure 7.6(b)) resemble that of a circular microwave waveguide. This is because the core diameter is of the same order of magnitude as the wavelength of the radiation launched into it. The number of modes a stepped index fibre can sustain goes sharply down as the diameter reduces (Appendix 7) until it becomes unity. (A further reduction of d would stop propagation altogether, unless λ were reduced too.) Hence this fibre's name.

The stepped index monomode fibre is composed of a 2–8 μm silica-based glass core surrounded by an equally silica-based glass cladding over 30 μm thick. (The cladding must be thick, not only for mechanical reasons, but also to support some of the energy propagation.) The NA is very small, usually ≤0.11. Fragile, difficult to align, to splice and to

this is nevertheless the best 'long distance runner' for high
dth communications. At the time of writing the longest trunk
ction in the UK (London–Birmingham, 201 km) uses this kind of
and so does the longest *repeaterless* experimental link (British
com Labs, Martlesham, 1982 – 102 km).

7.3.3 The graded index fibre

This represents a fascinating development. Its working relies on the same
physical phenomenon that makes mirages appear to desert travellers.
When the refractive index of a transparent medium varies gradually, light
travels in curved paths. In the desert, the temperature gradient of the hot
air can make a remote oasis or town appear to be within easy reach. You
don't have to go to the Sahara to experience this kind of delusion; Figure
7.7 shows that the bending of light rays in unevenly heated layers of air can
play tricks with a perfectly English landscape.

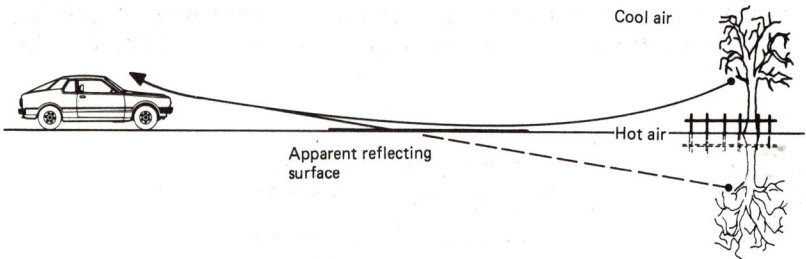

Figure 7.7 The bending of rays through inhomogeneous media

In a graded index fibre, a gradual reduction of the refractive index is
deliberately introduced into the glass core, with the effect shown to the
right of Figure 7.6(c). The most clever thing about this fibre is the law of
variation of n; as we move away from its axis towards the periphery, n
reduces in strict mathematical accordance with the relation

$$n(r) = n_{max} \left[1 - k \left(\frac{r}{d/2} \right)^2 \right]$$

(see Appendix 8) in which r is the distance from the axis and k the relative
variation of the refractive index $(n_{max} - n_{min})/n_{max}$, making all light paths
(modes) into sinusoids of the same period, p. (For mathematical proof see,
for example, Ref. 218, p. 34 or Ref. 130, appendix 1.) Rays with different
angles of incidence travel with (very nearly) the same longitudinal speed,
thereby *reducing modal dispersion almost to zero*. (Remembering from
Section 2.1.2 that $v \propto 1/n$, we can see how the outer rays, while making
longer journeys, catch up the inner ones, by travelling more speedily in the
faster parts of the medium.) Here, light no longer travels in straight lines,
not even in broken, zigzaging lines as in stepped index light guides. The
'meandering ray' of p. 89 suddenly falls into place!

So then the graded index fibre combines the advantages of good light
gathering compatibility with very low modal dispersion, i.e. it provides at

the same time large NAs and broad modulation frequency bandwidths. The only penalty for this valuable combination of virtues is high cost.

7.3.4 Fibre optic cables

Who in his right mind would dream of interconnecting two distant telephone exchanges by several kilometres of unsupported, tenuous and bare glass thread? For such interconnections, cables are needed, and it takes rather more than a fibre to make a cable. Despite the fact that the structures so far described include a 'cladding', strengthening members, stress relief components, void fillers, antiabrasion jackets and weatherproof covering are all vital adjuncts to the lightguide proper before it can be laid in ducts, buried in soil or hung overhead. Cable engineers came to the rescue of physicists in deciding on the choice of materials and manufacturing techniques. Together, they evolved several types of fully fledged *optical cable*, now on the market. To go to great length in describing them all is beyond the scope of this book. (The reader with a specific interest in cable technology is referred to references 39, 129, 131 and 132, the latter containing a rich biography of its own.) Instead, a few

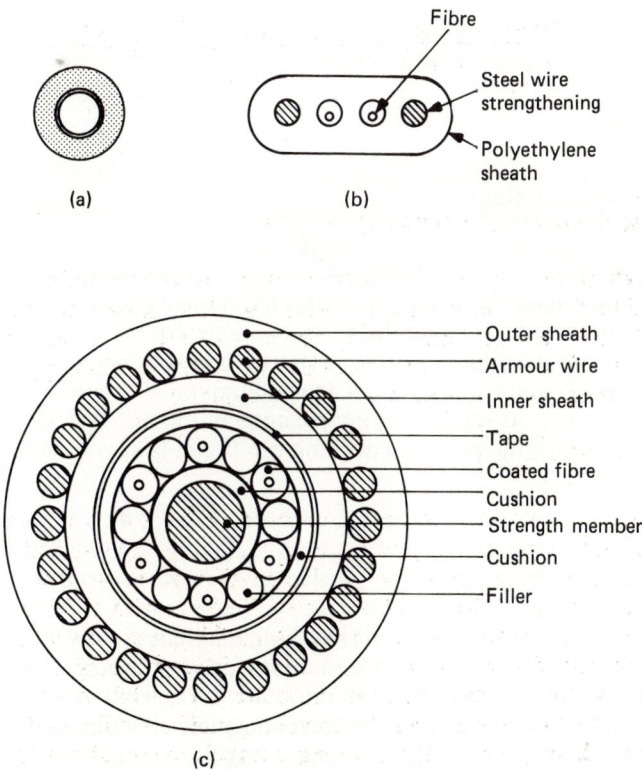

Figure 7.8 Examples of fibre optic cables. (a) Simple stepped index cable. Acrylic 1 mm core clad with a fluorine-containing polymer. Polyethylene jacket. ESKA® CH4001 Mitsubishi (Hytran). (b) Strengthened flat duplex cable (BICC). (c) Multiple heavy duty, fully fledged optical cable

examples of commercially available structures are shown in Figure 7.8 to illustrate the trends, namely:

1. Short-distance low-capacity indoor linkages. Instrumentation, speech, video. The OD of the jacket is 3 mm.
2. Twin fibre cables. The Hytron RS type is intended for indoor links only while the BICC wire-reinforced flat cable can be used for short underground ducted distribution, despite the absence of a moisture barrier [131, 132].
3. Conceptual representation of the generalised structure of a long-range, heavy-duty multiple fibre cable. The various members would be normally helically wound [39].
4. A specific variety of (3) for submarine use, OD = 26 mm. An armoured version of this cable was laid by the British Post Office in Loch Fyne in 1980 for trials in seabed conditions similar to those of the North Sea (see Section 7.7 and Ref. 133).
5. High-capacity 'multi-ribbon' structure. The ribbons are twisted within the cavity of the protective structure.
6. There are at least two types of aerial cable which are suitable adjuncts for HT power transport systems. The NKF (Nederlandische Kabel Fabrik) version is praised for being metal free [135]. The Fibral cable (CEGB and BICC) is of great interest as it embodies the first opto-fibre in the world *inside* an overhead earth wire application. It was installed between Fawley and Nursling (Hants, UK) in 1982 (Figure 7.22), and is 21 km long [134].

7.3.5   A velocity greater than c

The universally accepted symbol for the speed of light in vacuum is c. The numerical value of c is very nearly 300000 km/s ($c = 299782$ km/s). We remember that, when not in vacuum, unconstrained monochromatic light travels with a velocity $v = c/n$, n being the refractive index of the medium. Thus, in air, where $n = 1.000293$, v differs from c by less than 0.03%. 300000 km/s really is an astonishing speed! It would make a hypothetical traveller circle the earth on its equator in approximately 150 ms. We could have said the *most* astonishing speed, as nothing can travel faster. Indeed, Einstein's equation

$$m = m_0 \frac{1}{\sqrt{\left(1 - \dfrac{v^2}{c^2}\right)}} \tag{7.3}$$

shows at a glance that the effective mass of anything that moves grows fast with v, reaching infinity for $v = c$. (The mass doubles for $v = 0.86602c$, and increases 10 times for $v = 0.99498c$ and 100 times for $v = 0.99995c$.) Yet, returning to our FO light guides, we find that a point of constant phase could progress along the z-axis with a speed greater than c. Indeed, focusing our attention on the point of intersection of a wavefront F with the fibre axis z (Figure 7.9, point K) we find that it 'moves' to a new location K' within the same interval of time the photons need to move along the ray from R to R'.

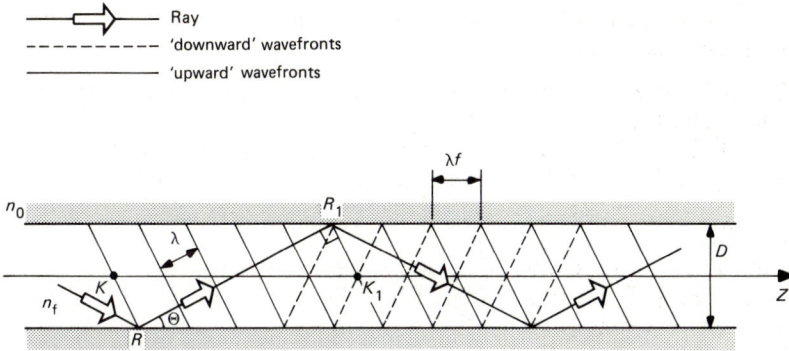

Figure 7.9 A velocity greater than c

The more oblique the wavefront, i.e. the more zigzaging the ray path, the faster the motion of this *constant phase point, K*. The *phase velocity*, V_p, is:

$$V_p = \frac{V}{\cos \theta_f} \qquad (7.4)$$

or

$$V_p = \frac{c}{n_f} \times \frac{1}{\cos \theta_f} \qquad (7.5)$$

For sufficiently shallow wavefronts (provided the ray injected into the fibre still lies within the funnel of admittance) and high propagation modes, $\cos \theta_f$ could become smaller than $1/n$ thus making $V_p > c$.

However, Einstein's equation lives on. Phase velocity is what its name says it is and no more – the speed of points or planes of constant phase. There is neither transport of matter nor transport of energy at this speed. In fact, the greater the phase velocity, the smaller the *signal* velocity in the fibre. Figure 7.9 shows that:

$$V_s = V \times \cos \theta_f \qquad (7.6)$$

This is the *effective* velocity of light in the guide, inevitably smaller than v (and hence, than c), the reduction resulting from the zigzaging travelling manners of the captive ray. From Equations 7.4 and 7.6 we find that:

$$V_s \times V_p = V^2 = \text{const.} \qquad (7.7)$$

showing that V_s goes down as V_p goes up. The signal velocity V_s corresponds to the so-called *group* velocity, usually denoted by V_g. The reasons for this name stem from motion of modulated waves in dispersive media. The inquisitive reader is referred to Appendix 8. For the time being, let us just retain the fact that there is a certain velocity greater than c.

7.4 Why fibres?

What is it that made fibre optic communications grow into a major industry within less than a decade? Why are the world's major telephone networks 'de-copperised' (proportionately speaking; for example, British Telecom reckons that half of its trunking will be FO cabled by the end of the 1990s)? Why use a quadruple conversion of acoustic vibrations for a telephone conversation (acoustic into electric, into luminous, into electric, into acoustic) when a double one does the job? After a near-century of absolute reign, the copper wire finds itself so seriously threatened by the glass fibre because in this application the propagation properties of light in glass proved to be superior to those of electricity in copper. (In much the same way the propagation of electricity in copper had proved, once upon a time, superior to that of sound in air, thus giving birth to the telephone.) A glance at Figure 7.10 shows convincingly the savings in *size and weight*

Figure 7.10 Six optical fibres in reinforced cable compare in signal carrying capacity to the 900 pairs of conventional copper telephone wires (Courtesy of Telefocus; a British Telecom photograph)

brought about by the new technique. Directly ensuing are labour savings in the area of cable transportation and cable laying. Lightwave communications can be likened to carrier telephony (or radio). A lightwave with a wavelength $\lambda = 0.9$ μm has a frequency

$$f = \frac{c}{\lambda} = \frac{3 \times 10^8 \text{ m}}{0.9 \times 10^6 \text{ sm}} \cong 3.3 \times 10^{14} \text{ Hz or } 330\,000 \text{ GHz}$$

Potentially, the *information carrying capacity* is five orders of magnitude greater than for coaxial cables. Even today's practical figures ($\sim$ 100 GHz km) [130] give an improvement factor of between 5 and 20! It is easy to calculate the tremendous number of telephone conversations and/or TV programs that a bandwidth can accommodate. The next advantage – a result of extensive R & D work – is *low attenuation per unit length*.

To begin with, fibre optic communications was a scientist's curiosity. This was no longer so in 1970, when a 20 dB/km attenuation was achieved, replacing the 1000 dB/km figure of the mid-1960s. With today's attenuations of under 1 dB/km the distance between repeaters (DBR) is around 30 km, i.e. some 5–15 times less than for copper cables. For countries such as the UK, this means that all repeaters of an intercity trunking can be housed *in existing buildings*. Next comes an absolutely unique advantage of fibre optics communications – the *immunity to electromagnetic interference* (EMI) that makes fibre optics communications unbeatable in electrified railway tunnels, along HT power lines and in other electrically noisy environments. The *absence of earthloops* is also worth noticing. Next comes the *safety* (no sparks) aspect. (Think of mines, explosive atmospheres in chemical plants, etc.) Penultimately, comes commercial and military *security* (no surreptitious tapping), and last but not least the *abundance of glass constituents in nature* – there is plenty of sand on the beaches. This is a factor of some importance in view of the rising prices and possible shortages of many world commodities of which copper is one.

All the 'pluses' printed in italics add up to tip the glass/copper economic balance in favour of glass, making the fibre optic technique into a winner. The one and only disadvantage is the splicing, jointing and connecting difficulty encountered with optical fibres. Here, when compared with soldered or wrapped joints, plugs, sockets and connectors of copper leads and cables, fibre optics has to admit a rather embarrassing inferiority. Perhaps our ingenuity will help us to conquer this last frontier in the not too distant future.

7.5   Other components of a fibre optic link

It takes more than a cable to make a link. While it is obvious that a transmitter (TX) and a receiver (RX) at each end of the FO cable are required to bring into existence an FO link, the need for specialised *passive* optical components in a linkage may, at first, escape attention. Not only do we need TX-to-cable and cable-to-RX connectors; more often than not the overall linkage lengths will be greater than the longest available one-piece cable, thus creating the need for cable-to-cable splicing and/or joining. Remembering that fibre diameters are often only 50 μm and sometimes only 4–8 μm, we can appreciate that this is an area in which some non-trivial mechanical problems may arise. This joining and connecting difficulty is, in fact, the one disadvantage of fibre optics technology.

7.5.1   TXs and their light sources

A transmitting head (TX) will usually be composed of a source of light, a means of modulating it, some optics for coupling it to the fibre and a

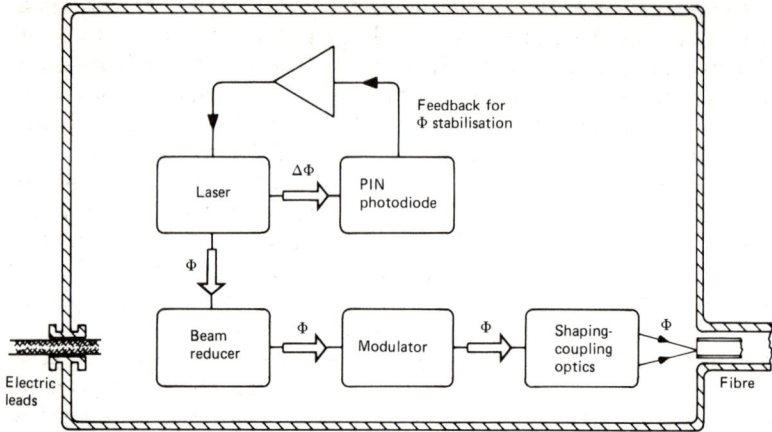

Figure 7.11 Schematic of TX using an externally modulated laser and coupling optics (cooling not shown)

housing. Only two kinds of light source are in use at present: LEDs (amply described in Chapter 5) and semiconductor lasers, also called diode lasers (to be described in Chapter 9). Two more sources are being considered, the $Nd^3 + YAG$ and the NLP solid-state laser [130], for long-wave communications. The simplest possible TX consists of a pigtailed LED in a single package using internal (current) modulation. A complex TX may contain several components: an externally (light) modulated laser, a feedback light path and circuit for Φ stabilisation, a beam-shaping lens, a cooling device and, of course, a housing (Figure 7.11). (Some relatively complex TXs look simple, thanks to integration. See the closing paragraphs of Section 7.5.2.) The coupling by lens of Figure 7.11 is sometimes improved by the use of a Selfoc rod (see Section 14.1) or simplified by the use of a quasi-spherical glass 'blob' ending on the fibre, acting as a crude lens. Here is how the two light sources, the LEDs and semiconductor lasers, compare. The LED is simpler, more reliable, has a longer life, works quite well without forced cooling and is cheaper. (High-radiance (HR) high-frequency (HF) diodes are an exception to the rule.) The laser, however, has a much higher peak radiant power (up to 3–5 W as opposed to ~ 0.2 W for the LED), a slightly higher mean power (say 5–10 mW against just 1 mW for the LED), possesses temporal and spatial coherence (see Section 10.2.1) and can be modulated up to (if not over) 1 GHz. The source/fibre coupling efficiency ranges from some 5% (for an LED plus a plano-convex lens) to say 60% (for a semiconductor laser plus a Selfoc rod). The typical power in a link is thus of the order of 1 mW. This is perhaps why power levels along a link path are usually expressed in dB m, i.e. referred to 1 mW taken as the 'zero dB m' reference.

7.5.2 RXs and their light receivers

Like the TX, the RX (receiving head) will usually have four main constituent parts: a fibre-to-receiver optical coupling, a light receiver proper, a preamplifier/demodulator and, obviously, a housing. Two kinds of receiver dominate the scene: the PIN photodiode (Section 6.1.4(a)) and the avalanche photodiode (APD) (Section 6.2.1). Both are semiconductor devices of the silicon type. Of these two, the PIN photodiode is not only cheaper but also much simpler to use and, hence, probably more reliable. The APD, on the other hand, has a much higher SNR and a much higher responsivity to offer. Long-wave systems (1.05–1.7 μm) tend to use the more recently introduced InGaAs, and non-amplifying photodiodes (Section 6.1.4(e)). Here again, the pigtailed semiconductor in a single package constitutes the simplest RX, while a complex one may contain, in addition to the photoreceiver, a beam-shaping lens or Selfoc rod or a microsphere, some fairly complex circuitry, and often a noise-reducing cooler.

Numerous firms have introduced on the market completely integrated RXs (black box type). Some contain hybrid and others monolithic (silicon) electronics. In the latter case the photoreceiver has become part of the integrated circuit. These RX boxes are destined mostly for extra-short (<10 m) or short (100–5000 m) linkages. As ready-to-use system 'building bricks', these terminals can present attractive engineering solutions to information transmission problems, either in instrumentation or in data processing (process control, servo and signal links aboard ship, aircraft or road vehicles, distributed computer networks). They are especially handy when they come in TX–RX pairs.

7.5.3 Splicers and connectors

The one technological problem in the bright picture of lightwave communications is that of jointing and connecting fibres together. Figure 7.12 shows the kinds of misalignment likely to occur at fibre optic joints and connections and the resulting losses. Here, the two fibres are of the same type and diameter, and their butt ends in perfect condition; their faces exactly perpendicular to the fibre axis and polished to the highest possible degree. In other words, the light transmission conditions are highly favourable but for the alignment geometry. The losses can, nevertheless, run into several decibels, so that a single imperfect connection may easily squander many more precious microwatts than a whole kilometre of a modern cable. (The reader is invited to have a close look at all three graphs of Figure 7.12(a) to become fully aware of how little misalignment it takes to cause a 3 dB loss in a 50 μm, 0.5 NA fibre system. Yet the commonly accepted maximum for a connection is only 1.0 dB.) For what happens when the butt end surfaces are rough or oblique, refer to Ref. 61. The residual 0.5 dB loss is almost entirely due to Fresnel (reflection) losses, present whenever there is an air gap between two glass surfaces.

In the text to follow the words 'joint' and 'jointing' will be used for *permanent* cable extensions while the terms 'connector' and 'connecting' will denote *demountable* matings. Jointing involves cutting, polishing and, usually, fusing. Both techniques are very young and still very fluid, so the following text gives only a few pointers to the present state of their evolution.

(a) Loss due to end separation

(b) Loss due to lateral misalignment

(c) Loss due to angular tilt

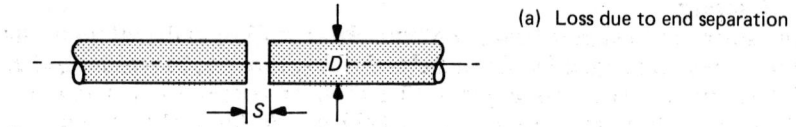

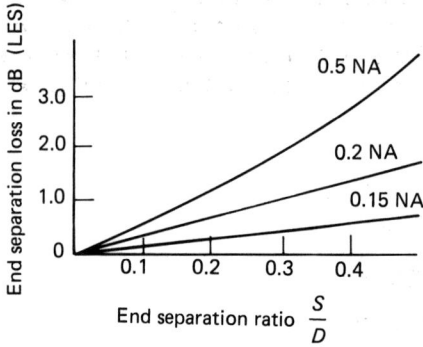

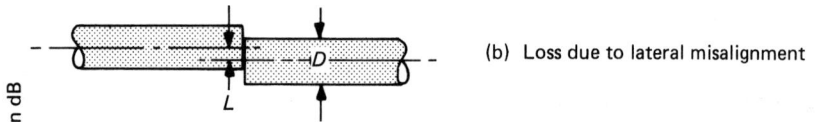

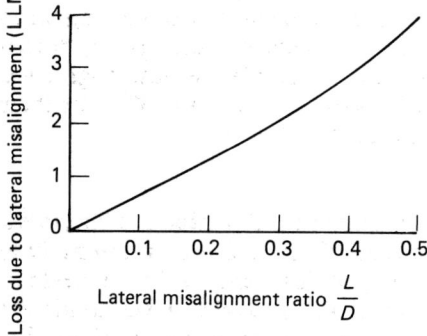

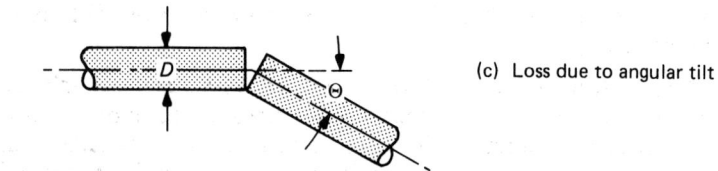

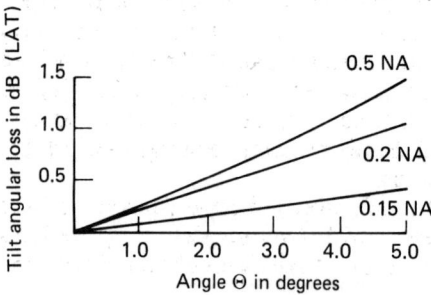

Figure 7.12 Jointing and connecting losses. The three basic types of misalignment

(a) Jointing

In the so-called 'fusion splicing' method, the two fibre ends, after cutting and polishing, are aligned with the help of a V-groove jig, brought together and heat fused. The gentle yet sufficient heat comes from a digitally controlled electric arc, battery fed in field usable kits. Microtorches or resistance heating can also be used (Figure 7.13). Surprisingly, thin as they are, single mode fibres can be fused/spliced too, thanks to a self-aligning process stemming from the surface tension of the molten material [129].

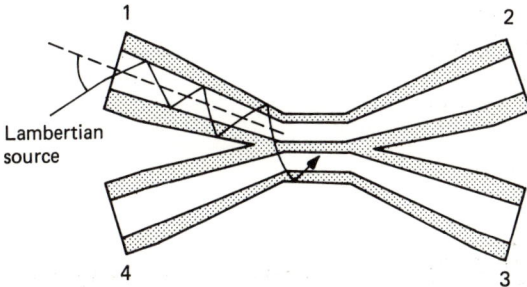

Figure 7.13 Fusion splicing method

Other splicing methods use precision machined heat-shrinkable snug tubes where the fibre ends meet, curved hollow enclosures and epoxy, 'three-roller' centring sets in combination with shrinkable sleeves, V-grooved platelets, etc. (see Ref. 129 and also Refs 44, 61, 130 and 137). For their protection, splices are often housed in junction boxes.

(b) Connecting

Repeatability of performance at successive re-matings, mechanical strength (relying on other than the fibre itself), exclusion of or insensitivity to dirt, and wearability are the main qualities required of connectors. Considering these together with the economic constraints gives an idea of the challenge facing FO connector designers. Of the profusion of possible solutions, all but a few rest heavily on *precision engineering*. The three main categories are:

1. *Ferrule types* A ferrule is a hollow, usually metallic, pin, not unlike a hypodermic needle. The connector comprises three parts – two identical ferrules terminating the fibre ends and a central alignment sleeve. Some ferrules are conical, mating with correspondingly shaped sleeves. The fibre is first secured to the ferrule by means of epoxy resin, then either cleaved or flush cut and polished. The fibre-to-fibre interface can consist of either a silicone resin pad (Figure 7.14(a)) or simply an air gap (Figure 7.14(b)). In some types of assemblies ferrules float a little within a fixed part of the plug (spring loading) to which the cable sheath is secured by crimping.

    Originally developed for multimode fibres, ferrule connectors have been tried – not without success – with monomodes [140]. Perhaps the most interesting monomode results have been obtained with the

(a)

(b)

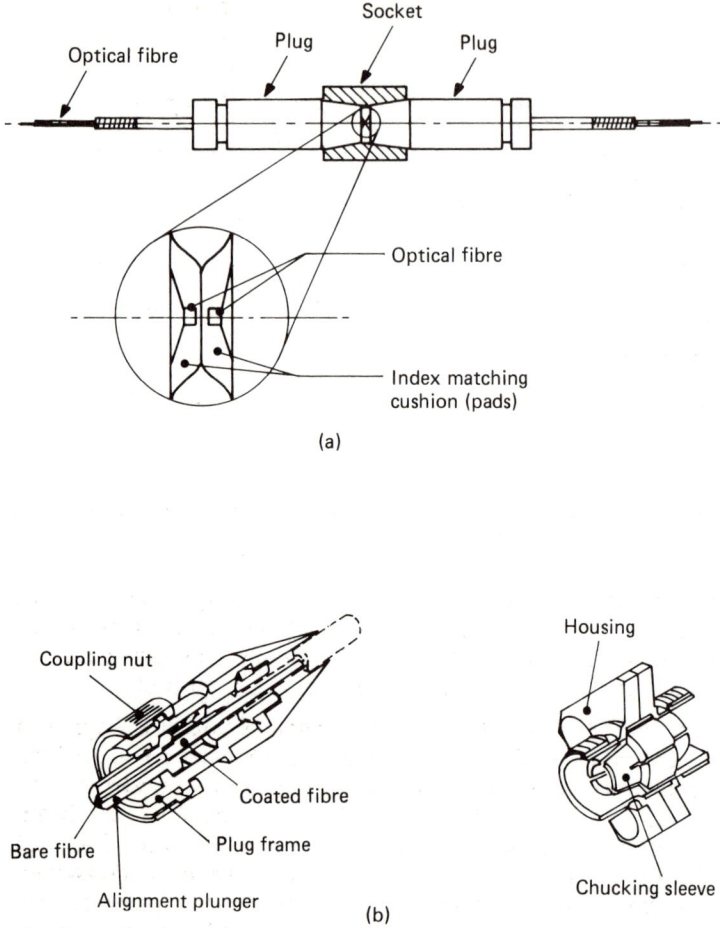

Figure 7.14 (a) Bell Laboratories ferrule connector (insertion loss < 1.0 dB); (b) Nippon Electric ferrule connector which uses a glass capillary for hosting the fibre and has a sprung chucking centre sleeve (insertion loss < 0.5 dB)

'double-eccentric' ferrule/plug arrangement, in which a once-for-ever rotational positioning of the fibres in their plugs is optimised for light transmission efficiency.

2. *Kinematic designs*   The shape and size of the empty space created by pressing together three identical circles is uniquely determined by their diameter D (Figure 7.15(a)). It follows that the diameter d of the much smaller circle enscribed in this space is also determined by D.

    The three-roller fibre centring set mentioned above relies on this principle for accurately positioning and holding together in a near perfect connection vis-à-vis the two fibre ends of the joint. A kinematic connector uses two sets of closely pressed ballbearing spheres for holding and accurately mating the two fibre ends (Figure 7.15(b)) [44, 137]. (For the original proposal see Ref. 141.)

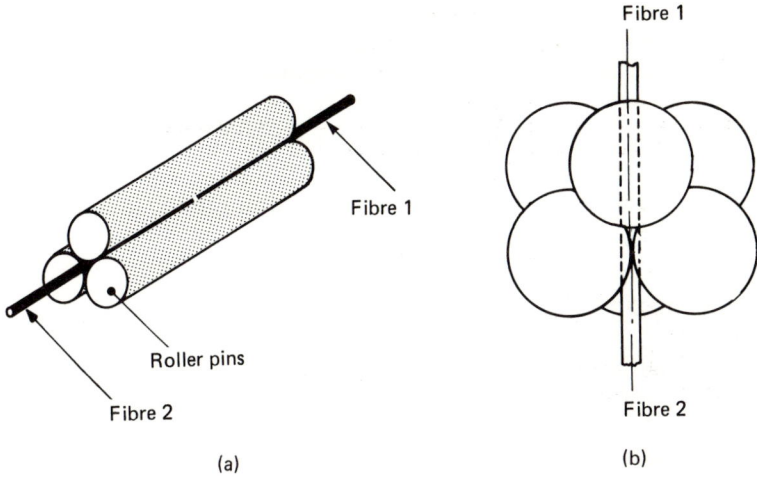

Fibre 1

Fibre 1

Roller pins

Fibre 2

Fibre 2

(a)

(b)

Figure 7.15 Fibre locating by three-roller set for shrink sleeve jointing and by two interlocked three-ball sets for re-matable connecting

In both cases 1 and 2 the designers have astutely shifted the onus of precision machining out of fibre optic communications and into existing old established technologies with excellent facilities for the mass production of closely toleranced items. The cost advantage gained is obvious.

3. *Lensed, so-called 'expanded beam' connectors* In all of the types of connector described so far, the slightest speck of dirt on one of the mating surfaces, or in their interface, can cause a serious obstruction to the propagation of the narrow light beam. Expanding the beam upstream and re-reducing it before it strikes the receiving fibre end should obviate this difficulty. This is exactly what the 'expanded beam' connectors do. The Mossman [137] version uses plano-convex lenses to collimate the incoming (to the connector) beam and decollimate the outgoing one. Each fibre end, supported by a glass capillary and accurately centred within the cylindrical holder, is abutted in the focal plane of the lens (Figure 7.16). It is true that the insertion losses of such a connector are higher than those of a *very clean* ferrule or kinematic connector, but here, at least, a speck of dust will *not* disrupt a communication. The expanded beam connectors must, surely, have a

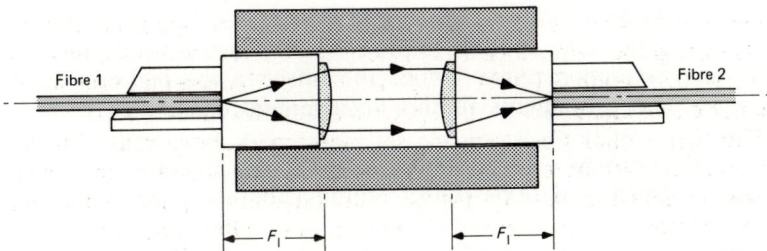

Fibre 1

Fibre 2

F_1

F_1

Figure 7.16 Expanded-beam connector

great future in polluted environments: mines, factories, building sites, military operations terrain, etc. Collimation offers the additional advantage of a high tolerance to end-separation variation (compare Figure 7.9).

Before closing this section, let us say that some jointing and connecting techniques use combinations of two or more of the above described (and other) methods, and that there is still great scope for ingenuity in this field.

7.6   Fibre optics communication systems

It takes more than a link to make a network. Even a plain A to B duplex installation calls for additional types of hardware, if one is not to build it, wastefully, as just a duplication of the basic simplex. A genuine data bus or other communication network will require even more additions to the jointing and connecting devices of a link – multiport components for dividing or merging light beams, en route access points for monitoring, signalling or feedback purposes, path switches, etc. A tremendous variety of such components exists and very many have been, and still are being, proposed by research workers. In just one of the published papers [143], the description of no fewer than 40 'couplers' has been given. Since the situation here seems even more fluid than in jointing and connecting technology, just a few examples of existing solutions will be given, hopefully sufficient to arouse awareness of the problems involved. Specialised reading should do the rest.

7.6.1   Access point components*

The fundamental requirement of this type of component is to be such that its failure will not endanger the functioning of the highway as a whole. A further requirement is that a tapping should rob no more than a few percent of the transmitted power. The simplest access point is obtained by way of 'ratio-splitting' of the on-line optical power. This can be easily (if not a little crudely) achieved with a *multifibre cable*: part of the bundle is simply split off and a portion of Φ with it (see Figure 7.17).

Increasing the splitting ratio to unity will degenerate the access point into a Y-coupler usable for RX/TX (or other) power division or power merging. Bundle fibres are used, however, for very short distances only (< 100 m). Moreover, the power-splitting method is very lossy.

Figure 7.18 shows a four-port *monofibre* splitter based on the principles of integrated optics (see Section 13.3). Here we have a trunk line with injection and extraction ports. Notice the offset in the highway cut of Figure 7.18. The proponents of this arrangement claim high coupling efficiency.

Figure 7.19 concerns a very interesting acess coupler for single mono-mode fibres. The undulations shown literally liberate a fraction of the

---

*This section is based on Ref. 62.

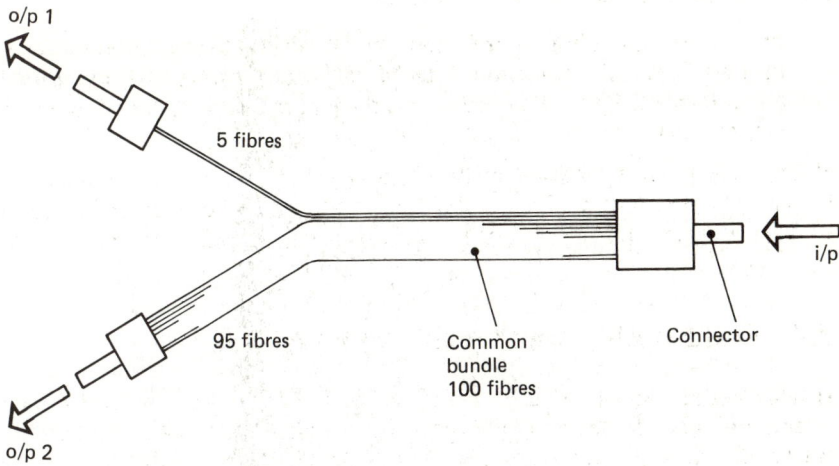

Figure 7.17 Simple fibre bundle Y-splitter

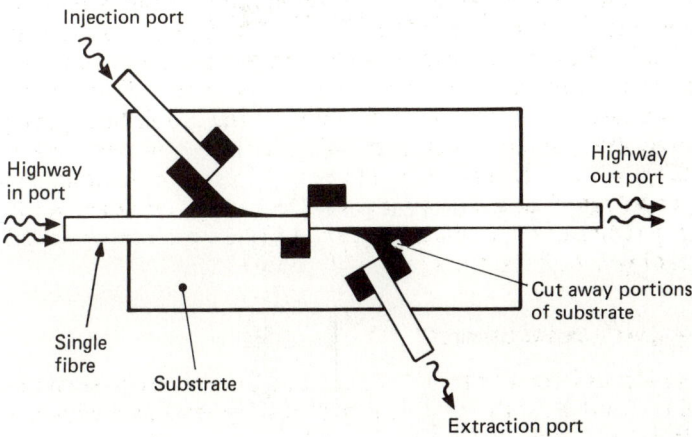

Figure 7.18 Low insertion loss access system based on integrated optics technology. Side fibres (not shown) extract (or inject) radiation from (or into) the main track

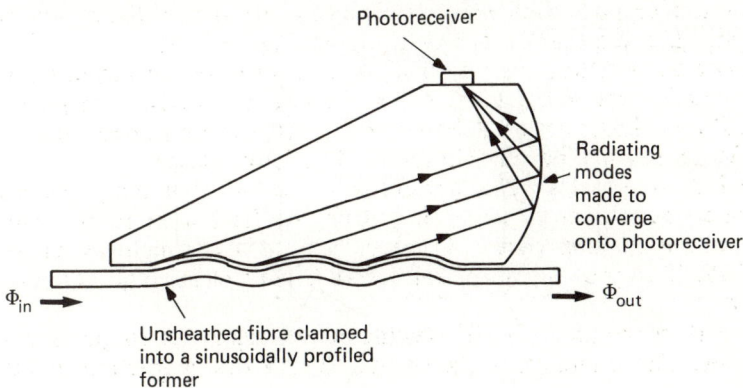

Figure 7.19 Developmental variable ratio coupler. The undulated section is approximately 1 mm long

captive ray's power which the lens directs onto the monitoring photodetector. The tapping ratio is even mechanically adjustable! Certainly worth noticing is the fact that the highway fibre *remains* uncut.

7.6.2 Networking components

Imagine that you have to connect n terminals into a network so that any one of them can communicate with any one of the remaining $(n - 1)$. Many networking options are open to you, the basic two being the ring (loop) network and the star network. The information stream is so formatted that the address code of the destination heads the data which thus, while 'talking' to all terminals, is being listened to by the desired one only. Figure 7.20 shows the arrangement of a *loop network* for $n = 5$. It will be seen that the interconnections are by means of a bus line and five Ts. A T is no other

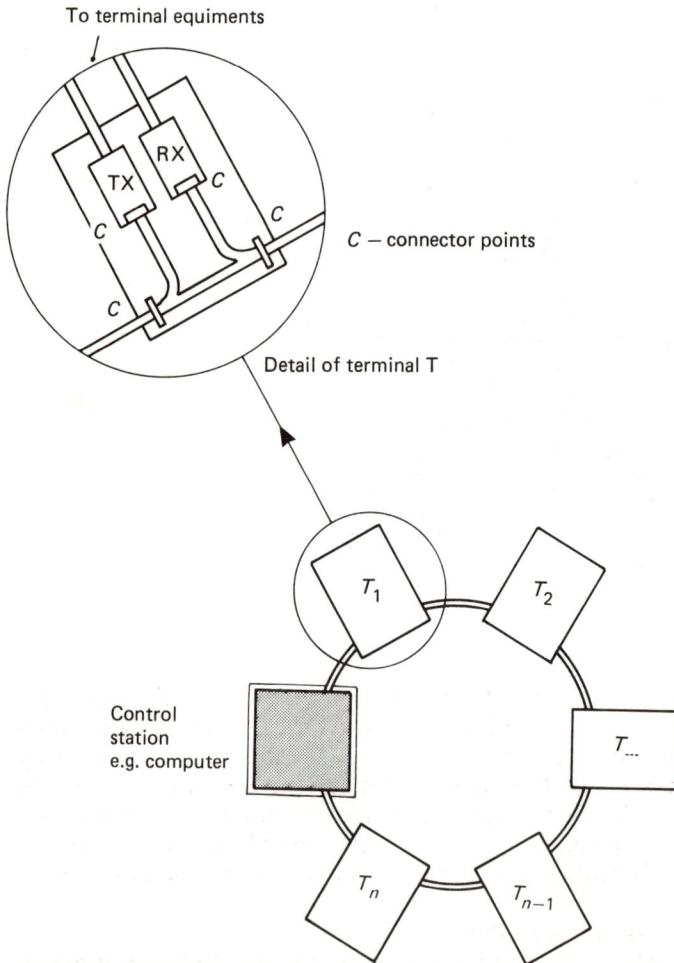

Figure 7.20 Loop networking

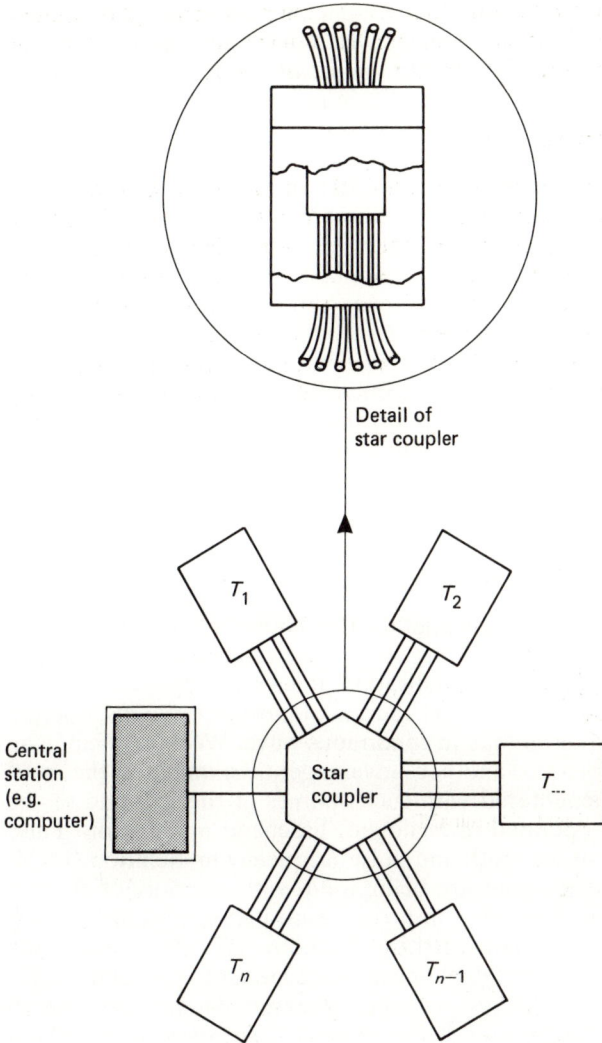

Figure 7.21 Star networking

than a four-port (not three-port!) coupler of one kind or another (typically as per Figure 7.21 but see also Figure 7.18). Thus, this configuration does not require any passive components not yet described. Note that every point C introduces losses. The *star configuration* (Figure 7.21) is, in several ways, superior to the ring [65], in particular by its inherent equivalence of all the terminals. It calls, however, on a specialised optical component – the star coupler. Several versions exist, of which two will be described.

(a) The reflective star coupler
In this coupler [65, p. 221], the inputs and outputs of all the terminals are abutted onto a glass rod which is, in fact, a short piece of thick light

guide fibre (Figure 7.21). The opposite endface of the rod is polished and gold plated. The fibre cladding with its TIR effects and the gold mirror together form a light trap, so that after one or two reflections the rays of the input port return to the rod/bundle interface and irradiate all the output ports. This is because of the NA-governed fan-out of light inside the rod. (The device is suitable for multimode FO only.) It can be seen that this geometry introduces an inherent 3 dB loss due to the presence of both the TX and RX ports at the interface. Adding to it the power splitting and other losses, the overall coupling efficiency of a 20-port star lies in the 20 dB region.

(b) The transmissive star coupler

Because the TX and RX ports of a transmissive star coupler are abutted to separate interfaces (Figure 7.21), the inherent losses are at least 3 dB smaller. In the NTT version [143, p.36] the mixer rod of the reflective coupler is replaced by a slab waveguide and the mirror by the second interface. (The transmissive star coupler is also suitable for multimode FO only.)

7.6.3 Modulation multiplexing formats

Nearly all FO transmissions are digital. The choice of this modulation technique in optical communications does more than just to reflect this decade's general trend towards digitalisation: it acknowledges the fact that both LEDs and laser diodes, the light sources of most FO systems, operate *more efficiently* when pulsed than in continuous wave. Working them in an on/off fashion presents yet another advantage: it eliminates the non-linearity problems encountered with such sources at the low end of the dynamic range when operated in analogue, baseband modulation. Pulse code, pulse position, pulse width and pulse frequency modulation (PCM, PPM, PWM and PFM) systems are being used, with a variety of channel densities. The diversity prevailing in private (intra-company) networks is too great to be dealt with here. British Telecom (BT, the organisation responsible for the UK public telephone network) tends to concentrate, for trunk lines, on the 140 Mbit/s bit rate, accommodating 1920 speech channels per carrier. Modulation is mostly eight-level sigma delta. Older BT installations use 8 Mbit/s (120 speech channels) and 34 Mbit/s (480 speech channels) [145] and a few short-haul ancillary 2 Mbit/s links. The newest systems use 565 and even 650 Mbit/s. 1 Gbit/s (and over) pulse rates have been experimented with (see Section 7.7).

BT's channel multiplexing is purely electronic (as opposed to optical, by wavelength mixing), by time division. Worth mentioning are Japanese and US experiments with optical multiplexing, made possible by the availability of a wide range of wavelength offered by semiconductor heterojunction light sources. Chromatic de-multiplexing is either by colour filters [143, p.34] or by integrated optics devices using diffraction [143, pp. 34, 40; and see Section 13.3].

One is tempted to speculate that the ways and combination of chromatic and electronic multiplexing techniques could, someday, extend channel capacity to genuinely fantastic figures.

7.7   State of the art

In view of the time it takes to write, produce and distribute a book, any attempt at describing in it the state of the art of such a fast and bold technological invasion as that of fibre optics is a brave attempt. It is likely that, by the time this text reaches the reader, the conquered territory will have doubled or trebled. It is hoped that this last section will catch readers' interest sufficiently to motivate them to keep updating this information through the technical press. The periodicals listed in Appendix 13 may help in this task.

7.7.1   UK achievements

Two excellent review articles [147, 149] reveal that, at least in Europe, the UK is the leader. The London–Birmingham link is British Telecom's pride. This, the longest (201 km) public multichannel link in the world, uses a long-wave ($\lambda = 1.3$ μm) carrier, has a useful bandwidth of 140 Mbit/s accommodating 1920 duplex speech channels and boasts a DBR 25% longer than that of other existing FO links. BT has several other firsts, among which are an experimental *one-piece* (no repeaters) 31.5 km monomode fibre transmitting a bandwidth of 650 Mbit/s (8000 telephone channels) and a 102 km repeaterless 140 Mbit/s link [150, 156]. BT's activities in the FO field are not characterised, however, by an obsession with record breaking. They have a coordinated long-term plan of action to have, by 1990, 100 000 km of fibre installed, interconnecting all the country's major cities and covering half of the UK public network. To date, the first stage of this plan is accomplished and the second well under way. Currently in exploitation is a fibre length of some 3500 km (exceeding twice the Lands End to John O'Groats distance). Some 6500 km more are being installed. Most routes (26) use or will use a 34 Mbit/s bit rate on a 1.3 μm carrier with a DBR of 10–13 km. Next come 8 Mbit/s short routes (24), with repeaters (when necessary), housed in existing buildings. Finally, come high-capacity 140 Mbit/s long-distance ($\simeq 60$ km) systems (6) with a DBR of 8 km. This DBR is four times greater than with coaxial cables. The 140 Mbit/s links of the second phase of the plan are to have their DBR raised to $\simeq 30$ km. Monomode 1.55 μm operation without buried repeaters is to characterise the third generation of BT's links. BT is also installing video channels for hire to end users for CCTV, CATV and networks for teleconferencing (confravision). In this respect, the town of Milton Keynes deserves a special mention, as it is rapidly becoming an experimental ground for a miniature Biarritz, France's 'fibred city' (see below). The Milton Keynes scheme provides a 3 km CATV cable sprawling out into an 18-branch star distributing multiprogramme television inside private homes. Interactive terminals are to provide the 18 users with other services too.

Another UK venture worth mentioning is the Fawley–Nursling (Hampshire) (24 km) land cable in which fibre optics are integrated with metal in the so-called 'earth wire' of a 400 kV EHT Central Electricity Generating Board (CEGB) line. Fibre optics are used for the transmission of vital monitoring information and control signals in a high EMI environment

Figure 7.22 The composite earth-wire crowning the Fawley–Nursling 400 kV power line (Courtesy of CEGB)

[134, 147]. A 1000 km ring, mostly equipped with conductors of this kind, is being considered by CEGB. The composite opto-electric conductor and the 400 kV pylons carrying it above the power lines are shown in Figure 7.22. Narrow-band data links are also planned along BR railway lines.

Finally, at least with regard to inland installations, London may well get an extra-modern communication network for its business laid in ducts built in the 19th century. The plan contrives to convert and re-utilise a long-disused 280 km network of pipes which once ducted high-pressure water to the City's lifts and to its famous hinged Tower Bridge.

Turning now to international FO links, by far the most interesting project now underway is TAT-8 – the next Transatlantic Automatic Telephone link. This will be the first non-electric capacity extension of the 6500 km London–New York route the linking the UK (and thereby Continental Europe) with the USA. Preparatory trials have already begun. A 9 km trial loop submerged in Loch Fyne, Scotland (not Loch Ness) in 1980 is performing satisfactorily [153–155; see also Section 7.3.4] but owing to the prescribed length of service for submarine equipment prior to approval, TAT-8 has not yet come into service. The cable under test has a 6000 speech channel capacity. The economic attraction of fibre optics for TAT-8 stems mostly from the large value of the DBR, planned to be 50 km or more.

Figure 7.23 Submarine Loch Fyne cable (Courtesy of Telefocus; a British Telecom photograph)

7.7.2 International pointers

Much of the USA's telephone trunking is high density. As many ducting systems are narrow gauge, FO's main promoting factor is its high information per unit cross-section capability. In some practical situations this capability is nearly 150 times greater than that of a copper cable with, for example, a two-fibre optical cable carrying as much traffic as a 20000 copper wire cable [149, p.17]. Particularly large savings can be made on long-haul high-density trunkings.

Two FO highways, the North-East Corridor and the Pacific FO Project are such cases. The North-East Corridor will run from Cambridge (Massachusetts) to Mosley (Virginia) via Washington DC, an impressive 1300 km, and carry 100000 simultaneous telephone calls. The system will be wholly digital, inclusive of electronic 'superswitches'. The Pacific Project is for an 830 km run along the Californian coast. Altogether, the American FO giants (Corning Glass, Bell Labs, ATT, Western Electric), 'helped' by the aggressively expansionist Japanese FO industry, intend to interconnect most of the USA's metropolitan areas by the 1990s. Of this vast network, they planned to have some 3700 route-km operational by the end of 1984.

Across the border, the Canadians planned for a 3400 km highway (for Saskatchewan Telephone) [160]. Decidely, things American are big!

Turning now to Japan, we come more to a country of high technological achievement [161] than to a land of spectacular installations. Nippon Electric Co (NEC), Nippon Telephone and Telegraph Co (NTT), Fujitsu Ltd and Nippon Sheet Glass Co Ltd, rank among the world's most outstanding organisations in the field of FO research, development and component production. Hundreds of projects around the globe use NEC FO products. Systems are being catered for as well, however – 34, 90 and 140 Mbit/s networks have been installed and long-haul 270 and 400 Mbit/s networks are now being planned. NTT intended to install 10000 km of fibre per year. Fujitsu, in addition to doing work on 100 and 400 Mbit/s long-haul systems, is engaged in valuable developments on long-life TX sources and on single-mode transmission networks. Most systems appear to have been installed by the Japanese outside their country, unless the scarcity of information regarding FO links on their own soil is responsible for creating such an impression. One Japanese town, however, is very special for us – Higashi Ikoma (see below).

For France, we turn our pointer to the beautiful Atlantic resort of the Basque coast, Biarritz, which is in the course of becoming the captive ray's city. An FO network is to provide 5000 homes of the 30000 inhabitants with full duplex high-quality 625 videophones. In addition, the home terminals are to supply two simultaneous national or local TV programmes (from a choice of 15), pictorial information by request from an 'image library', 12 stereo radio programmes, and other narrow-band information such as the French 'on screen' telephone directory listings, 20000 km of fibre will form the network, consisting of a central dispatch/switching centre and four satellite nodes. Picture transmission is analogue FM. All other fibre-propagated information is digital or digitalised, with time division multiplexing. Switching is analogue for pictorial channels, digital for narrow bands. The TV terminals are interactive, i.e. subscriber channel selection, while controlled from home, is effected at the main centre. The operational bandwidth is approximately 150 MHz. Graded index 50/125 μm fibre has been chosen for the project, with $A = 3.5$–4.0 dB/km and 2, 10 or as many as 70 fibres per cable. More than 10000 km east of Biarritz, in Japan, lies another FO city: Higashi Ikoma [160, p.37]. In the years to come these two large-scale yet compact live experiments will yield a host of invaluable information on the operational and exploitational aspects of the captive ray technology.

For further reading on this chapter, see Refs 44, 61, 129, 130 and 137. For the convenience of readers seeking a full treatment of fibre optics communications, the principal books on this subject have been listed in the Bibliography, where they are denoted by a dagger, †.

Chapter 8

The liberated ray or fibreless optical communications

8.1 Free space optical communications from 3500 years ago to the present

Around the year 1200 BC, Troy fell to the Greeks. This was known in Argos, hundreds of miles and a sea away, only hours later. At a time when the speed of information transfer was the same as the speed of transportation and the fastest means of travel was on horseback, this was surely an extraordinary feat. According to Aeschylus [284, *Agamemnon*, lines 10–11] optical transmission took place. Although he doesn't tell us how he knew this, his character Clytaemnestra explains the link of beacons [284, *Agamemnon*, lines 281–318]. Two and a half thousand years after Troy, in England, in AD 1588, another outstanding use was made of the same method of signal transmission. The Spanish Armada had just been sighted. Within 30 hours 70 000 men were mustered, most of them within 24 hours [70]. At about the same time, John Dee's Moonbeam Telegraph, the invention of an English astrologer, was used in the Austro-Hungarian Empire [47, chapter 17]. Habsburg Emperor Rudolf II (1550–1612), then in Prague, learned of his Hungarian troops' victory over the Turks *minutes* after they captured the Turk fortress of Győr. From Győr to Prague, a 400 km distance, this 'most important fresh news from the front' was transmitted by Dee's Moonbeam Telegraph using no more than ten relay stations.

To me, the history of Free Space Optical Communications begins with Alexander Graham Bell's photophone. Troy, the Spanish Armada and the Turkish fortress are prehistory not so much because of their legendary character as because of the limited information contents (*one* 'bit') these transmissions handled: no more in fact than did hill-to-hill fires lit from times immemorial to announce, if not simply to confirm, the readily expected 'big' events, such as the birth of a son to the country's ruler. On 19 February 1880, however, A. G. Bell converted *speech* to modulated light (Figure 8.1) [66]. A week later he wrote: "I have heard articulate speech produced by sunlight! I have heard a ray of sun laugh and cough and sing." Free Space Optical Communications (FSOCs) were born!

Five weeks later, Bell and Tainer transmitted speech over 82 m. By 1 April they extended the range to 213 m, on a site in Washington DC where today a plaque commemorates Bell's achievement, 'the first wireless

Figure 8.1 Photophone transmitter which uses sunlight reflected from a voice-modulated thin mirror (Courtesy of Bell Laboratories)

telephone in the History of the World'. Marconi's radio transmission didn't come until 15 years later.

Modern FSOC began with the emergence of semiconductor LEDs in the early 1960s.

8.2   The nature of free space optical communications

8.2.1   Where fibre won't go

We know from Chapter 7 that FO cables are much thinner and very much lighter than electric cables of an equivalent information carrying capacity. There are situations, however, in which – thin as they may be – FO cables are still too obtrusive. Take the case of a linkage crossing a railway line, a waterway, or simply a busy city road. Even the thinnest of silken threads would be too much here, without indulging in highly objectionable tunnel digging. There is no need for that, however. Simply take a fibre link, throw away the fibre, add a couple of lenses and you have a fibreless optical link (Figure 8.2) – conceptually speaking, at least.

Unobtrusiveness is not the only argument in favour of FSOCs. Through their intrinsic cheapness they have the potential of being of help to isolated communities. Goods and skills do not represent all the problems of such communities. Take the case of a village isolated by virtue of its situation, e.g. on a small island or in the middle of marshland or rugged mountainous terrain. For the inhabitants, the ability of getting in touch with the nearest

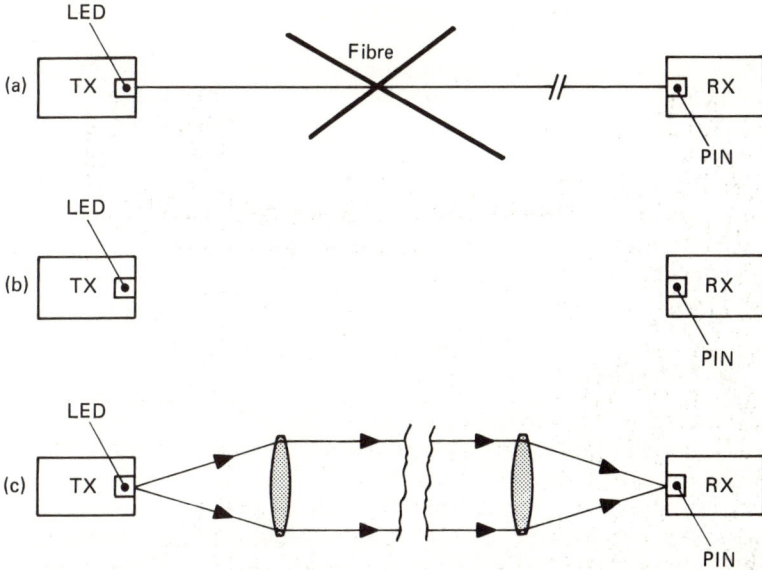

Figure 8.2 Throw away the fibre, add a couple of lenses and you have an FSOC link

hospital, school or other service shared by a wider area can be of great, if not vital, importance at times of crisis. In many geographic and other local conditions FSOCs should enable them to do so for a lesser initial outlay and a smaller running cost than any other type of linkage.

8.2.2 The fascination of FSOCs

Free space optical communications have a Marconian fascination about them. This wire-less, fibre-less, radio-less and nearly always licence-less means of transmitting speech, music, pictures or remote control signals by way of harmless and totally invisible radiations, often for a modest capital outlay, is of a kind to spur some day a new, intense amateur activity. Its fibrelessness raises the question: Is FSOC to fibre optic communications what radio is to telephony? The answer, at present, is a qualified no, because of FSOCs' inaptitude to traverse or circumvent material obstacles, the poor way they have of coping with atmospheric hazards and, above all, their highly pronounced directionality. However, FSOCs do share some of their shortcomings with microwave *radio* links, though admittedly the latter suffer from them to a lesser extent.

8.2.3 The liberated ray

The act of dispensing with the fibre makes the expression 'the captive ray' lose its applicability. The radiation is now simply *guided*. In FSOC, light waves are directed rather than ruthlessly imprisoned in the narrowest of tunnels. Contrasting them with fibre optic communications invokes irresistibly the notion of 'the liberated ray'.

8.3   The hardware of a simplex FSOC link

The change of the propagating medium from an optical fibre to fresh air need not necessarily affect the basic electronics of a lightwave link. In Figure 8.2(a) and (c), provided the between-terminals insertion losses remain the same (this *could*, but does not have to, entail an alteration of range), the modulating, driving, demodulating and amplifying electronics of the TX and RX boxes do not need changing. (However, ancillary circuits may be required, e.g. for alignment.) The truly big difference in hardware concerns the means of launching (TX) and of collecting (RX) the information-carrying radiation.

8.3.1   The TX antenna

Let us assume that the radiation generator is an LED. With neither fibre nor lens (Figure 8.2(b)) the linkage would cover a range of no more than a few centimetres, certainly no more than a few metres. Indeed, the irradiance I produced by the source at the receiver would decay rapidly, in accordance with the known $I_{RX} \propto 1/r^2$ law, which is to say that in the case of an LED producing 1 mW/mm^2 at a distance $r = 1$ mm, I_{RX} would dwindle to a mere 1 nW/mm^2 at 1 m. This would be accompanied by a corresponding fall of the SNR. Clearly, then, a beam-shaping device is required: the TX antenna (TXA).

A single-element biconvex lens represents the simplest TXA. In this connection, it will be remembered that there is no need, here, to correct the chromatic or the coma aberrations. Nor is astigmatism to be feared. At the same time, such a simple lens is one of real practical value. More elaborate TXAs are examined further. For the wavelengths considered so far (0.85–1.55 µm), all currently used optical glasses and practically most optical plastic materials are suitable.

(a)   Collimation

The object of the TXA lens is to produce as parallel a beam of radiation as possible. This beam shaping, termed *collimation*, is fundamental to FSOCs. Basically, collimation is the reverse of the operation illustrated in Figure 8.3(b) and (c). If we reverse the arrows – a perfectly legitimate beam-tracing operation – we replace the focusing of a perfectly parallel beam of light to a single point by the formation of a perfectly parallel beam of light from an idealised perfectly point-like source. (In both cases the phenomenon of diffraction spoils this splendid simplicity a little. Diffraction will be treated in Chapter 10.) Real emitters, however small, are not point sources but physical objects with finite dimensions. Each emitting point produces its own parallel beam (as shown in Figure 8.4 for points K_1 and K_2). This results in a beam spread, as illustrated. P–N junction emitters offer the advantage of being two-dimensional light sources. With volume emitters (e.g. filament bulbs) collimation would have far greater beam spreads, as it would be impossible to locate *all* the emitting points in the focal plane of the lens.

The above discussion makes us realise that genuinely parallel beams do *not* exist in practice. A little reasoning and a few sketches will show that it

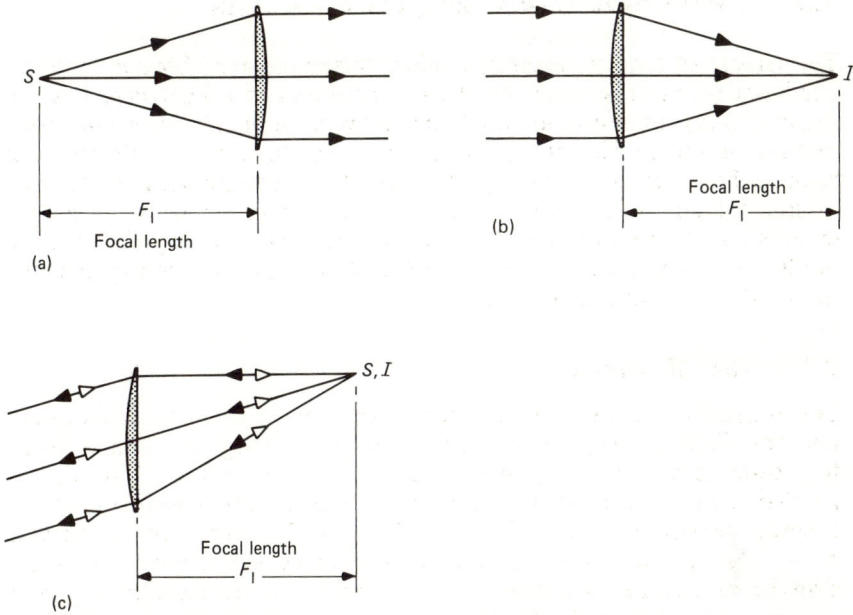

Figure 8.3 (a) Collimation shapes an idealised point-like source into a parallel beam of light. (b) Focalisation refocuses it into a small image, I. (c) Oblique ray emitter and receiver situations (obliqueness exaggerated)

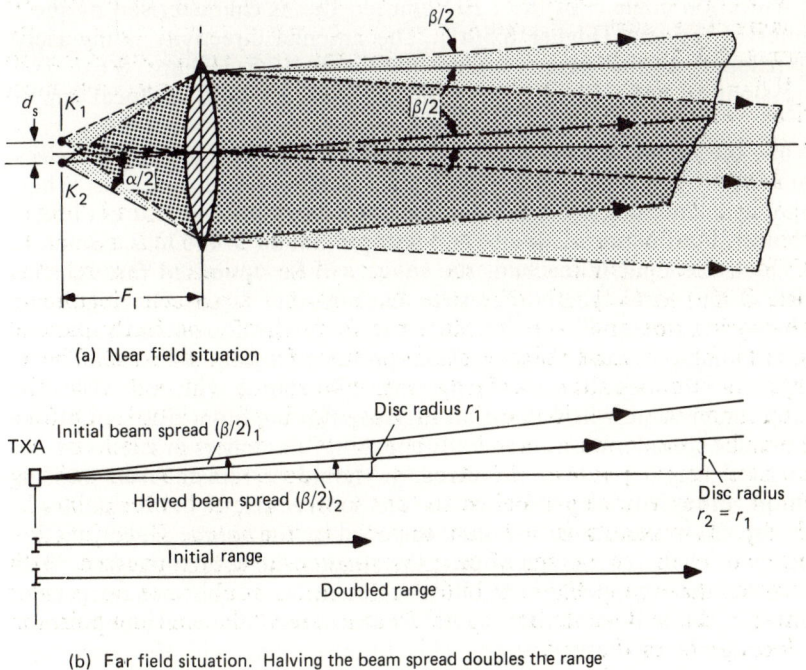

(a) Near field situation

(b) Far field situation. Halving the beam spread doubles the range

Figure 8.4 Beam spread $\beta/2$ is easily derived from dimensions d_s and f_L

is equally impossible to produce with a TXA lens a truly *convergent* beam at an appreciable distance from the source: no matter what we do, the far field beam will always end up diverging. Collimation remains the best bet in our endeavours for range. The half-angle of the beam spread (also called the divergence half-angle), $\beta/2$, is a very important design parameter. Its value is easily derived from a simple geometrical relation between d_s, the diameter of the source (usually circular), and the focal length F_L of the lens (Figure 8.4).

$$\tan \frac{\beta}{2} = \frac{d_s}{2F_L} \tag{8.1}$$

$\beta/2$ is usually expressed in milliradians. This is very practical as this angle is sufficiently small to be replaced by its tangent. We thus have

$$\beta/2 = 500 \frac{d_s}{F_L} \tag{8.2}$$

It is useful to remember that:

$1° = 17.5$ mrad

$\left(\text{from } 1° = \dfrac{2\pi}{360} \ 1000 \text{ mrad}\right)$

(b) Flux collection angle and directivity: the importance of brightness

The collection angle α of the LED emitted flux is characterised by the f number of the lens (Figure 8.5(a)). The antenna directivity is inversely proportional to the beam spread half-angle, $\beta/2$ (Figure 8.6). The expressions $f_\# = F_L/D_L$ and $\beta/2 = d_s/2F_L$ both contains the focal length F_L.

While the f number can remain the same for a whole range of lens sizes (Figure 8.6(b)), larger lenses will have smaller beam spreads. When designing an antenna, the lens has to be optimised for range. Larger lenses are better, but bulkiness, weight and cost put a limit to the lens diameter, D_L. Once this limit is reached, the range will be optimised through the choice of the focal length, F_L, remembering that larger F_L values are conducive to greater directivity but also to a reduction of the radiation collection angle, α, and thereby of the power collecting efficiency. Thus, we perceive the need for an engineering compromise with regard to F_L. The optimum value will depend on the shape of the polar diagram of the LED under consideration. Generally speaking, f numbers in excess of 2.0 (long focal lengths) will be preferred. It should be remembered that, in addition, directivity also depends on the source size, d_s, and that for a given directivity the range will be influenced by the source power, P_{TOT}. We can see that, for a given power, the *smaller* the source the *better* the directivity, and thereby the greater the range. We begin to appreciate, intuitively, the important role played by the source *radiance* (see Section 4.2) in the struggle for range. This intuitive feel will be confirmed rationally in a further section, where it will be shown that, leaving out atmospheric effects, it is this parameter that matters for range and *not* the

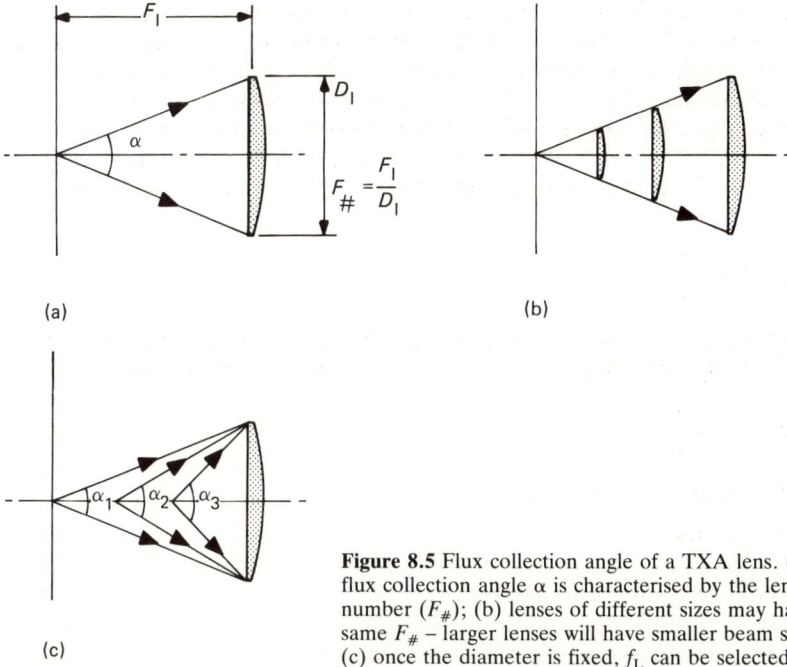

(a)

(b)

(c)

Figure 8.5 Flux collection angle of a TXA lens. (a) The flux collection angle α is characterised by the lens F number ($F_\#$); (b) lenses of different sizes may have the same $F_\#$ – larger lenses will have smaller beam spreads; (c) once the diameter is fixed, f_L can be selected

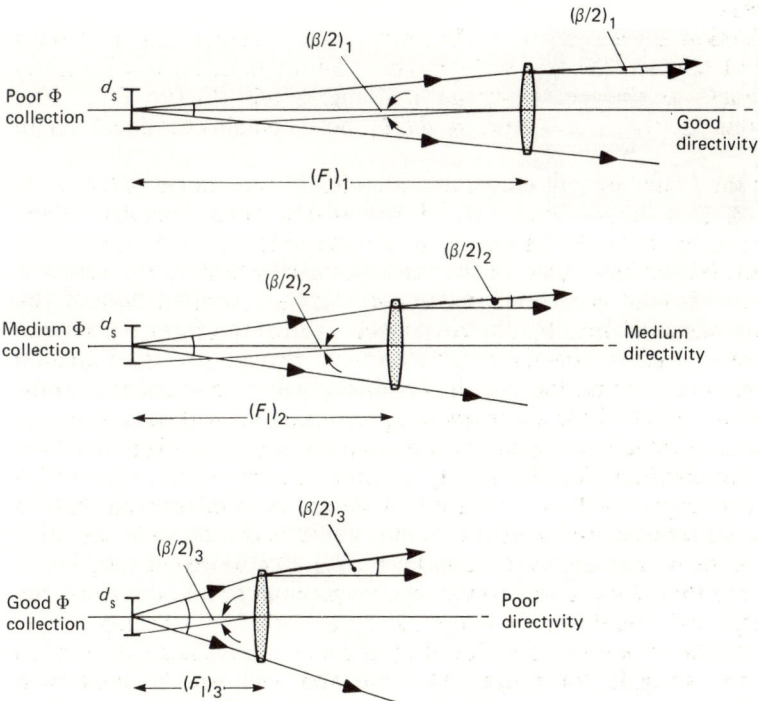

Figure 8.6 Collection angle and directivity

total source power, P_{TOT}, taken in isolation (i.e. with disregard to source size). More will be said about TXA optimisation in the section that deals with the range of a linkage as a whole.

It is important always to be aware of the fact that the representation of angles frequently used in drawings (including those in this book and most polar radiation patterns featured in catalogues) can be very deceptive for what in fact concerns *three-dimensional* geometries. Appendix 3 includes the conversion of flat angles α into space angles θ. The deception is greatest in connection with small increments. A modest increment $\triangle(\alpha/2)$ can yield a very substantial increase $\triangle\theta$ of large space angles, as can be seen from Table A3.1 and Figure A3.5. For instance, while the increase of the flat angle by one degree from 4° to 5° produces $\triangle\theta$ of 8.6 msr, adding 1° to 89° produces a $\triangle\theta$ of 110 msr. The effect of $\triangle\theta$ upon radiation collection may thus be much greater than what one would be led to assume by looking at its two-dimensional representation $\triangle(\alpha/2)$.

(c) Lensless TXAs
Collimation can be obtained by the use of mirrors instead of lenses [2, pp. 385–394]. This is not completely without problems, though. Surface mirrors do not lend themselves well to scratch-free cleaning, while the glass thickness of second surface reflectors introduces refraction. Spherical mirrors are poorer collimators than paraboloidal ones but can be much cheaper. In either case, the now reversed light source and its supporting structure obstruct some of the outgoing radiation. Collimation adjustments

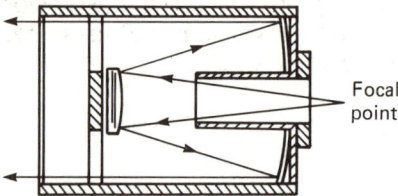

Figure 8.7 Example of Cassegrain optics

are more difficult here than in the case of lensed TXAs as the source is located awkwardly *inside* the mirror structure. A combination of two mirrors, one parabolic, the other hyperbolic, called the *Cassegrain* objective, does not suffer from this predicament (Figure 8.7) and has the advantage of being compact for the focal length it offers. The Cassegrain is, however, expensive to fabricate.

(d) Mirror-and-lens systems
The combination of refraction and reflection can be of interest. Objectives using second surface mirrors with a non-uniform thickness of the glass portion are called *catadioptric* [2, pp. 394–398]. Best known is the Mangin objective. Little information on the use of catadioptric TXAs is available.

(e) General
Apart from the light source and the collimating optics, a TXA emitting 'gun' will usually house some electronics. Its inside walls should be non-reflective (e.g. metal treated or painted to an absorbent matt black

finish) in order to avoid the formation of radiation squandering sidelobes. In addition to the gun a TXA will carry an alignment telescope, for sighting the RXA, often mounted on the outside of the housing. The housing itself must be sturdy, strong and vibration damped, for the sake of pointing stability. In outdoor sitings, weatherproof enclosures with de-misting facilities are a must.

8.3.2   The RX antenna

There is little difference between TXA and RXA architectures. Where there was an LED or a laser emitter there is a PIN or an APD receiver; where there was a small piece of driver electronics there is a PCB-preamplifier; where there was collimation optics there is, obviously, decollimation optics. Regarding size, the belief is that, in the case of a 'Point A to Point B' link, cost minimisation requires the TX and RX optics to be of the same diameter. The one antenna element particular to RXAs is the IR band pass filter, intended for cutting out unwanted radiation, in the interest of the SNR. As in the case of the TXA, a single-element biconvex lens represents the simplest, yet often adequate, optics. The $\beta/2$ angle becomes the *angle of admittance*, $\gamma/2$. The focal length f_L will be more often than not the same as that of the TXA lens, for simplicity and economy, but sometimes smaller to increase $\gamma/2$ and thereby ease alignment. In RXAs, too, mirrors can replace lenses, and catadioptric arrangements can be used. The photoreceiver area will usually be some 2–5 times larger than the TX radiating area, with a compromise on junction capacitance, noise, flux collection capability and admittance angle being reached, always bearing in mind that optical blur will cause the diameter of the received 'spot', d_r, to be larger than d_s [2].

8.4   The FSOC link as a whole

8.4.1   Alignment problems

In the interest of range and security, the TXA ought to have as small a divergence half angle, $\beta/2$, as possible. The acceptance half angle, $\gamma/2$, of the RXA, while larger than $\beta/2$, should nevertheless remain sufficiently small to ensure a good signal-to-noise ratio. The penalty for the high directivity of these terminal equipments appears in the shape of alignment problems. In fixed and mobile installations alike, the supporting structures must be stable and 'deadened' (damped) against mechanical oscillation, despite their firmness. Mobile links require, in addition, the provision of a pan-and-tilt mechanism with judiciously chosen pivot axes, being usually expected to be erected and aligned speedily. Such designs will allow for both terminals to be manipulated simultaneously with maximum speed during the alignment operation. One such design is described in Ref. 5. With just one exception [163], all equipments known to the writer use alignment telescopes, usually permanently attached to, if not built into, the TXA and RXA heads. Telescopes suffice for the initial establishment of communication in the field. The final trimming is best done by maximising the strength of the received signal.

8.4.2 Factors affecting range

Why is there a limit to the distance a liberated ray optical communications channel will carry information? In fibre optics, the captive ray is contained *within* the physical boundaries of its guide – the fibre. With this spatial confinement of the radiation, range limitations stem exclusively from attenuation and modal dispersion caused by the medium. In free space optical communications this is not so. There is relatively little attenuation in the atmosphere and next to none above it, and modal dispersion can be ignored in all but a few rare situations. The major reason why an FSOC link cannot carry information right into infinity (above the atmosphere) lies in the unbeatable *beam divergence* problem. However hard we try to collimate it, the TXA radiation will just not be corsetted into a *truly* tubular shape (Section 8.3.1(a)). Refusing to keep its diameter constant, the beam keeps widening as the TXA–RXA distance grows (Figure 8.8), thus continually weakening the irradiance at the RXA aperture. Beam divergence is the primary enemy. It can be shown that the range, R, is inversely proportional to the beam spread half angle $\beta/2$ (Appendix 9).

$$R \propto \frac{1}{\beta/2} \tag{8.3}$$

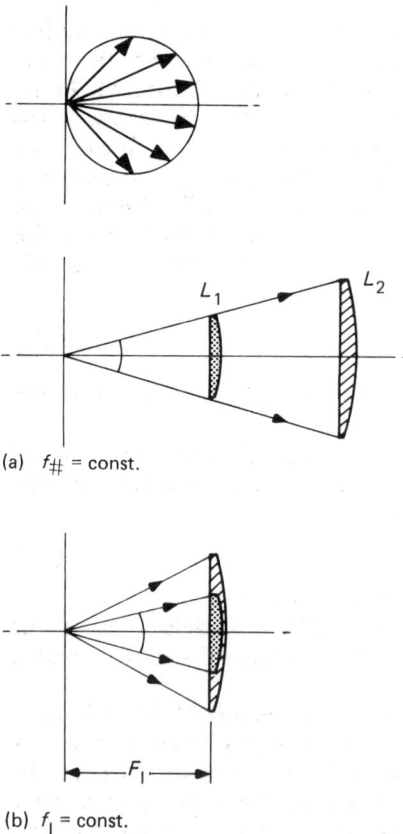

(a) $f_\# = \text{const.}$

(b) $f_1 = \text{const.}$

Figure 8.8 While changing the lens diameter of a design we can keep constant either (a) the f number or (b) the focal length, but not both

What are the other factors affecting the attainable range? Consider first the transmitter *power*. As in any communication link, the ultimate limit to range is set by the signal-to-noise ratio at the receiver. As the irradiance at a given distance is proportional to the total transmitted power, $(P_{TX})_T$, everything else being equal, the signal strength, $(P_{RX})_T$, will be proportional to it. It can be shown, however (see Appendix 9), that doubling the transmitter power pays less dividends than halving its divergence. Indeed, it works out that the range R is proportional to the square root of $(P_{TX})_T$ only:

$$R \propto \sqrt{(P_{TX})_T} \tag{8.4}$$

Let us now look at the influence of the lens size. We have already considered the irradiance at the RXA input aperture. Obviously, the larger the admission aperture, the greater the intercepted fraction of the total interceptable flux. It follows that the signal power, $(P_{RX})_T$, is proportional to the clear lens surface area, A_{RXA}, and hence to the square of its diameter, d_{RXA}. As for *range*, it can be shown that, everything else being equal (e.e.b.e.), it *is directly proportional* to d_{RXA} (see Appendix 9):

$$R \propto d_{RXA} \tag{8.5}$$

A moment's reflection will show that the e.e.b.e. condition requires clarification when the lens diameter undergoes a change: we can keep constant either $f_\#$ or the focal length f_L, but not both (Figure 8.8). In the above case of the RX lens, it was tacitly assumed that the lens *size* was being varied while its *shape* remained unaltered ($f_\# =$ const.). If the same assumption is made for the TX lens, every increase of d_{TXA} results in a similar increase of f_L and hence in a similar decrease of β/2. This is one of the two reasons underlying the truth already contained in relation 8.3. We can thus state that, everything else being equal, the *range* is *directly proportional* to d_{TXA}:

$$R \propto d_{TXA} \tag{8.6}$$

bearing clearly in mind that Expression 8.3 is an implication of Expression 8.6.

As, in most links, TX and RX lenses of equal diameter represent the most economical solution, the range R becomes heavily influenced by their size. Namely $d_{TXA} = d_{RXA} = d$ leads to:

$$R \propto d^2 \tag{8.7}$$

Increasing the lens size pays handsome returns indeed.

The treatment of d variations becomes a little more delicate when the e.e.b.e. condition is to imply a constant f_L. The more inquisitive reader may find Appendix 3 of some help here.

The type of reasoning used to explore the influence of β/2, $(P_{TXA})_T$, d_{TXA} and d_{RXA} can be applied to other engineering parameters of the TX and RX terminals, such as f_L and $f_\#$ of both antennas, NEP of the photoreceiver, link bandwidth BWD. Some of these parameters have *multiple* effects, not all of which are linear, making things a little complicated. At this stage, we sum up the above discussion by giving a *scaling* relation for the range:

$$R \propto \frac{d^2, \ \sqrt{(P_{TX})}}{\beta/2, \ \sqrt{NEP}, \ \sqrt[4]{BWD}} \tag{8.8}$$

to show the influence exerted by the design parameters α, P_{TX}, $\beta/2$, NEP and BWD on the range of an *atmosphere-free* link. The presence of commas in relation 8.8 indicates that it is *not* to be treated as a formula for calculating r, as *each* parameter is supposed to be varied alone, e.e.b.e.

The form of the actual range equation varies substantially from publication to publication. So do its limitations. I prefer an easy four-step calculation method to a single, but obtuse all-embracing formula. Expounded in Appendix 9 and demonstrated by way of a worked example in Appendix 10, the method is based on the relation:

$$\log R + \frac{AR}{20} \log R_{FS} \tag{8.9}$$

which, though not solved for R, easily yields a numerical solution for a given set of conditions, after only three to four quick iterative trials. R_{FS} is the range of an atmosphere-free link given by:

$$R_{FS} = R_o 10^{I_{dB}/20} \tag{8.10}$$

in which R_o is the 'back distance' of a virtual source and has the value:

$$R_o = \frac{d_{TX}}{\beta} \tag{8.11}$$

I_{dB} is the permissible total insertion loss of the link, and A, in Equation 8.9, the atmospheric attenuation in dB/km.

(a) Atmospheric effects

It has been stated earlier that there is relatively little attenuation in the atmosphere and next to none above it. Table 8.1 gives some figures; with their support we can establish to what extent real-life atmospheric attenuation reduces the range of an FSOC link.

For our calculations we shall assume a beam divergence of 0.3 mrad ($\beta/2 = 0.15$ mrad), a TX starting power of 1 mW (LED), a TXA optical efficiency $\simeq 50\%$, an RXA collecting efficiency $\simeq 100\%$ (coated lenses), a

Table 8.1

Typical atmospheric attenuation A (dB/km)	Conditions	Human eye visibility (km)
0.04	Extremely clear	50–150
0.45	Very clear	2–50
1.25	Clear	10–20
2.6	Light haze	4–10
10	Haze	2–4
13	Fog	1–2

single voice channel bandwidth of 3 kHz and a receiver NEP = 2×10^{-14} W/Hz$^{1/2}$ (as per Section 6.11.6).

From these calculations (detailed in Appendix 10) we find that:

1. The permissible path loss is 76.8 dB (allowing 12 dB as an 'intelligibility margin' SNR).
2. The range, in atmosphere-free space (attenuation $A = 0$ dB/km), is an amazing 3230 km.

If we now assume an atmospheric attenuation of 0.2 dB/km, a US Naval Research Lab figure for $\lambda = 0.9$ μm and a relative air humidity of 66% [68], the range reduces to 137 km. This is still a highly respectable figure. With $A = 2.6$ dB/km (Table 8.1, line 4, Light haze) it reduces further to 17.5 km.

The first and even the second range of figures in this calculation must be viewed, however, with great caution, for more than one reason. Firstly, A undergoes large diurnal and, in some geographic locations, very large seasonal variations. Secondly, the clearest of weathers in the driest of climates can have its drawbacks too: hot air turbulence and still air beam bending, both caused by the variation of the refractive index with temperature (see mirages in Section 7.3.3). With the very small divergence $\beta/2$ assumed, hot air turbulence may cause the beam to perform a bumble-bee dance in the receiver plane, around and over the RXA, causing heavy fading. Refractive bending can throw the beam off target altogether. Thirdly, it must be remembered that *optical* noise has been left out of the calculation. In fact, the atoms and molecules making up air (not only nitrogen and oxygen but also water vapour and carbon dioxide), and the often present foreign bodies such as pollen, insect swarms and industrial pollutants can, in addition to causing scattering and absorption of signal energy responsible for the already mentioned atmospheric attenuation, act as strong scatter centres for sun rays and generate optical noise. Fourthly, for distance much above 50–75 km (depending on the elevation of the terminals), earth curvature can come into play and cause the horizon to intercept the beam on its TXA–RXA journey.

8.4.3 Ranges achieved so far

Despite the various adverse effects a range of 50 km has been covered by an experimental FSOC link as early as in 1963 [165] and a number of commercial equipments have been marketed since the late 1970s for distances in excess of 20 km, with some claiming a possible 64 km [67, 166]. In my view, above the atmosphere, i.e. in space, links using powerful lasers with expander optics (see Chapter 10) should be capable of reaching through distances of thousands, maybe tens of thousands, of kilometres, for narrow-band data transmission.

It would be wrong to close this section without mentioning interesting, as well as useful, achievements at the other end of the scale: wire-less, radio-less and licence-free telephone extensions, remote TV controls and toys, operating over 5–10 m, all relying for their indirectionality on wide-beam TXAs and RXAs, and on photons bouncing off the walls.

8.4.4   Tolerable BERs and SNRs

With the continually growing penetration of digital techniques into communications, it was inevitable that the operational reliability of a linkage would eventually be expressed in terms of the relative number of digits lost or gained in the course of transmission through the presence of noise. The Bit Error Rate (BER) represents this number statistically.

Pulse modulation is of a particular interest in optical communications, mainly because the instantaneous output power attainable with LEDs and lasers can easily exceed their mean power by 10 dB or more, and thus increase the range. (This argument must be viewed with caution as in some installations part or all of the power benefit can be lost through the demands of greater receiver bandwidth. In many practical cases, though, there is an overall range benefit. See Ref. 168.) Hence, the importance of the BER concept to us. Speech intelligibility and BER can be numerically related. For instance, with a BER = 10^{-2}, 90% of the words of an English verbal message will be understood [167]. Mathematically minded communication engineers have succeeded in elegantly linking the BER not only to this intelligibility, but also to the somewhat older signal-to-noise ratio. The subject has received some coverage in Ref. 60 (p. 9).

8.5   Examples of FSOC equipment

To give a practical notion of the size, shape and fixing of FSOC equipment, illustrations of a few embodiments of their terminals will now be given. Some show that we have come some way since Bell's days. Figure 8.9 shows the OE indoor 'no-cables' telephone system mentioned in Section 8.4.3. An outdoor one-way speech link was designed and built by the writer in 1964, to demonstrate the possibilities offered by the then two-year-old industrial GaAs IRED in support of an IEE Lecture [270]. It was followed, a few years later, by a similar looking semiduplex (press-to-talk) link of a similar range using a common optical head for both the transmitting and receiving function, as well as a common TX/RX GaAs junction [48]. This semiduplex was probably one of the very first

Figure 8.9 Siemens cordless telephone extension. Omnidirectionality is obtained through the use of backscatter from walls (Courtesy of Siemens AG)

Figure 8.10 Interlaser System (Courtesy of Modular Technology Ltd)

Figure 8.11 Long-range TXA/RXA pair intended for experimental work on atmospheric propagation at University College London

publicly shown [INEL, Basle, Ref. 179] application of two-way (emitter/receiver) devices. The video transmission equipment developed in 1975 under the name of 'Leelink' was quite advanced for its time – it accommodated three outgoing and one return channel [271].

The outgoing beam of the master terminal carried one video and two audio channels (programme image and sound plus service speech), while the single incoming channel of the incoming beam carried service speech. The full duplex service speech channels also hosted an alignment tone. The terminal featured a 'quick alignment' ($\sim$ 40 s) optohead geometry [5], invaluable in mobile links. Figure 8.10 shows one of the many commercially available (at the time of writing) long-range equipments that use lasers. It is a simplex, so that full duplication is required for duplex communication. Finally, Figure 8.11 shows a long-range TXA and RXA pair intended for experimental work on atmospheric propagation (University College, London).

8.6   FSOCs and their competitors

Section 8.2 considered the value of FSOCs in situations where fibres won't go. The enumeration of such situations there amounts to pointing out the superiority of FSOCs in *special* circumstances. However great the fascination the liberated ray exerts on our imagination through its possibilities in such applications – and great it certainly is – it mustn't blind us to the existence of a mighty competitor: the microwave link. Microwaves will also jump rivers, cross railway tracks and spare us digging up the high street to convey data, speech, music or pictures from one side to the other and, hence, are more of a competitor to FSOCs than FOCs. How do these two cableless means of communication compare? Microwave (MW) links score better in long-range jobs, especially when high reliability is of concern, because the propagation of the much longer wavelength radiation used here is less subject to severe rises of atmospheric attenuation or to fadings caused by fog, rain, snow and heat. In other words, MW links have lower down-time figures. For short and moderate range (say 10 m to 1 km) applications in moderate climates, AGC will cope with occasional 10 dB or 20 dB surges of atmospheric attenuation. There, FSOCs should win because of:

1. Lower equipment weight and size (unobtrusiveness, portability)
2. Lower power consumption permitting battery operation
3. Freedom from licence requirements
4. Greater ease of installation and alignments
5. Greater information carrying capacity (bandwidth)
6. Greater security
7. Better immunity to EMI
8. Lower cost (when fully developed)

In short-range cases (e.g. TV shows and interviewing) UHF radio scores a superiority, despite its high EMI susceptibility and licensing requirement, because of its lesser directionality.

For long distances it may well be that a technique based on the use of a wavelength lying between the micrometre and the centimetre range of FSOCs and MWs respectively will provide the best solution to cableless communications.

Chapter 9

The dangerous and the non-dangerous laser

9.1 Why the laser is such a special light source

The laser is a very special light source because the light it produces is *coherent* (Figure 9.1). As such, it can be easily shaped into highly parallel beams. Highly parallel beams, in turn, can be easily focused into extremely small spots to produce enormous irradiances. No other light sources can produce highly parallel beams at significant power levels. In wavefront parlance, flatness, parallelism and front equidistance (i.e. uniqueness of wavelength) of the radiation are far better here than in light of any other origin, all because of the extremely high degree of coherence. Coherence concerns more than the frequency of the radiation; not only is laser light monochromatic but the *phase* of all its constituent photons is also the same. They work in unison, which makes them so effective (splendid proof of the value of cooperation).

While present in radiowaves, this phase coherence is absent from the light produced by an LED, even when monochromatic radiation is extracted from it by equipping it with an optical notch-filter. Monochromatic it may well be, but not coherent. There, wavefronts are produced in 'packets', by randomly born photons and, since the various births are unrelated, the result is a non-synchronised light emission. As for incandescent sources, where photons have vastly different energies as well as random times of emission, the electronics engineer may view them as enormous noise generators leading to the thought – 'not the best of devices for telecommunication usage'. Before more is said on this important subject of coherence (Section 10.1) let us look at the way lasers work.

9.2 The way lasers work

It has just been implied that, in lasers, the various photon births *are* related. The photon is a wavetrain. Many an electronics engineer will associate the idea of time-related wavetrains with that of an oscillator, their source. The laser *is* an oscillator, a *light oscillator*, despite the fact that its acronym (LASER: Light Amplification by Stimulated Emission of Radiation) alludes to an amplifier. Of course, it is thanks to high gain amplification and positive feedback that oscillations are established, here

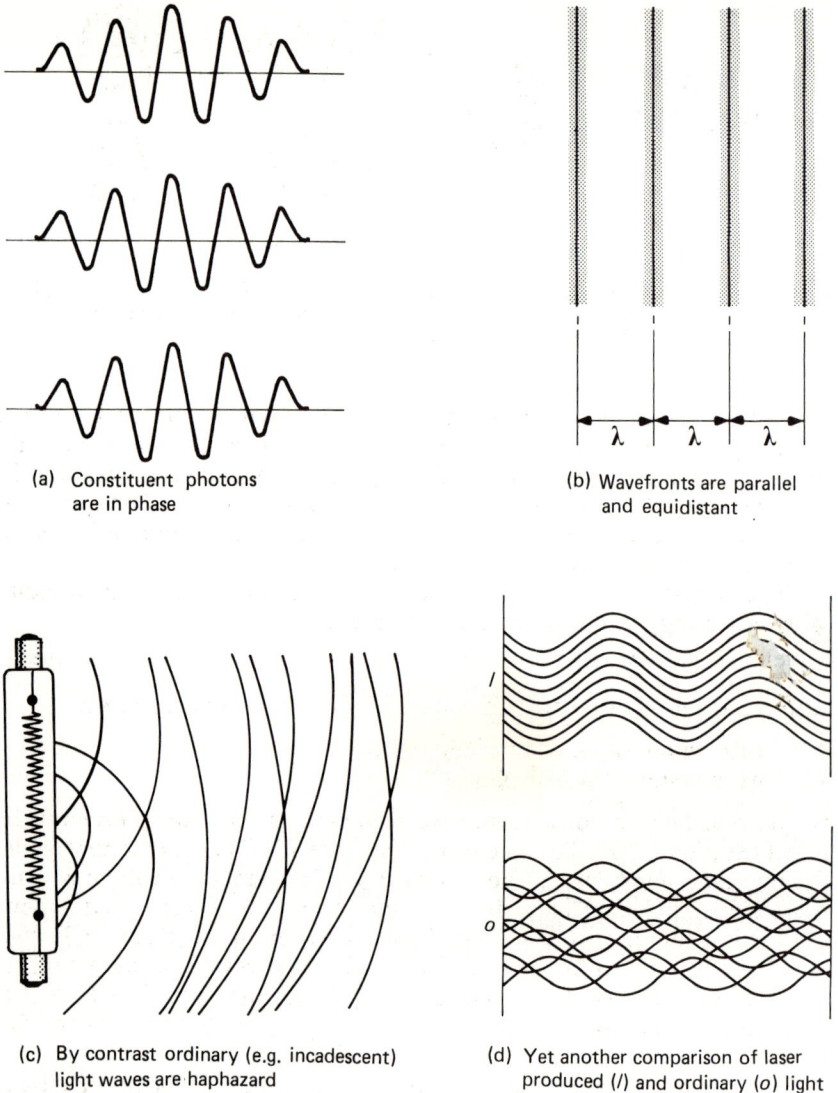

(a) Constituent photons
are in phase

(b) Wavefronts are parallel
and equidistant

(c) By contrast ordinary (e.g. incandescent)
light waves are haphazard

(d) Yet another comparison of laser
produced (*l*) and ordinary (*o*) light

Figure 9.1 Laser light has a high degree of coherence. Constituent photons are in phase, wavefronts are in step – parallel and equidistant

like elsewhere, but neither a semiconductor nor a HeNe laser is just an amplifier, since both put out oscillations with none being supplied to them. Their only feed is electrical d.c. power source: d.c. in, 500000 GHz out! At the heart of the light-modification process lies Einstein's stimulated emission of photons. Einstein predicted in 1917 that, under certain circumstances, an *incident photon* will generate another one, of exactly the same energy and hence the same frequency (Figure 9.2). This two-for-one process is, of course, equivalent to amplification and to a high one, as the

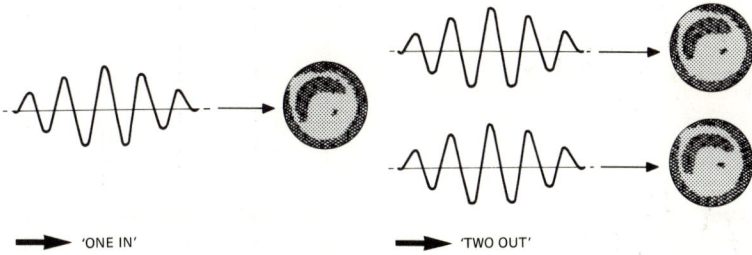

'ONE IN'   'TWO OUT'

Figure 9.2 Figurative representation of Einstein's stimulated emission of photons taking place in 'lasing' media. Both photons, old and new, have identical energy (and hence frequency), phase, polarisation and direction of propagation

two-for-one mechanism can go on reproducing itself. Einstein added that in this type of emission both photons, old and new, will be *in phase*, will have the same polarisation and will propagate in the same direction. His entirely theoretical prediction was made 43 years before the first laser was made, at a time when electronics was unheard of, transistors didn't exist and vacuum valves were still a novelty.

Einstein stated that the impact of a photon of frequency f on an atom with two energy levels differing by the amount:

$$\Delta E = E_2 - E_1 = hf \qquad (9.1)$$

where h is the Planck constant can have one of two consequences:

1. The creation of an electron–hole pair
2. The emission of a new photon

In the first type of consequence we recognise the familiar photoelectric effect (Section 3.3.2). The electron moves upwards (in energy terms) while the photon is annihilated. The second type of consequence will be new to most electronics engineers. In it, the electron *loses* energy, and a new photon is created. This is the *stimulated* photon emission (Figure 9.3), the very basis of laser action. Note again that either process requires that there be a kind of attunement between the energy level difference $E_2 - E_1$ of the atom and the photon energy/frequency. Unless $f = \Delta E/h$, nothing will happen.

Of fundamental importance was Einstein's statement of the probabilistic character of the reaction: the probability of (1) and/or (2) taking place in a

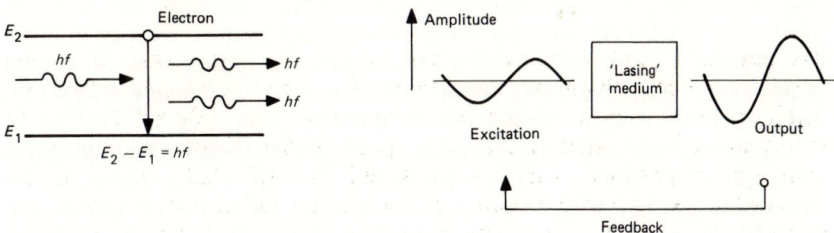

Figure 9.3 In stimulated emission of photons, the $E_2 - E_1 = hf$ attunement is essential

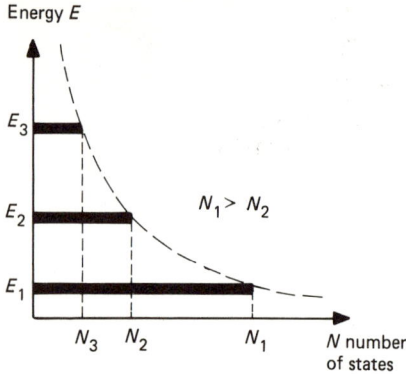

Figure 9.4 Normal distribution of states

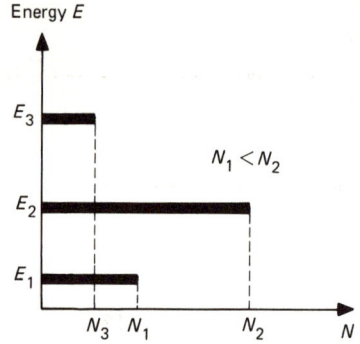

Figure 9.5 Population inversion of states (through pumping)

material is linked to the specific number of active sites N_1 (energy level E_1) and N_2 (energy level E_2). If we are used to thinking of an electron moving upwards in its energy level upon the incidence of a photon, it is because in all materials in their normal (ground) state the lower energy levels in atoms are more crowded than the higher ones, in accordance with the general trend in nature of energy minimisation (Figure 9.4). This makes the upward transitions immensely more likely to happen.

The likelihood of the inverse, *downward* transition, can be artificially increased by increasing the number of electrons in the upper energy level. Atoms in such a condition are in the *excited* state. In strongly excited materials the number N_2 can become larger than N_1. This abnormality is rather imaginatively described as *population inversion* (Figure 9.5), with the rich outnumbering the poor!

The excited atoms are in a meta-stable state and hence always keen to revert to the ground condition. The ones that do so 'unprovoked' cause *spontaneous* light emission. This is, for instance, the case of atoms in an LED or a fluorescent tube. Such reversions occur at random. In strongly excited, i.e. population inverted, conditions, however, inversions are caused by an incident photon. Attuned to the energy surplus $\Delta E = E_2 - E_1$ of the raised electrons by its own energy, $\Delta E = hf$, the incident photon topples the electron which, upon dropping to the E_1 state, instantly loses its energy surplus ΔE to the newborn photon. This is *stimulated* light emission. In strongly inverted populations the stimulation will be a self-perpetuating and a much amplifying mechanism, the two-for-one reaction leading on to four-for-two, eight-for-four reactions, and so forth, until the quest for equilibrium in the material has gone far enough to make E_2 electrons short in supply. (The process is reminiscent of carrier multiplication in avalanche photodiodes, and, to a lesser extent, of electron multiplication in photomultiplier tubes.)

The stimulated reversions are *not* random: all the photons belonging to a group of downward ($E_2 \rightarrow E_1$) transitions are *strictly in step*. To sustain the in-step photon generation one must sustain the population inversion, i.e. keep topping up the E_2 level with excited electrons. This operation is called *pumping*. Laser pumping can be done continuously or 'in bangs'. In lasers,

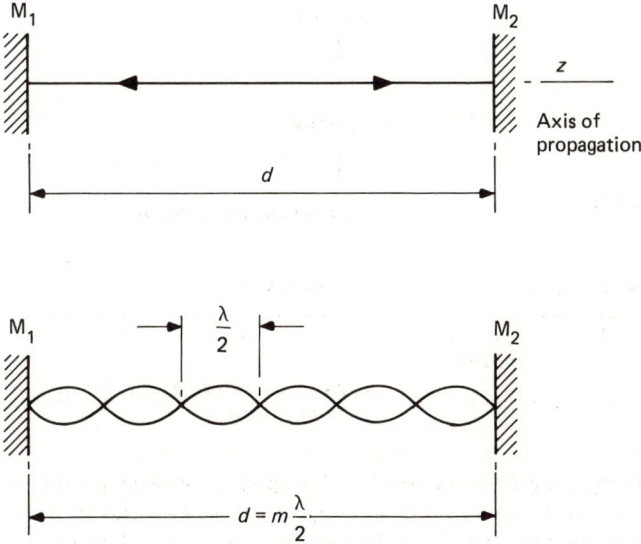

Figure 9.6 Mirrors are placed at both extremities of the lasing medium. The presence of mirrors favours wavelengths for which $m\lambda/2$ fills neatly the inter-mirror distance

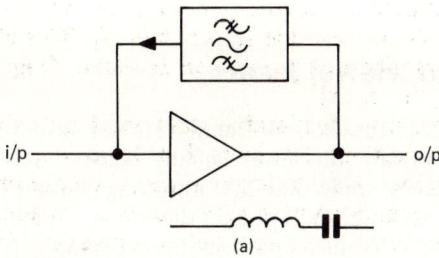

M_1 – Fully reflective mirror
M_2 – Partly transmitting mirror
E – Enclosure

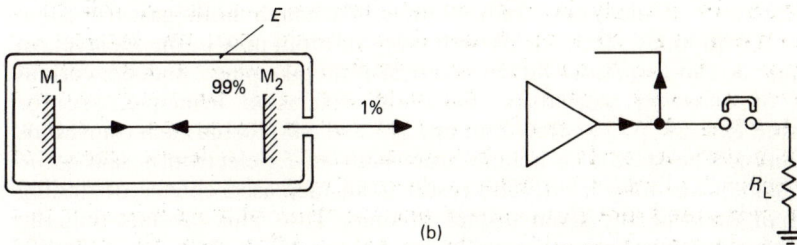

Figure 9.7 (a) In a laser cavity we recognise the optical analogue of the selective positive feedback in an electronic amplifier. (b) Useful output power is obtained by 'bleeding off' the resonator

mirrors are placed at the boundaries of the active medium at 90° to the chosen direction of light propagation. This has the following, most important effects:

1. It causes photon flights to make multiple passes in the medium, thus increasing the chances of photon/electron encounters (positive feedback!).
2. It favours the light production in the z-direction (spatial coherence).
3. It creates conditions for the establishment of standing waves between the mirrors by favouring the generation of such wavelengths, λ, for which an integer number m of half wavelength $m\lambda/2$ fills neatly the mirror-to-mirror distance d, which becomes the *periodicity* defining element (frequency coherence).

We recognise the optical analogue of *selective* positive feedback in an electronic amplifier, made into an oscillator (Figure 9.7(a)). Neither the oscillator nor the laser would be of much use without an output. In both cases the useful output power is obtained by 'bleeding off' the resonator (Figure 9.7(b)). In a laser, this is done by making one of its end mirrors partly reflective and partly transmittive. In a HeNe laser, 1%, typically, of the oscillating power is allowed to escape from the resonant cavity by the use of a fractionally transparent mirror at the output end. Laser action can be obtained in solids (especially, though by far not exclusively, in semiconductors), gases and liquids. Pumping can be electrical, optical or chemical (for further reading see Refs 75, 23). Two of the most frequently encountered lasers will be described now; the HeNe and the GaAs laser.

9.3   Helium–neon (HeNe) lasers

The most widely used is the helium–neon (HeNe) laser. Inexpensive, reliable and easy to operate, this laser usually comes in the shape of a 30–50 mm tubular housing some 300–500 mm long, linked by cable to a power supply unit the size of a small pocket dictionary. A range of models deliver from 0.5 to 50 mW of pleasing bright red beams.

The working of a HeNe laser (Figures 9.8 and 9.9) is based on a gas-to-gas energy transfer: the helium does the pumping, the neon does the 'lasing'. Energy-wise, a three-level process is involved (Figure 9.10). The gas mixture is ionised by the application of some 2 kV d.c. (10 kV starting) to the two end electrodes of the laser. Although both gases become ionised, with many electrons of both types of atom now in the E_3 level, in neon alone can a *radiative* downward transition take place. Indeed, certain rules of quantum mechanics (the Pauli Exclusion Principle, see for example Ref. 29, p. 9 or Ref. 280, pp. 515 and 548) forbid such transitions in helium. The state of a helium atom with an E_3 electron is said to be *meta-stable*, i.e. long lived. The result is an increased chance of finding many of them at any given instant of time. This, plus the fact that the mixture contains some 10 times more helium than neon atoms makes collisions between excited helium atoms of the above state, and ground state neon atoms, highly probable. It is these collisions that cause the gas-to-gas energy transfer. Called *inelastic*, they make the helium atoms

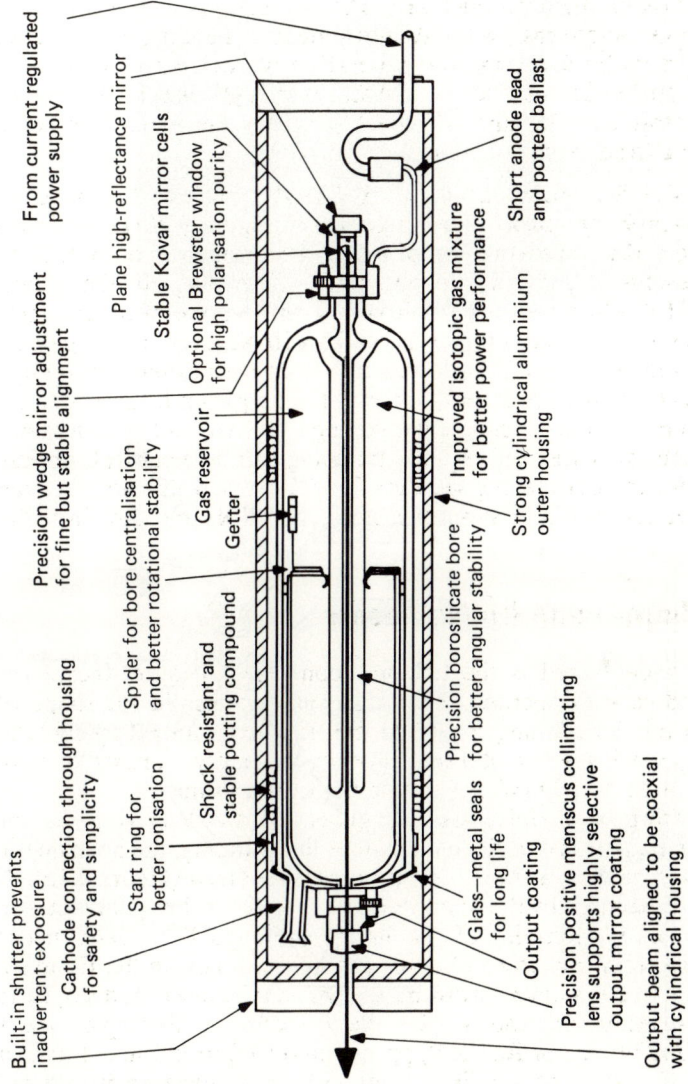

Built-in shutter prevents inadvertent exposure

Cathode connection through housing for safety and simplicity

Start ring for better ionisation

Shock resistant and stable potting compound

Precision positive meniscus collimating lens supports highly selective output mirror coating

Output beam aligned to be coaxial with cylindrical housing

Precision wedge mirror adjustment for fine but stable alignment

Spider for bore centralisation and better rotational stability

Gas reservoir

Getter

Precision borosilicate bore for better angular stability

Glass—metal seals for long life

Output coating

From current regulated power supply

Plane high-reflectance mirror

Stable Kovar mirror cells

Optional Brewster window for high polarisation purity

Improved isotopic gas mixture for better power performance

Strong cylindrical aluminium outer housing

Short anode lead and potted ballast

Figure 9.8 Internal structure of a 5 mW commercially available HeNe laser (Courtesy of Melles Griot)

142

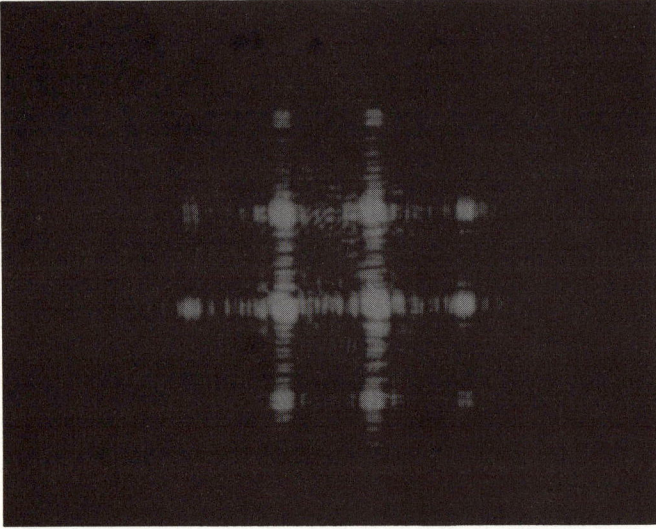

Figure 9.9 One of the stationary patterns the laser can generate

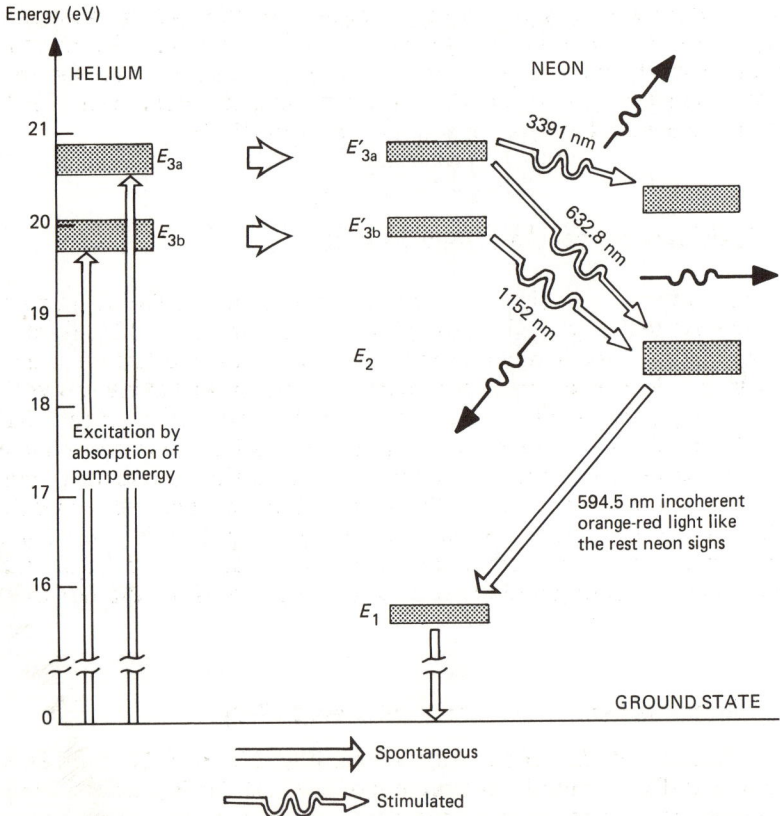

Figure 9.10 The working of a HeNe laser is based on a three-level energy transfer. Helium does the pumping, neon the lasing

into absolute losers and neon atoms into absolute winners: the former drops back to the E_1 state, the latter rises to E_3. We see straight away that the collisions cause a population inversion in neon, one of the three necessary conditions for lasing. The two other conditions, photonic stimulation and optical selective positive feedback, are satisfied as follows. Permitted radiative $E_3 \rightarrow E_2$ downward electron transitions in neon cause spontaneous emission whose photons can trigger stimulated emission and avalanching, while mirrors at both laser ends spaced at $d = m\lambda/2$ ensure the selective feedback. Figure 9.11 shows the cross-section of a commercially produced HeNe laser. Before closing, it should be mentioned that the emission line structure of the helium–neon mixture is more complex than in the above description, with more than one allowed radiative transition in neon and four main lasing wavelengths, three of which are in the infrared. The predominantly exploited one, however, is the immensely popular visible $\lambda = 632.8$ bright red radiation resulting from the above described process.

9.3.1   The HeNe laser and the electronics engineer

HeNe lasers are driven by special-purpose power supplies which produce a starting and a maintenance voltage, both of a few kilovolts. As internal (electronic) modulation is not possible there is really nothing for the electronics engineer to do, apart from connecting the laser to its EHT supply and plugging the latter into the mains socket!

9.4   Gallium arsenide (GaAs) lasers

If not the most widely used, this is the smallest of all lasers. The active part of the semiconductor chip often occupies a volume of a mere 1/10000 of a cubic millimetre – and sometimes much less! It is also the laser that lies nearest to the electronics engineer's heart, as its packaging and its mode of use are highly reminiscent of transistors, diodes, and the now familiar LED. Within this small chip of crystalline GaAs lie all the constituent parts required for laser action: the seat of population inversion, the replenishing 'pump', the resonance chamber with its two mirrors for positive feedback and the generator of stimulator photons! As the output of this little light factory can be of several milliwatts, it is no wonder that a large and efficient heat sink is required to keep its temperature within the required limits.

9.4.1   Double heterojunction laser diodes (DHLD)

These lasers are of the stripe geometry variety. Figure 9.11 represents diagrammatically the multilayer structure of one such device, while Figure 9.12 shows the energy levels of the active part of its transverse cross-section. Before proceeding, the reader may find it helpful to refer back to Sections 3.2.2 and 3.3.1 (especially to Figures 3.6 and 3.8) with regard to

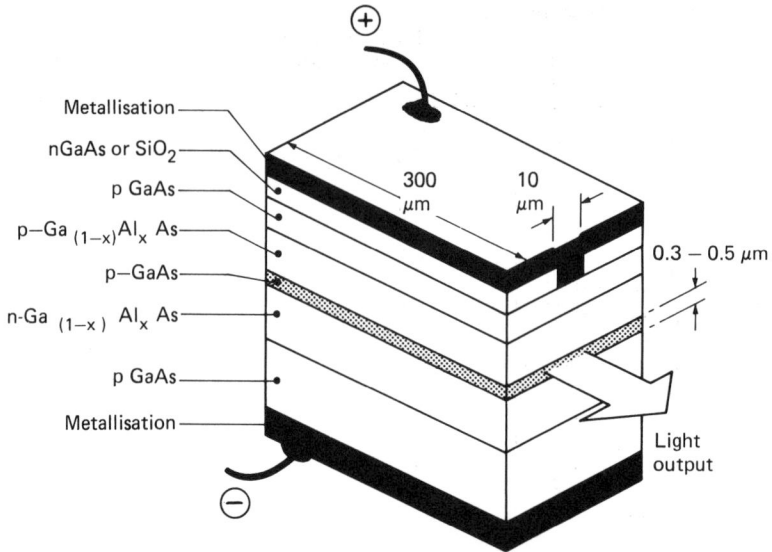

Figure 9.11 The laser chip has a layer structure (double heterojunction stripe geometry)

electron injection in FQM operated PN junctions, and to Sections 5.3.1–5.3.3 (especially Figures 5.7 and 5.8), with regard to the use of heterojunctions for electrical and optical confinement.

The idea behind the use of stripe geometry and heterojunctions is to make everything functionally important happen within a very small volume of material. If the I_F current of the laser diode could be made to flow through a very narrow channel, large current densities would be achieved with low threshold currents, I_{th} (the smallest value of I_F for which laser action takes place). This has been achieved and I_{th} values of some 30–200 mA are commonplace today. If electron–hole recombination could be restricted to a small volume of material, large photon densities would increase the chances of stimulated emission. Here, volumes of some 300 μm^3 have been achieved. Finally, if light could be ducted in a stripe of material sufficiently thin to behave like a waveguide, its propagation/amplification would be quasi-unidirectional, yielding a further enhancement to laser action. In the example below (and many other DHLDs) a stripe of material 0.1 μm thick (a fraction of the wavelength) is being used.

In practice, the active chip has a layer structure (Figure 9.11), the various layers being obtained by epitaxial growth (a continuation of the lattice structure of the crystal) on a GaAs substrate and doping. In some of them, aluminium atoms have been added to the lattice as a replacement for gallium. The ternary composition is referred to as $Ga_{(1-x)}Al_xAs$, with x representing the mixing proportions, both ternaries being doped in the usual way for P or N polarity. The addition of aluminium increases the energy gap of the material. In the middle of the structure lies the *active* p layer, the all-important small volume of material within which radiative carrier recombinations and light amplification takes place. The injection of

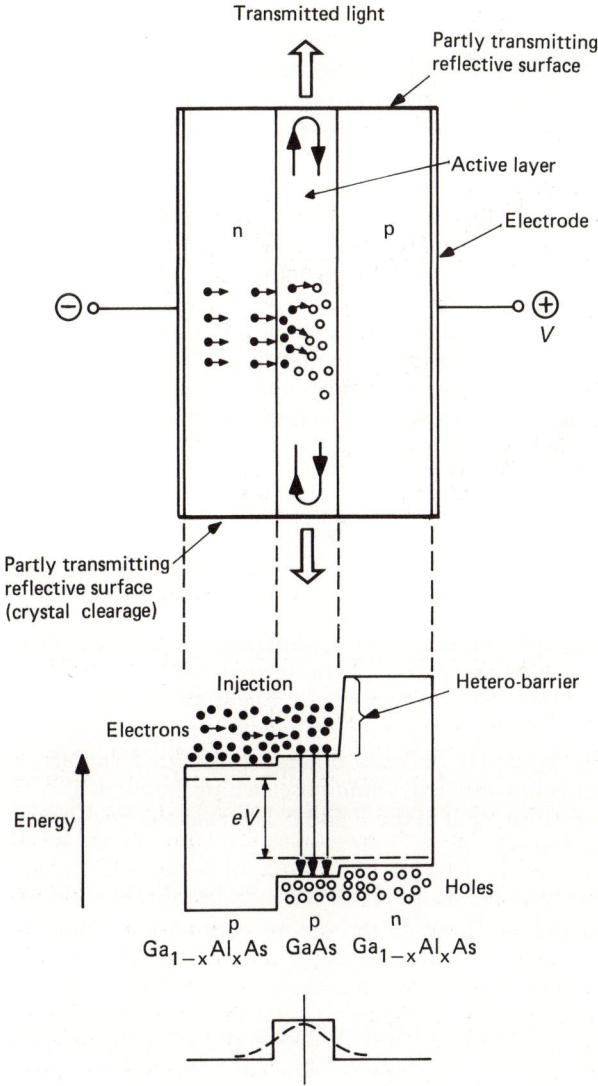

Figure 9.12 Energy level diagram for a forward-biased DHLD of the type shown in Figure 9.11

electrons into it, from the n layer, is explained by way of the energy level diagram of Figure 9.12. This electron flow corresponds to the FQM diode action of the device (see Section 3.3). It is this injection that produces the population inversion in the active layer p. Some of the injected electrons recombine there, some of them radiatively, producing photons of the wavelength $\lambda = hc/E_g$, the LED way, explained in Section 3.3.1. These photons are unrelated to each other and hence the resulting radiation is of the *spontaneous* type. If the current density is increased sufficiently, the population inversion reaches a level at which *stimulated* radiation can take place, the extra electrons making up, continually, for recombination and other losses. This is pumping. The stripe geometry of Figure 9.11 helps to achieve adequate current densities with I_{th} seldom more than 50 mA. This allows continuous wave operation at room temperatures. (Older GaAs lasers required an I_F of several amps, forced cooling and low duty pulse operation.)

For light amplification to exist, here, as in the HeNe laser, two parallel mirrors at the extremities of the long axis of the device provide positive feedback. The bouncing forwards and backwards of the radiation increases the chances of creating additional coherent photons. The inter-mirror space is, of course, the resonance chamber of the laser. The mirrors of this laser are simply shiny crystalline planes of the chip, obtained by cleaving. They are partly transparent in a natural way and thus allow some of the coherent light to escape and become the useful end product of the laser. (An interface between two materials with refractive indices n_1 and n_2 has a reflectance factor for perpendicular rays

$$r = \left(\frac{n_2 - n_1}{n_2 + n_1} \right)^2$$

For GaAs, $n_1 = 1$, $n_2 \simeq 3.5$ and $r \simeq 0.3$.) The heterojunctions play a treble role:

1. They assure a horizontal (Figure 9.11) confinement to the light flux, thanks to the lower refractive index of the ternary layers below and above the p layer, which they thus turn into a light guide.
2. They prevent carrier injection to penetrate beyond the heterobarriers P–p and N–p (see potential barriers in Figure 9.12). This carrier confinement prevents low-efficiency luminescence in P and N.
3. They reduce photon absorption in the P and N layers as the energy gap is greater there than in p.

The three 'horizontal' confinements are completed by two 'vertical' ones:

1. The current confinement through the stripe geometry of the top electrode.
2. The end mirrors.

Together, they restrict the volume of this photon-generating plant to an incredible half of a millionth of a millimetre cube. It produces 10 mW CW of coherent light.

The above described device operates in the near infrared, with λ in the 0.8–0.9 μm region. The newer long-wave devices for optical communications ($\lambda = 1.3$–1.55 μm) use $In_{(1-x)} Ga_x As_z P_{(1-z)}$ and InP layers.

9.4.2 Using GaAs lasers

To the electronics engineer the injection laser is a diode. The DHLD variety requires an I_F value that produces a voltage drop V_F of around 2–3 V. As CW operation at room temperature is possible, circuit design presents very few problems. (In all semiconductor lasers PSU transients must be avoided, as they are likely to cause catastrophic device failure.) At I_F values below the threshold value, I_{th}, the device behaves like an LED. A typical DHLD without stripe geometry requires an I_F value of 1–2 A and must be pulsed, with a duty ratio of only a few, say five, per cent. The peak light output ϕ_{pk} of such a device is 200–300 mW. The single heterojunction laser diode (SHLD) will deliver an amazing ϕ_{pk} of 10 W at $I_F = 40$ A. (Multiple chip arrays are available beaming a frightening 300 W of coherent luminous power.) The penalty for such a higher performance is an extremely low permissible duty ratio – typically a mere 0.01% with a maximum pulse length of 200 ns. A device of this type calls for an elaborate circuit design and a thoughtful layout, followed by a neat, caring implementation. Figures 9.13–9.16 give examples of drive circuits and ϕ versus I_F characteristics for all three types. Notice the linear regions. In Figure 9.16, note the feedback photodiode. This is used for ϕ stabilisation to counteract temperature and ageing effects. In some modern devices such a photodiode is integrated into the laser package. Pigtailing and thermoelectric (Peltier) cooling are other sophistications sometimes available (at a price).

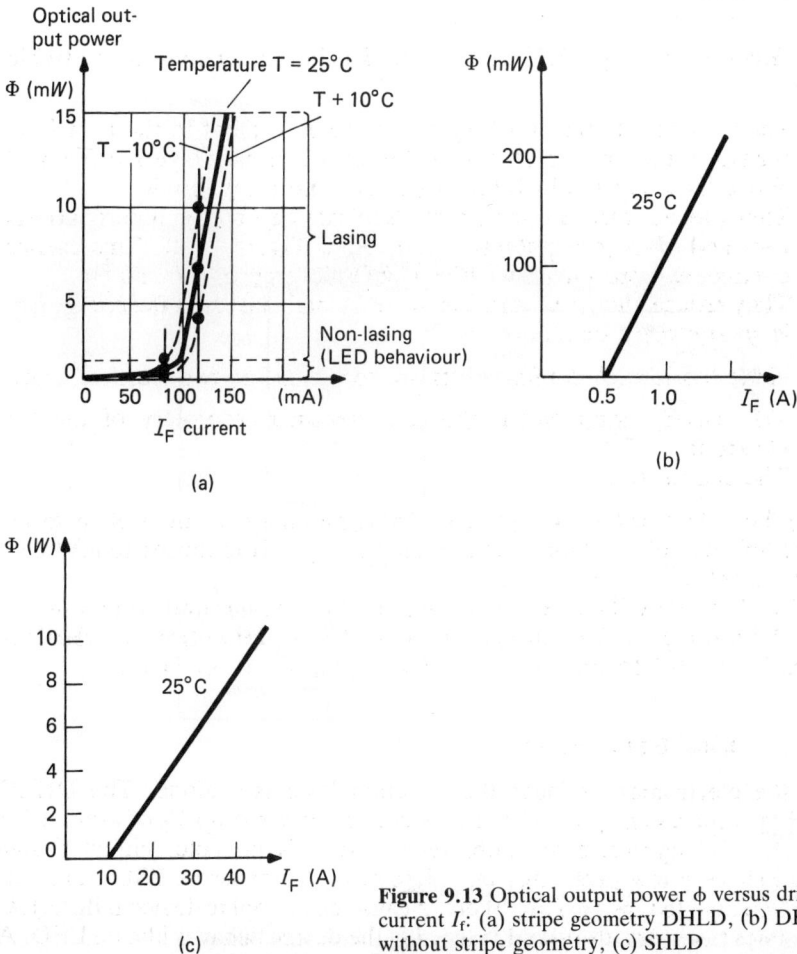

Figure 9.13 Optical output power ϕ versus drive current I_f: (a) stripe geometry DHLD, (b) DHLD without stripe geometry, (c) SHLD

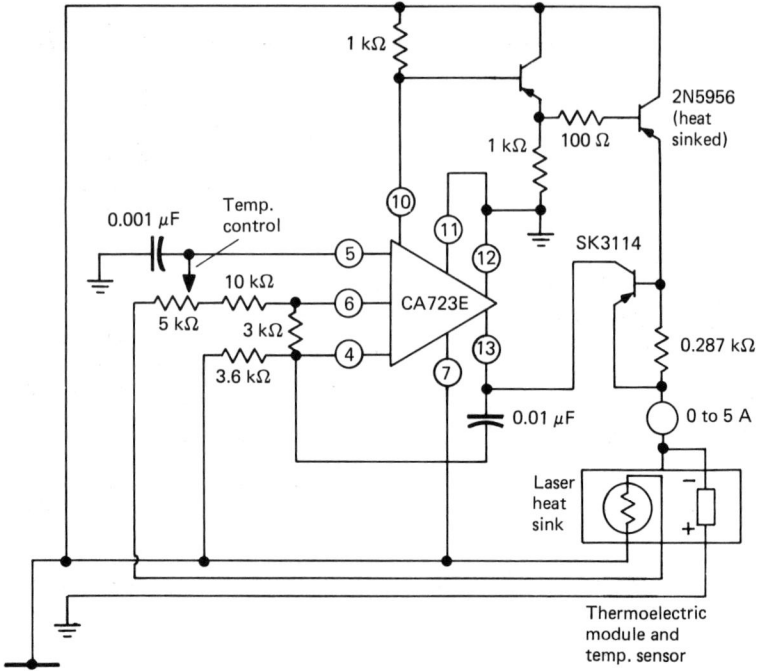

DRIVE SECTION

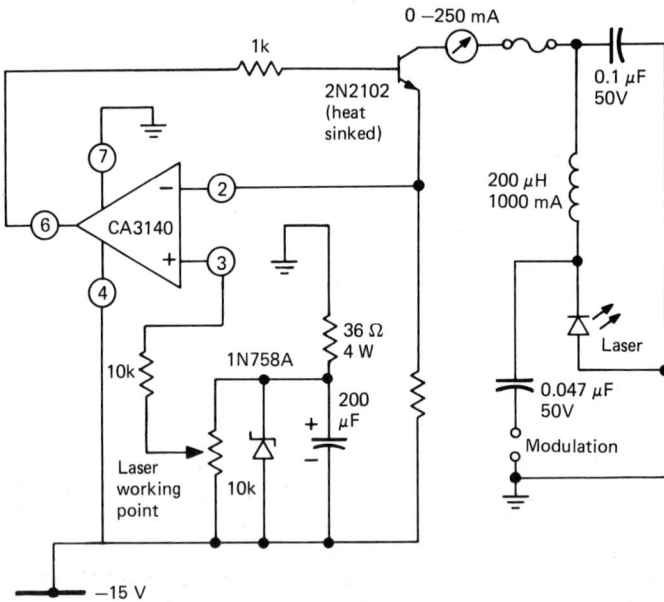

TEMPERATURE CONTROL SECTION

Figure 9.14 Drive circuits for a DHLD with temperature stabilisation

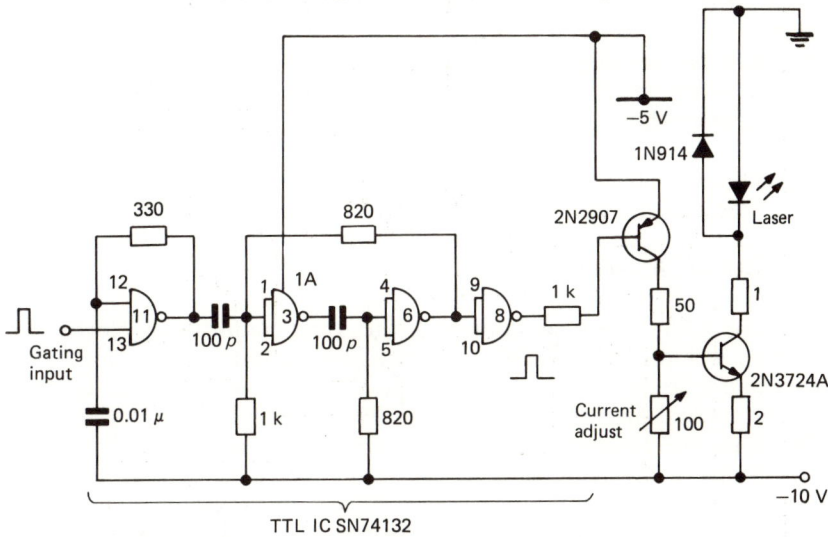

Figure 9.15 Digital (TTL) drive circuit for DHLD without stripe geometry produces typically 1 A, 50 ns pulses at 2 MHz repetition rate

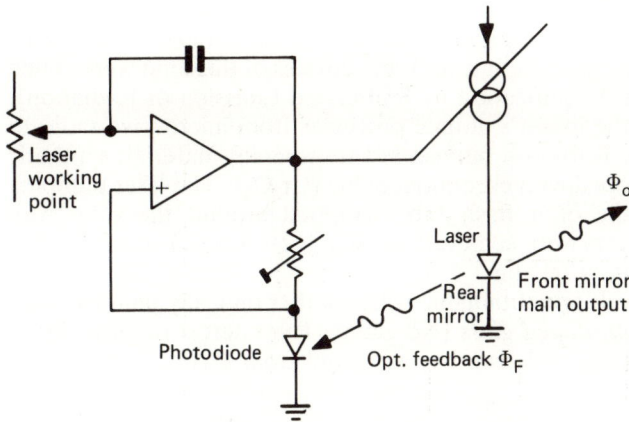

Figure 9.16 The principle of using a feedback photodiode to stabilise the optical output ϕ

The two great attractions of the injection laser to the electronics engineer are *direct* modulation of the light output (by I_F) and modest voltage requirements – PSUs of usually less than 30 V rising only occasionally (avalanche transistor drivers) up to 120 V.

From the optical point of view, the injection laser has a rather poor temporal and spatial coherence (see Section 10.2.1), with its numerous longitudinal and transverse modes. (However, a recently introduced (by RCA) and still very new CDH-LOC device claims to operate in a single longitudinal and single transverse mode.) Its polar diagram (Figure 10.46) reveals different parallel and perpendicular (to the junction plane) beam spreads.

It would not be proper to close the section on the utilisation of GaAs lasers without a word of warning on their safety. Let the innocent look of these transistor-like little cans not deceive you – the radiation can be dangerous, especially to the eye. Remember that it is invisible. Read the manufacturer's safety notes (and comply!).

9.5   Other lasers

There are many types of laser that the great majority of electronics engineers will probably never use. With laser action possible in solids, liquids and gases, the variety is enormous. The best known are mentioned here for general awareness of the subject. Of these, the carbon dioxide (CO_2) and the neodymium yttrium aluminium garnet (Nd-YAG) are probably the most widely used, not counting the HeNe and GaAs types. The power range is enormous too: from 0.5 mW to 1.5 kW (CW) and from 5 W to 300 TW (peak) (1 TW (terawatt) = 10^{12} W). In contrast, the wavelength range seems rather restricted (from some 0.2 μm to just under 11 μm), although some research in the X-ray band is under way.

9.5.1   Solid-state lasers

The *ruby* laser (Figure 9.17) is undoubtedly the oldest. It existed before its name, as when T. Maiman built it, in 1960, devices of this kind were called *masers* (Microwave Amplification by Stimulated Emission of Radiation). Maiman extended the maser's atomic processes from microwave radio to optical frequencies. Ruby is a pinkish-red transparent material; synthetic sapphire (Al_2O_3) doped with chromium oxide (Cr_2O_3). The laser pumping is optical, by means of a flash tube wrapped around the ruby rod. Megawatt peak powers were achieved as early as 1965 (see Table 9.1). The ruby laser emits at λ = 694 nm.

Similar power levels are obtainable with another optically pumped solid material: *neodymium-doped glass* (Nd-glass). The interest of glass lies in the fact that very large rods can be fabricated from it, i.e. diameters of

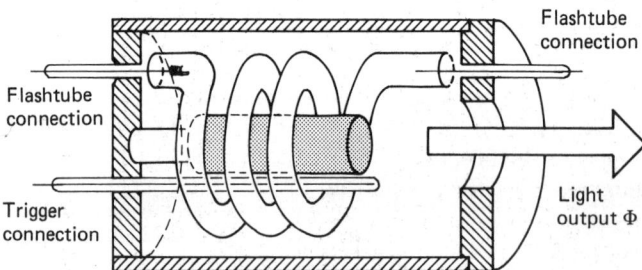

Figure 9.17 In Maiman's laser a flash tube is coiled round a ruby rod of typically 1–1.5 cm diameter

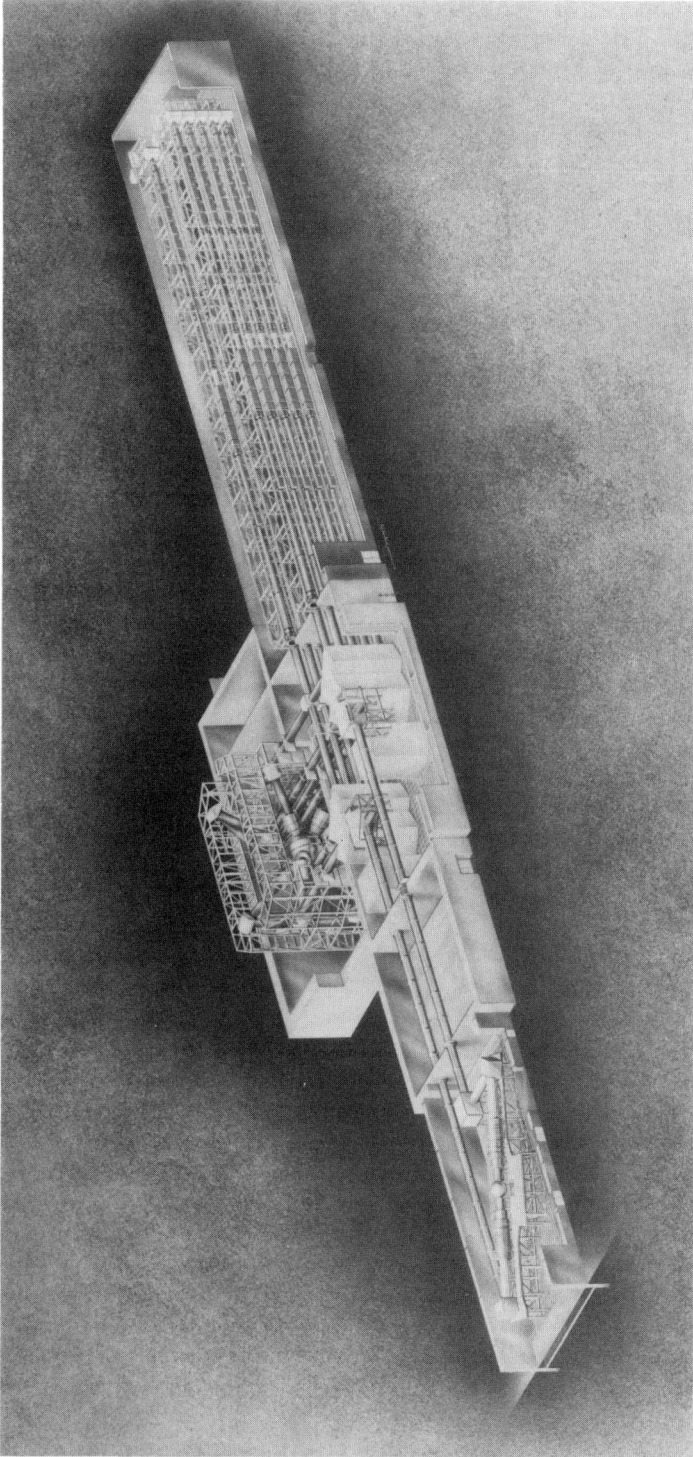

Figure 9.18 Nova laser – the target frame is six storeys tall (Courtesy of University of California, Lawrence Livermore National Laboratory, and the US Department of Energy)

several centimetres and lengths of over a metre. Using arrays of large rods, mind-boggling powers can be attained. The Shiva laser (Livermore Lab., USA) has been designed to deliver 30 TW pulses. It is already superseded by the Nova 200–300 TW (Figure 9.18) [75]. (To put this in perspective, the author of Ref. 75, J. Holmes, quotes the total instalment power of the CGEB, the UK national grid, as 0.06 TW.) The Nova's colossal target bay doors weigh 320 tonnes, its target frame stands 6 storeys tall, and the total length of its structural steel members is over 17 km. At the other extreme, glass rod diameters have been so reduced that they become 'lasing fibres'. The attraction of diameter reduction lies in easier cooling. All Nd-glass lasers emit at $\lambda = 1.06$ μm.

Another member of the solid-state family is the popular yttrium aluminium garnet laser. This crystalline material is also neodymium-doped for lasing; hence its name, Nd-YAG. Like the Nd-glass laser, the Nd-YAG also emits at 1.06 μm, although other wavelengths in the 1.052–1.338 μm range can be obtained with remarkable gain (see Section 10.2.2(a) for longitudinal modes), and harmonics in both the visible and the ultraviolet are sometimes used. (See Section 14.3 for non-linear optics.)

9.5.2   Ion and metal vapour lasers

The lasers examined so far produce either infrared or red radiation. Photographic films and papers are usually more sensitive in the blue-green part of the spectrum. Here, the *argon, krypton* and *helium cadmium* lasers fit the bill. Their uses, however, extend far beyond photographic applications. The first two are often referred to as *ion lasers* because their active element is an ionised gas. The third (Figure 9.19) belongs to the *metal vapour* category of lasers as cadmium acts in it in a vapour state. Between them, these three lasers produce a large number of wavelengths right across the visible spectrum [173] with krypton featuring an extremely powerful (4 W) red line at $\lambda = 647$ nm. Wavelength selection is by dispersive prism colour separation. The pumping of ion lasers is very similar to that of HeNe lasers.

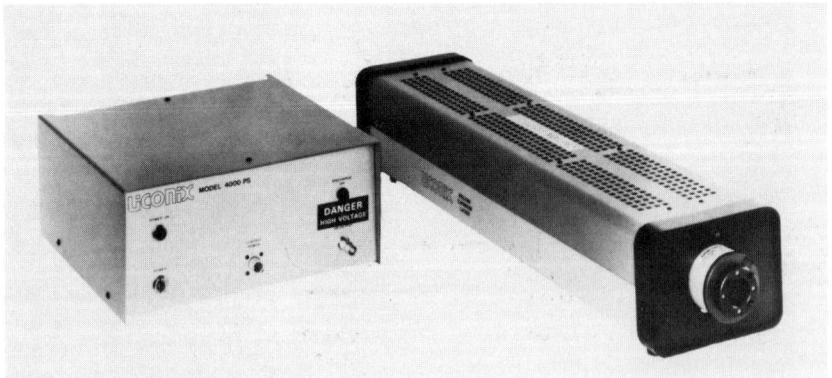

Figure 9.19 10 mW helium cadmium (blue) laser (Courtesy of Laser Lines Ltd)

9.5.3   Carbon dioxide lasers

The carbon dioxide (CO_2) laser deserves a special mention for more than one reason: it covers an enormous range of CW, as well as pulsed, powers (see Table 9.1); it operates in parts of the infrared spectrum that other lasers cannot reach and where applications are plentiful; and it is intrinsically more efficient than atomic gas lasers. In fact, CO_2 lasers form a family of devices with various internal structures ('sealed', 'axial flow', 'transverse flow', 'waveguide'), linked by a common working principle. It is not surprising that such versatility should have led to great commercial success.

All CO_2 lasers use a mixture of carbon dioxide, nitrogen and helium as active medium in which helium and *not* carbon dioxide is the majority component ($\simeq 80\%$). Their name, which may appear unfair to helium, stems from the carbon dioxide's role as light emitter. Nitrogen does the pumping of carbon dioxide (to cause population inversion), while helium does the mopping up (depopulation) of the intermediate energy level of the lasing mechanism. CO_2 lasers belong to the *molecular* category. Here, the energy levels involved in laser action are the quantised amounts of *vibrational* (V) and *rotational* (R) energy of the O–C–O molecule (Figure 9.20) and not the electronic energy levels of atomic gases [7, p. 266; 178; 181, p.219]. V and R levels are much more closely spaced than electronic levels (tens rather than thousands of meV), so that the resultant transitions produce much lazier (longer wave) photons. The resultant basic wavelengths are 9.6 and 10.6 μm with many lines in between. Sealed tube structures generate powers of less than 100 W. One inexpensive bench

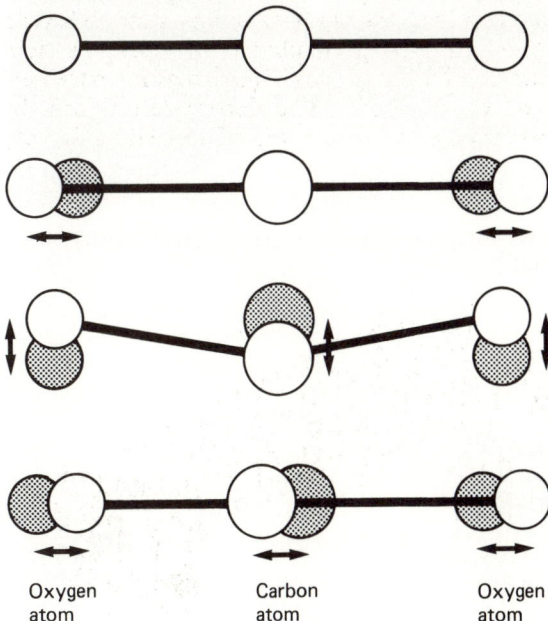

| Oxygen | Carbon | Oxygen |
| atom | atom | atom |

Figure 9.20 In CO_2 lasers, quantised amounts of vibrational and rotational energy of the O–C–O molecule are responsible for the laser action

model intended mainly for fusing optical fibres (Type 3038H, Hughes Aircraft Co [178, 179]) delivers only 1 W.

(a) The CO_2 waveguide type

These lasers (see Figure 9.21, for example) use the *waveguide* technique, which is to say that the inner tube is so narrow that it constitutes a dielectric waveguide. Making the gas mixture flow through the laser tube solves much of its cooling problems. The gas can be recycled. In structures in which it is not, the supply need not be embarrassingly large, owing to the use of small pressures. A longitudinal gas flow is suitable for powers of up

Figure 9.21 Ferranti CM3044 (4 W) waveguide CO_2 laser engraving glass (Courtesy of Ferranti Industrial Electronics Ltd)

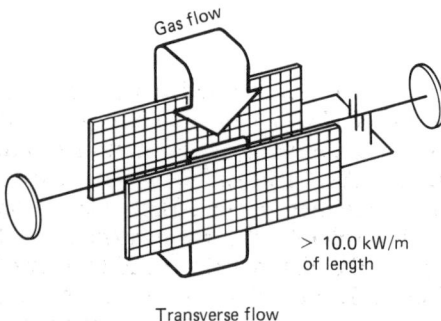

Figure 9.22 Basic structure of a transverse flow CO_2 laser

to a few hundred watts. For powers in the kilowatt range transverse flow structures (Figure 9.22) are used. Here, gas flow, electrical field and light flux are mutually orthogonal.

(b)   The CO_2 TEA type

The pressure within the enclosures of most so far described CO_2 lasers is only a small fraction of an atmosphere. TEA lasers are those Transversally Excited Lasers that use gases at Atmospheric or near atmospheric pressures (Figure 9.23). They are capable of producing very powerful (kilowatt) very short (nanosecond or microsecond) pulses.

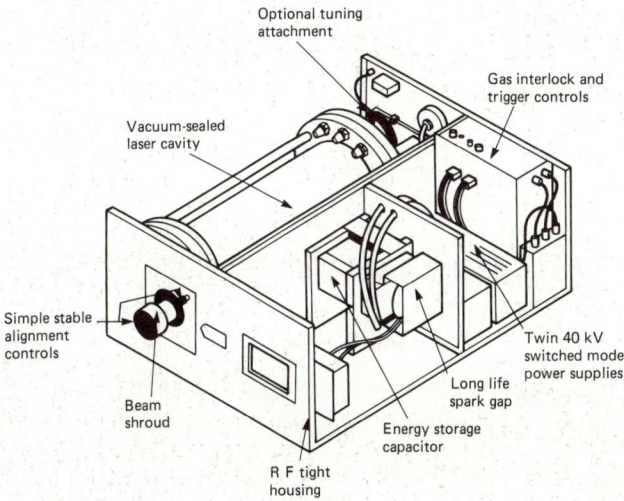

Figure 9.23 Internal construction of a TEA laser (Courtesy of Lumonics Inc)

To sum up, CO_2 lasers, starting with a benchtop 1 W model and ending with the 15 kW transverse flow system, a real 'radiation factory', cover a truly formidable power range.

9.5.4   The excimer laser

An *excimer* laser is a kind of molecular laser using for active medium a gas which is stable with its molecules electronically excited. In other words, its normal condition is much unlike that of other gases, whose molecules are stable in a ground state, e.g. XeCl, ArF, ArCl and KrCl. The wavelength of excimer radiation is usually short ultraviolet or blue. High conversion efficiency in the ultraviolet range is probably the main attraction of excimer lasers. One of the detractions is the high reactivity and toxicity of excimer gases. Powers are as per Table 9.1. Basic designs are quite similar to those of CO_2 TEA lasers. Applications are, as yet, mostly scientific, with a potential use for atomic fusion work.

9.5.5   The liquid (dye) laser

In *dye* laser apparatus, lasing takes place in a liquid (Figure 9.24). Here, the active medium is an organic dye hosted by a solvent, also organic, circulated in a long capillary or a compact flat glass vessel. The solution is pumped optically, by means of one of the previously described laser varieties, e.g. Nd-YAG, ruby or krypton. The chief interest of dye lasers lies in their tunability, obtainable in both the CW and pulsed regimes. Tuning, for a given dye, is by wavelength selection through a Fabry–Perot etalon or a dispersive prism, or by replacing one of the laser mirrors with a fine, tilted grating. (A Fabry–Perot etalon consists basically of two strictly parallel semi-reflective plates spaced an integer number of half-waves $\lambda/2$

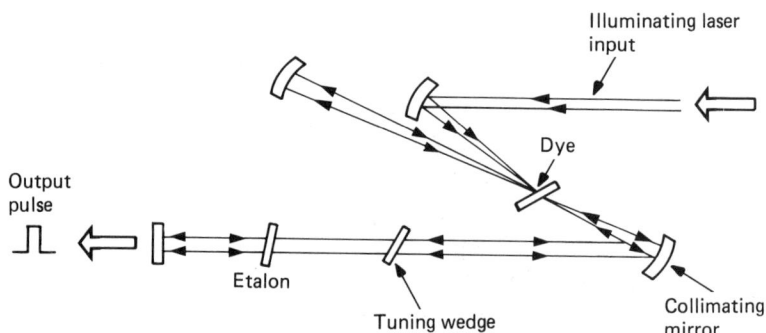

Figure 9.24 Schematic of a dye laser

away. Constructive interference favours strongly the transmission of this particular wavelength. See also Section 14.3.2(a) with Figure 14.7 and, for fuller treatment, Ref. 17, p.307 and Ref. 192, pp. 21, 240.) Dye selection can yield a further range extension. Certain lasers can be tuned within the incredible range of 0.6–0.73 μm, with a good conversion efficiency. The overall wavelength range for dye lasers stretches from the ultraviolet (190 nm) through the whole of the visible to the near infrared (1.1 μm). Table 9.1 shows that the range of obtainable powers is at most as formidable as that of CO_2 lasers.

Industrial applications of dye lasers are few. Prevailing non-industrial applications are in medicine (UV) and physics and chemistry research.

9.5.6   The free electron laser

Stanford University's John Madey evolved the idea of the free electron laser (FEL) (Figure 9.25) in 1971. By 1977 he had his hardware working [180]. Despite its cost and almost monumental bulkiness, the system is pledged to a bright future because of its enormous applications potential.

Today, the FEL is the only *laser* still true to its name: a light *amplifier* in which the input (not the pump) is a light beam. This comes from another, say a CO_2, laser. Powering this amplifier is not the ordinary PSU we know, but a high-energy electron gun aiming a megaelectron volt beam into it.

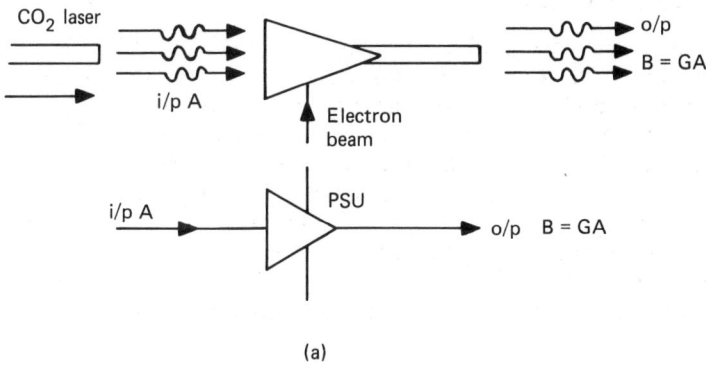

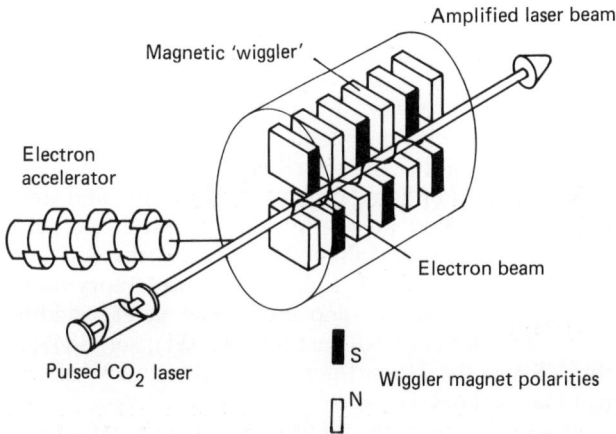

Figure 9.25 The free electron laser. (a) FEL and its electronic analogue; (b) schematic diagram of the magnetic 'wiggler'

(The 'gun' is, in fact, a large particle accelerator. With this and the external laser the FEL can hardly be called a stand-alone unit!) Inside the FEL, a row of alternating NS/SN magnets creates a wiggling magnetic field through which the electrons slalom at Einsteinian speeds, amidst the photon cloud of the incident light beam. Oscillating electrons create electromagnetic radiation that is monochromatic, its frequency depending on the electron velocity, the magnet spacing and the transverse magnetic field [181, p. 262]. It is the interaction between this radiation and the incident coherent light – acting as a seed – that produces the energy transfer from the electronic to the photonic beams, resulting in light amplification. As in other lasers, two end part-mirrors provide the necessary positive feedback.

When considering the applications potential of FEL, remember that its tunability offers an enormous range of λ, theoretically from short ultra-violet through the visible spectrum to millimetric waves. Therapeutic and surgical uses of various kinds could be obtained with *one* instead of many lasers. Cancer treatment raises new hopes, with highly selective cell killing without damaging the neighbouring cells (through extra fine spot size control selection). Hopes for the future include the so-called 'photoradiation' therapy. In this, the patient is injected with dyes capable of lasing. Selective saturation of diseased tissues takes place. Irradiation by a tunable laser stimulates the dyes, killing the saturated (diseased) cells only. Chemical reaction too can be selectively provoked, by picking the molecules for specific reactions (e.g. synthesis of polymers). There is also talk of laser fusion and (alas!) weaponry applications. Added to all this is an indication that high conversion efficiencies (around 30%) and self-oscillation should be achievable. Thus, despite the high initial cost and its present bulkiness, the free electron laser has a most promising future.

9.6   The power range of lasers

A glance at Table 9.1 shows that some lasers generate kilowatts of CW power and that some others are capable of producing megawatts, gigawatts and even terawatts of pulsed peak power. We see immediately that they could not possibly be expected to be harmless, especially with the high degree of coherence and focusability characteristic of laser beams. The hazards to the human body and other living organisms are many, and strict safety regulations must be respected at all times by research workers, medical personnel and industrial users alike. (Lasers are divided into classes and should be labelled accordingly. Laboratory and factory doors should give due warnings [Ref. 182, p. 466 etc.]. The safety guiding authority in the UK is the British Standards Institution (BSI); see BS4803. The ruling body in the USA is the Department of Health and Welfare's Bureau of Radiological Health (BRH).)

On the other hand, to see in every laser a James Bond body carver or a death ray killer would be ridiculously wrong. Examples of perfect harmlessness (excluding, of course, cases of malicious or totally inconsiderate

Table 9.1 Power and energy range of lasers

Type	Mean power	Peak power	Single pulse energy
GaAlAs	1–5 mW (RCA claims 20 mW)	5 W	
HeNe	0.5–50 mW	0.5–50 mW	
HeCd	5–20 mW	5–20 mW	
Argon	<20 W	<20 W	
CO_2	1 W–1.5 kW	15 kW (up to 40 kW in 1 ns)	
YAG	150 W	20 mW	1.5 J
Nd-glass		100 MW–5 GW	1.5–100 J
Ruby		100 MW–5 GW	10 J
Nitrogen		200 kW	
Dye	2.5 mW	25 kW	
Excimers	20 W		100 mJ
'Shiva' (Nd-glass array)		30 TW	
'Nova' (Nd-glass array)		300 TW	200–300 kJ

misuse) are the Polypoint PL-505 lecturer's torch pointer (Polytec, El Toro, Calif., USA) using a 500 μW HeNe laser, and the supermarket bar code reader. In the often encountered 2–20 mW (CW) power range (GaAs and HeNe), it is advisable to wear protective goggles to guard against accidental direct (or, worse, exposure via collecting optics) beam exposure to the eye, the most vulnerable part of the body. The unaided eye can, however, be allowed to view the HeNe spots or patterns projected on a screen. When there is a case for wearing goggles, these should be carefully selected, not only for their spectral response but also for their attenuation. Excessive attenuation of goggles worn during work with *visible* lasers can be more dangerous than no goggles at all, as it makes the user forget, after a while, that the laser is on and, once this adjustment is completed, remove them without switching off or shutting off the beam. The effect is particularly bad in a dark or darkened room with the pupils of the eyes wide open for accommodation. Thus, an optical density that reduces the radiation to a safe level while leaving a vestigial spot is safer, in the writer's view. The deceptiveness of the *invisible* radiation from infrared lasers must always be borne in mind, even at moderate power levels. With high-power lasers, dangers other than to the eyesight exist; heat, fire, projectiles, high voltage, etc., and interlocks, barriers, protective gloves, etc., are sometimes stipulated (see Figure 9.26). Finally, some classes of lasers must be enclosed at all times while others require a full-time attendant.

Figure 9.26 High-power (15 kW) industrial laser. Note the protective enclosure and the operator's goggles (Courtesy of Avco Everret)

9.7   The jobs lasers do

It was easy to write about the applications of transistors in the mid-1950s. This was no longer so 20 years later. What the passing of two decades did to transistors, it also did to lasers. The famous (or infamous) description that they were given in the mid-1960s of 'a solution in search of a problem' has been proved very wrong. The fields in which lasers are used are so many, the applications so numerous, that to deal with them comprehensively now is almost as difficult a task as that of covering all the uses of semiconductors in electronics.

Some applications of lasers have been mentioned in the text so far, either in connection with a particular EO technology, e.g. optical communications for the semiconductor laser, or à propos a particular type of laser, e.g. medicine for dye lasers and the up-coming free electron laser or atomic fusion for the excimer laser. Some others, such as in civil engineering, metrology and optical sensors, will be treated in Chapter 12. Two groups of applications, manufacturing and health care will be reviewed now, and followed by a complementary listing. The more demanding reader is referred to a work by H. M. Muncheryan [176] with the extra inquisitive one being advised to treat him or herself to a systematic scrutiny of specialised magazines, such as *Lasers and Applications* (High Tech Publications, USA) and *Laser Focus* (Advanced Technology Publications, USA).

9.7.1   Applications in manufacturing

Figure 9.27 shows a CO_2 laser beam cutting a 5 mm thick stainless steel sheet. The beam itself is, of course, invisible. The plumes of sparks seen are caused by the expulsion of impurities from the cut. Laser tools can cut a variety of materials, from high carbon steel through titanium alloys, ceramics and reinforced ruber [87] to wood, cloth and cardboard. Computer programming of complex cutting patterns ensures flexibility, excellent repeatability after a period of disuse and high productivity. The latter results from high cutting speeds and the use of simultaneously operating multiple work stations. Other advantages are: low noise, dust, fume and vibration levels, the possibility of operating *through* a glass shield, no fraying (fabrics), the ease of starting a cut in the middle of a workpiece, and the elimination of the need for a wide range of cutting tools.

In soft materials (wood, cardboard, cloth) the cutting mechanism works through material vaporisation while in metals it relies on highly localised material melting. The ablated metal is removed by a gas stream. In some cases the gas is oxygen. Reacting exothermically with the metal, oxygen saves beam energy while increasing the cutting speed.

When it comes to industrial hole drilling, the laser proves of value mostly in working materials at both extremes of the hardness scale: diamonds and rubies on the one hand and polythene and rubber on the other. In the first case, laser beams cope much more swiftly than mechanical drills with the very high hardness: no more broken bits in cutting rubies for watches and instruments, fabricating dies for wire pulling or drilling ceramic substrates for semiconductor chips. In the second case, accurate aiming is easier when

Figure 9.27 A CO_2 laser cutting a 5 mm thick stainless steel sheet. The beam is invisible; the sparks seen are caused by the expulsion of liquid or solid particles from the cut

no mechanical contact with an exceedingly soft and shifty material is needed. Making holes in general metal work by laser, however, is not an economic proposition (yet), as the removal of large quantities of material involves the latent heat not only of melting, but also of evaporating the material. This can be very great indeed, especially for large holes. However, when small holes have to be made either in inaccessible places or at very accurate angles and depth, such as in a vane for a jet engine [87], laser drilling can be invaluable.

Laser welding has already found its way to the motor car industry, e.g. the Turin plant of Fiat. Most lasers welders use a deep penetration process, in which a kernel of molten and partly vaporised and ionised metal is first formed in mid-thickness of the material and then moved along the weld line, without the vapour even leaving the metal. The high efficiency of this process is due mostly to an extremely high degree of localisation of enormous amounts of heat. Weld depths some 12 times their width can be achieved. The weld is very pure, has a good crystallographic structure and is therefore as strong or stronger than the base metal. Other industries (shipbuilding, pipe manufacture, aerospace) have begun using the method which, thanks to recent progress, is gradually extending to light metals and alloys too, half-inch thick plates of which can be welded together. Laser welding comes into its own when access is difficult – optically guided beams get at otherwise inaccessible parts. Finally, a word about two of welding's remote cousins: microsoldering and microbrazing. For the first, an interesting combination of coaxially arranged CO_2 and HeNe laser beams produces precision positioned joints on microchips: the visible 1 mW HeNe laser does the guiding while a brief flash of the 20 W CO_2 laser does the soldering (no wire insulation stripping is necessary) [176, p. 118]. Microbrazing is still in the experimental stage. At Lawrance Livermore Labs, H. Witherell [185] reports that the best results are obtained with a Nd-YAG laser beaming onto a powder brazing alloy. As an example, submillimetre Cu–Ni tubes have been successfully brazed onto similarly thin stainless steel fittings (Figure 9.28). While this and other results are promising, much more R&D work remains to be done before microbrazing reaches the factory floor.

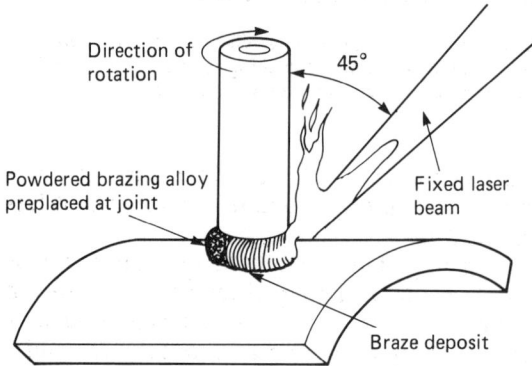

Figure 9.28 Submillimetre Cu-Ni tubes brazed on to stainless steel fittings

Alloying, annealing and surface treatment with laser heat is still a minority technology. In the same category lie surface hardening and surface cladding. The technology is used in cases in which advantage can be taken of the marvellous ease with which laser heat can be accurately confined either in space or in time, or both. In addition the processes are clean, cause no tool wear and can be easily intensity controlled. Highly localised action leads to such savings. In Figure 9.29, for example, a small portion of a hollow structure has to be hardened. Optical beam guiding does the trick, and in this instance on a mass production scale. Localised surface alloying too is a laser province. Small amounts of alloying additive are quickly molten on the surface and diffused into the main body of the part. The kind of forces mentioned in connection with the kernel in laser welding assist stirring and fast cooling produces a good microstructure [87].

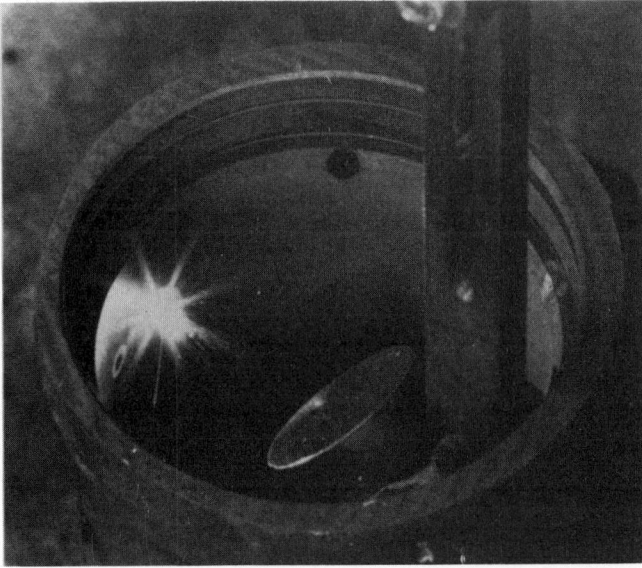

Figure 9.29 Partial surface hardening (inside a gear housing) at General Motors (USA). A 1 kW CO_2 laser is directed downwards by a lens on to a travelling mirror. Several vertical passes combine with mirror rotation to cover the required inner surface area

At least two methods of laser treatment of large surfaces are known – cladding (e.g. cobalt onto steel) and glazing, obtained by the use of fast quenching in hardening treatment.

In a class of its own is the purification, in depth, of semiconductor materials. 'Gettering' of unwanted impurities is obtained in a two-step process [187]. Other well publicised uses of lasers in the semiconductor industry include resistor trimming, quartz tuning, substrate (ceramic) drilling and die scribbling and separation.

In conclusion, the use of laser beams in manufacturing, although not massive, is highly diverse. Wherever applied it leads not only to higher yields but also to superior product quality.

9.7.2 Applications in medicine and surgery

Laser light can be concentrated into spots smaller than the great majority of the five thousand or so kinds of cells entering into the constitution of the human body. (The total number of cells in man is estimated to some 10^{10}. Among the smallest are the familiar red blood cells, some 8 μm across.) It is therefore not surprising that the laser has found applications not only in diagnosis but also in treatment and surgery. In diagnosis, laser-induced fluorescence seems to be playing an important role [181]. Through its use, bronchoscopy, known for some time in its simpler form (see Sections 2.2.8 and 12.3) is improved so that it helps to detect lung tumours in their early stages. The most recent diagnostic tool is the scanning ophthalmoscope which uses the eye's own lens to collect the back-scattered light from the low-power travelling spot projected onto the retina [189]. The image of the retina is electronically enlarged and thus made readily viewable on a TV monitor. Moreover, an alphanumeric pattern can be formed *directly* on the patient's retina during a frame scan by modulating the beam, and the patient's reaction to it observed. Another diagnostic laser tool is the laser Doppler velocimeter (see Section 12.7) which can measure the speed of various bodily fluids, e.g. blood. These are but a few examples.

Laser surgery has been known since the mid-1960s, when the first retinal lesions were being successfully repaired and retina detachments averted. Today, laser surgery is a vast field of activity covering gynaecology, otorhinolaryngology (e.g. tonsils removal), histology (e.g. drilling and cutting bone tissues), gastric bleeding, dermatology (e.g. removal of port wine birth marks and tattoos), and even circumcision of haemophiliac boys. The advantages of the laser scalpel (CO_2 lasers, usually) are: aiming accuracy, the ability to reach inaccessible places, much reduced bleeding (the laser scalpel attacks fewer cells than a steel knife and evaporates them quickly), near-absence of oedema, and reduction or suppression of pain. On the negative side are low cutting speed and safety-related problems.

Two types of development work are at present underway, in which the laser beam induces changes in cells, as opposed to destroying them. Microbeam techniques are using 0.5 μm argon laser spots to irradiate parts of cells, i.e. cell subsystems [188]. With this technique, genetic engineering does not seem far away as hereditary characteristics could be affected. In laser acupuncture, the thousand-year-old silver and gold needles are replaced by fine, micromanipulator-orientated laser beams.

9.7.3 Some other applications

The coherent light produced by lasers lends itself admirably to sensing and measuring of all kinds of physical parameters. Distance (within the submicrometre to multikilometre range), velocity (from micrometres per second to kilometres per second), temperature, pressure, frequency and electric current intensity, are but a few of these. The description of the ways lasers are used in metrology and sensing has been relegated to a later chapter, to be preceded by the treatment of beam engineering techniques and that of coherence itself. Suffice to say here that the methods employed rely often on interference and heterodyne frequency shifting, both familiar to the electronics expert. Surveying, civil engineering and shipbuilding, on the other hand, use the laser (HeNe) beam in a simpler way: as a very long, very light and very straight indeed piece of string or a very, very long chalk line. Lidar (an acronym akin to radar) is used mainly for probing the atmosphere (weather forecasting and pollution studies). Lettering, scribing, engraving (on metal, ceramic, wood!) and, of course, phototypesetting and plate-making (in the printing industries), are applications in which accurate laser spot creation is required. In phototypesetting and non-impact printing, operational speed is an asset. Laser beams and microprocessors combined make it possible to compose a full broadsheet page in under a minute, inclusive of screened halftone pictorial material. The entertainment industry, too, uses lasers. (A certain sales representative was heard reporting that his Company's sales of HeNe lasers to disco clubs exceeded those to professional users.) The video-disc machine using a semiconductor laser in its readout head is also an item serving this industry. Optical discs for digital storage in data processing use laser heads too, as does the compact disc player. A single 12-inch optical disc can store as many megabytes of digital data as 12 magnetic discs of a similar size. Holography (see Section 13.1), invented by Dennis Gabor in 1947, had not led to applications until lasers became available. Today, it is used in a multitude of ways including three-dimensional representation of objects, fingerprint identification and laser beam diffraction scanning.

Last, but certainly not least, of the human endeavours in which lasers play an important role is the entire field of scientific research. This encompasses physics, chemistry, biology (with genetic engineering) and various atomic studies, the most prominent of which is perhaps atomic fusion, with the hope it brings to mankind of becoming a new source of energy.

As for the military applications of lasers which might, alas, become more consequential to us than the discovery of new energy sources, I have neither the competence nor the inclination to write about them.

Chapter 10

Laser beam engineering

10.1 Exploiting the laser beam

Chapter 9 describes some of the uses of laser beams. Only a small minority of these can be performed by the beam directly out of the laser exit. Generally speaking, in order to be exploitable, the laser beam has to be shaped, in both space and time. This chapter is concerned with this shaping to which we give the name of *laser beam engineering*.

The most obvious of the operations needed are intensity modulation and deflection (perhaps by analogy with the cathode ray tube). Further comes static cross-section control, i.e. the operations of expanding or reducing the beam diameter, astigmatism correction and focusing the radiation to a tiny spot. Next comes static beam attenuation, beam bending, beam splitting and beam mixing. Further still we come to beam purifying techniques such as mode selection and noise removal, and this list may well not be exhaustive. Prior to the description of the principal beam engineering techniques, we shall look a little closer at the principal characteristics of the untreated laser beam.

10.2 The untreated laser beam

10.2.1 Coherence

Section 9.1 points out that laser light owes its uniqueness of character to a high degree of coherence. Electromagnetic radiation has *temporal* and *spatial* aspects of coherence. Both are important.

(a) Temporal coherence
Singleness of frequency combined with its stability is an early recognisable face of temporal coherence. It is represented in Figure 10.1(a) as a pure sine wave of constant amplitude with neither beginning nor end and in Figure 10.1(b) as a single, zero-thickness line in the time and frequency domains, respectively. Only outside programme broadcasting hours does the working radiotransmitter approach this type of temporal coherence.

An *absolute coherence* (idealised) of radiation precludes all kinds of modulation: here, as in radiotransmission, as soon as we vary the amplitude, A, or modulate the frequency, f, sidebands appear and the spectrum thickens (Figure 10.2).

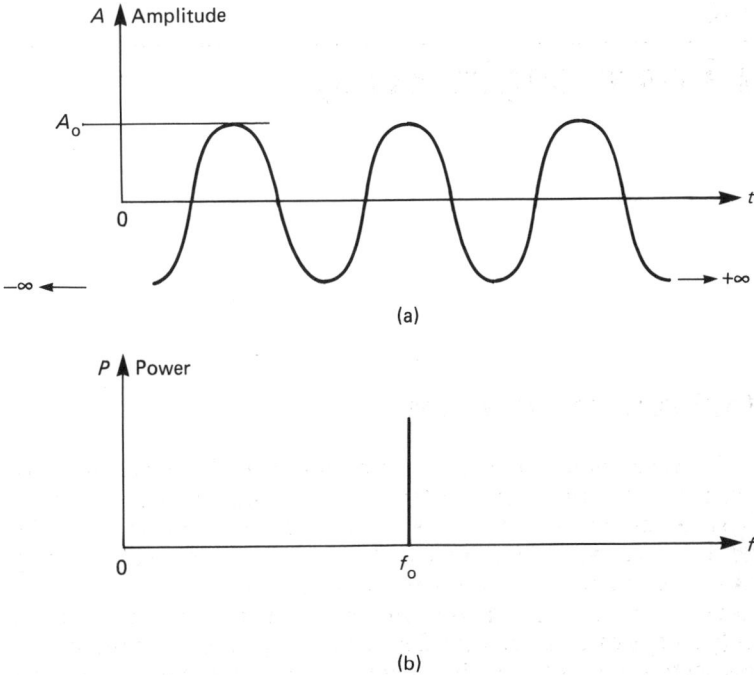

(a)

(b)

Figure 10.1 A pure sine wave of constant amplitude with neither beginning nor end has a zero-thickness single-line frequency spectrum

In practical lasers, the resonant cavity can accommodate more than one wavelength, λ, as more than one λ can satisfy the $1 = m\lambda/2$ condition (Section 10.2.3), giving birth to a range of *longitudinal modes*, spaced, for example, $\triangle\lambda = 6.7 \times 10^{-4}$ nm apart (Figure 10.3). As the system gain diminishes the modes away from λ_0 grow weaker and finally evanesce. To obtain a superior temporal coherence only one of the longitudinal modes has to be selected (by prism, grating, etalon, etc.) (Figure 10.4). The remaining single line can be as thin as 1.6×10^{-6} nm (1 MHz), that of a highly stabilised very high quality single mode CW HeNe laser can be as narrow as a few kilohertz, a spectral purity no non-lasing monochromatic source can approach. One of the reasons for which the line cannot be even thinner is a certain dimensional (longitudinal) instability of the laser cavity caused by vibrations and thermal expansion. To explain other reasons for line broadening one has to delve into physics of a kind that relies heavily on specialised mathematics. The interested reader is referred to good books on lasers (section 2.5 of Ref. 181, for example).

Monochromaticity, however, is no guarantee of temporal coherence. Indeed, sudden amplitude or phase jumps of a monochromatic wave reduce coherence (Figure 10.5). The length of a coherent wavetrain is called the *coherence length*, L_c. Note that, despite being a length, L_c is *linked to temporal* and not *spatial* coherence. The time, $t = L_c/c$, that this length of the wavetrain takes to travel is, naturally enough, the *coherence time*, t_c.

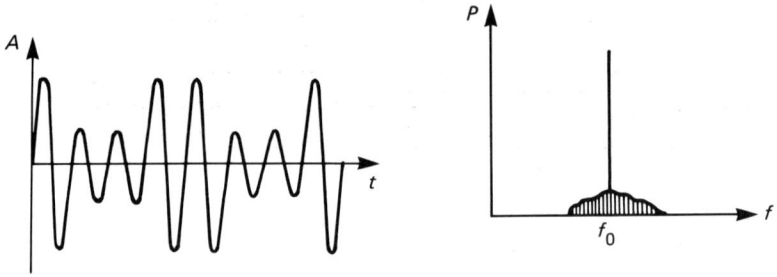

(a) The amplitude A only of a fixed frequency (f_0) signal is modulated

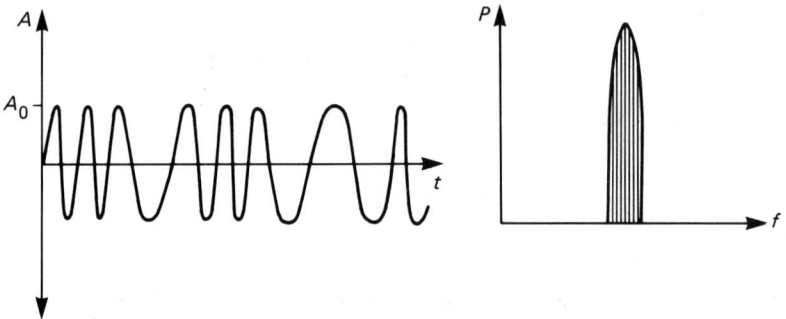

(b) The frequency only of a fixed amplitude (A_0) signal is modulated

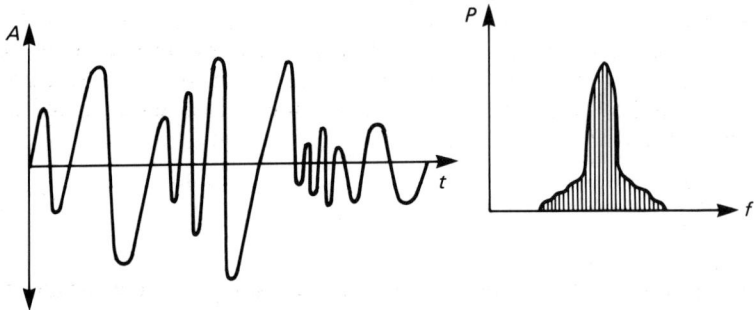

(c) Both amplitude and frequency are modulated

Figure 10.2 When the amplitude is varied or the frequency modulated, sidebands appear and the spectrum thickens

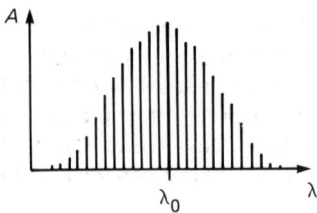

Figure 10.3 The resonant cavity can accommodate more than one wavelength

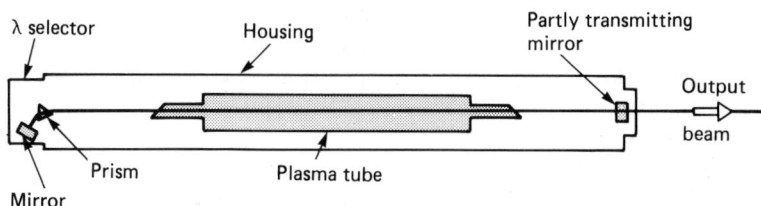

Figure 10.4 Selecting a single longitudinal mode by internal prism

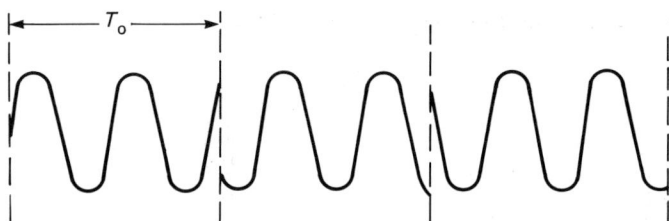

Figure 10.5 Sudden phase jumps of a monochromatic wave reduce coherence

The typical coherence length of a practical (hence, mostly of a multi-longitudinal-mode type) HeNe laser is some 20–30 cm, that of an ion laser 5–10 cm and that of a semiconductor laser less than 1 mm. On the other hand, the highly stabilised single mode CW laser mentioned earlier, with a linewidth of 1 kHz, would have $L_c = 300$ km. The numerical value of the coherence length, not always given by the manufacturer, can be easily calculated from 'line width' (the spectrum width FWHM, full width at half maximum), always to be found in a data sheet, using:

$$L_c = \left| \frac{\lambda^2}{\Delta\lambda} \right| \qquad (10.1)$$

which is derived in Appendix 11.

It will intuitively be perceived that no meaningful interferometric measurements can be made over distances exceeding the coherence length, L_c. The fringe fields of imperfectly coherent (practical) sources will have poor bright-to-dark contrast in experiments in which the path length difference approaches L_c. This contrast factor:

$$V = \frac{I_b - I_d}{I_b + I_d} \qquad (10.2)$$

with I_b being the maximum brightness and I_d the minimum brightness (dark fringes) is called the *visibility factor*. It gives a measure of the degree of coherence, $V = 1$ corresponding to perfect coherence and $V = 0$ to total incoherence. A Michelson interferometer (see textbooks on optics, e.g. Ref. 1 or 17) can therefore be used to measure L_c, as this will be the path difference for which V just reaches zero ($I_b = I_d$), i.e. the fringes are no longer discernible.

We are familiar, in electronics, with the correspondence between the duration of a wavetrain (e.g. a keyed carrier system) and the resulting frequency bandwidth. Here, also, the spectrum broadening caused by imperfect coherence is:

$$\Delta f_c \simeq \frac{1}{t_c} \tag{10.3}$$

More will be said on this subject in Chapter 13 in connection with the Fourier Transform and power spectra.

(b) Spatial coherence

Imagine two neighbouring laser beams, A and B, both of a high degree of temporal coherence. It is more than likely that they will fall short of being mutually coherent. Similarly, two neighbouring portions A_1 and A_2 of the cross-section of the same laser beam could be lacking in respect to mutual coherence. Points P_1 and P_2 would then be out of step: one moment on the same wavefront, the next on two separate wavefronts. No valid interference experiments would be possible with the radiations originated by these points. The type of coherence pertaining to the *transverse* similarity of beam parameters is called *spatial coherence*. The area of the beam cross-section over which interference can be obtained is the *coherence area*, A_c. Here, just as in the case of temporal coherence, the fringe visibility factor:

$$V = \frac{I_b - I_d}{I_b + I_d}$$

can be used as the yardstick of the degree of coherence.

Portions A_1 and A_2 could suffer from low temporal coherence and yet, within the area A_c, have a good spatial coherence. It can thus be seen that spatial and temporal coherence are independent of each other.

Spatial coherence is greatest (approaching perfection, with the conerence area A_c confused with the beam cross-section A_b, $A_c = A_b$) in lasers in which the optical cavity supports a single transverse mode, e.g. in a TEM_{00} HeNe laser (see Section 10.2.2(b)). It is lowest, with A_c representing a few per cent of A_b, in lasers with numerous transverse modes (e.g. some inexpensive varieties of semiconductor laser). Spatial coherence is completely destroyed (practically) by the passing of the beam through a ground glass plate or by reflecting it from a diffusing surface.

10.2.2 Oscillation modes

Remember that the laser is a light oscillator and that the space between its two end mirrors forms the resonating cavity. Unless special measures are taken, the cavity will accommodate several strings of standing waves. These are called *modes*.

(a) Longitudinal modes

Like an air whistle or an organ pipe, in which more than one tone can be sustained by the resonator length (Figure 10.6), a laser can sustain oscillation of more than just one frequency. As long as an integer number

M$_1$ M$_2$

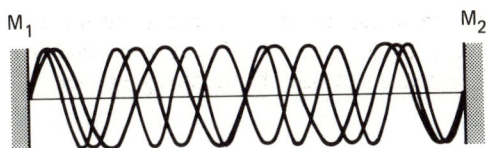

(a) Optical cavity, showing the simultaneous
accommodation of three modes

(b) Acoustic cavity
Three possible modes shown seperately

Figure 10.6 A laser can sustain more than one longitudinal oscillation

of half-waves fits neatly into the cavity length and provided there is
sufficient system gain, an oscillation may be sustained. Those frequencies
and wavelength at which stable oscillations of this kind actually exist in a
laser characterise its *longitudinal modes*.

Because the length of a laser cavity is several orders of magnitude
greater than the wavelength, the frequency difference between two
neighbouring longitudinal modes is very small. Indeed, if for a given mode,
n, there are k half-waves accommodating the stationary wave string, there
will be $k + 1$ of them in the next or $(n + 1)$th mode. The difference

$$\frac{(k-1)-k}{k}$$

or $1/k$ is indicative of the frequency differences Δf usually called *mode
spacing*. The mode spacing, being inversely proportional to k, is therefore
also inversely proportional to the cavity length, L. It works out that:

$$\Delta f = \frac{c}{2L} \tag{10.4}$$

This is no other than the reciprocal of the time t the radiation takes to make
a mirror-to-mirror return journey, $t = 2L/c$. (To be accurate, Expression
10.4 must be corrected for the speed of light in the laser medium:

$$\Delta f = \frac{c}{2Ln}$$

Small for gases, the correction can be very large for solids, e.g. for GaAs, where $n = 3.5$.) For a 50 cm long laser $2L = 1$ m. This represents 300 MHz for Δf and 3.3 ns for t. Shorter lasers are more wieldy. The degree of shortening a laser, however, has a limit as this operation not only expands Δf but also reduces the system gain.

Only a very few of the mathematically possible longitudinal modes exist in a laser, namely those that fit the energy level transition Equation 9.1. Influenced by the effects of the incessant motion of molecules (for gas lasers) and the thermal spread of electron energies (for semiconductor lasers) this condition yields a bell shaped gain curve acting as an envelope for the amplitudes of existing longitudinal modes (Figure 10.3). (See Section 12.7.1 for the Doppler effect.)

When a single mode operation is required, a prism or other 'colour' selective optical component is used, either inside or outside the cavity, to get rid of the unwanted modes. Internal 'line' selection yields higher output powers while external selection affords tunability (Figure 10.4).

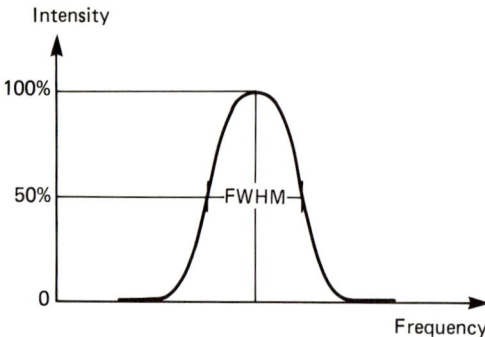

Figure 10.7 The FWHM concept of a single line frequency spectrum

The FWHM (Figure 10.7) of the remaining mode lies, typically, around 3 MHz (2.8×10^{-6} nm) for argon, 1 MHz (1.6×10^{-6} nm) for HeNe and 4000 MHz (0.01 nm) for GaAlAs. Compare these figures with the spectrum width of a typical monochromatic LED of approximately 30 nm and you will appreciate the monochromaticity of lasers.

(b) Transverse modes

With some HeNe lasers, pictures of great beauty can be obtained by simply projecting a slightly expanded beam on to a sheet of white typing paper in a darkened room. The enlarged beam cross-section reveals a *fragmented* structure. The distribution of irradiance shows a few bright zones, separated by dark 'nul' corridors, the whole pattern giving an impression of some hidden symmetry. The picture is characteristic of a higher transversal mode. The name TEM stems from Transverse Electric and Magnetic – specifying the position of the electric and magnetic field vectors, the

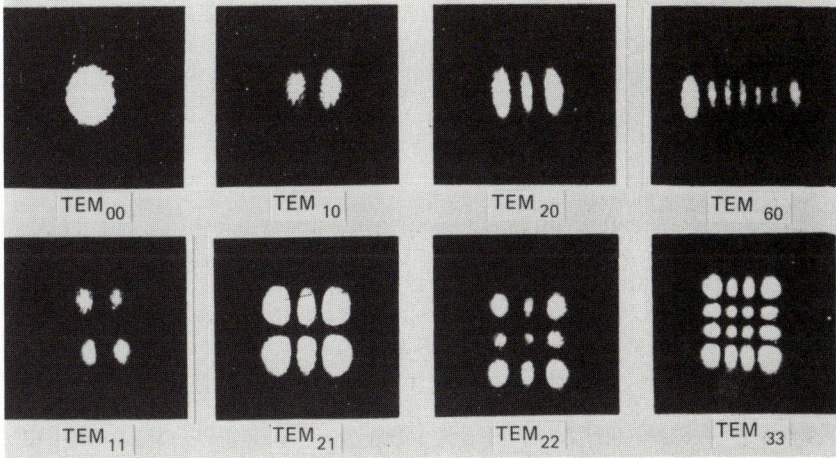

Figure 10.8 Demonstrating the way transversal laser modes are identified

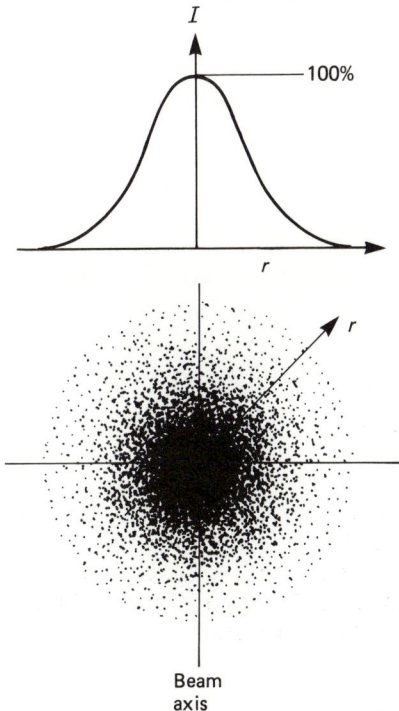

Figure 10.9 The Gaussian intensity distribution of a TEM_{00} beam giving rise to the irradiance shown below it

subscripts referring to the number of nul zones (the corridors) in the horizontal (first subscript) and vertical (second subscript) directions. Figure 10.8 should eliminate any doubts regarding the use of these subscripts. In many applications lasers operating in higher transversal modes are of little use, and the TEM_{00} or *fundamental mode* is required. Here, the projection gives a single spot, a circular bright zone, fading gradually away, outward from the centre (Figure 10.9). The bell-shaped curve showing the $I_r = f(r)$ variation (irradiance across a diametral line) is well defined, mathematically determined by an equation, due to Gauss (which can be likened to the Gaussian probability density function [37]) of the form:

$$F_{(x)} = \sqrt{\left(\frac{a}{\pi}\right)} e^{-ax^2} \qquad (10.5)$$

(where a is a constant). In its simplest form, this is $y = e^{-x^2}$. The TEM_{00} beam is therefore referred to as the Gaussian beam. More will be said about the Gaussian beam and its importance to the EO engineer in the following sections. Physically speaking, the existence of higher transverse modes is a result of the establishment of standing waves along a line at a small angle to the cavity axis, i.e. to the centreline working the laser and mirrors. Analogies with microwave cavity modes will be immediately perceived. The modal spacing Δf (frequency difference between two consecutive transversal modes) is much smaller than in the case of longitudinal frequency shifts, typically, for a HeNe laser, a few megahertz only. Nevertheless, temporal and spatial laser coherence suffer from their presence, as could perhaps be felt intuitively. Finally, on the subject of transverse modes and their notation, let us mention the existence in some lasers of cylindrical *circular symmetry* bright and dark zone configurations, which can result from 'of revolution' or concentric laser architectures. They are designated by an asterix, TEM*, such as the 'doughnut' mode TEM_{01}^* with a beam cross-section and a transverse intensity variation illustrated by Figure 10.10, which can be thought of as a combination of TEM_{01} and TEM_{10}.

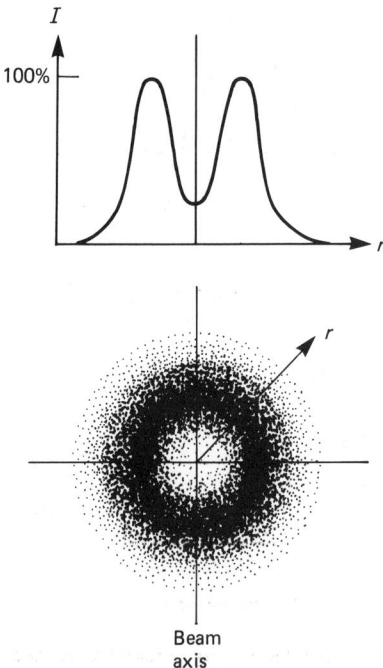

Figure 10.10 The 'doughnut' mode TEM*₀₁

10.2.3 The Gaussian beam – its diameter and divergence

The Gaussian beam, corresponding to the TEM_{00} or fundamental mode, is almost certainly the most useful light beam available to the EO engineer and probably the best behaved beam the optical designer has ever had at his or her disposal. Perfect spatial coherence, smallest possible longitudinal divergence, and softly, gracefully varying intensity profile are its valuable attributes. Here, most aberrations can be forgotten, because the rays are nearly always paraxial and the origin of radiation a virtually monochromatic point source. The TEM_{00} beam is widely used, especially when radiation has to be concentrated into as small a spot as possible or transmitted over as long a distance as possible.

(a) Beam diameter
How does one define the diameter of a beam, the cross-section of which, unlike that of a glass or copper rod, does not have a clearly delineated boundary (Figure 10.11)? An electronics engineer would be tempted to take for r the abscissa corresponding to a -3 dB point, a spectral analyst a FWHM point, while a pulse technique fanatic might insist on the use of the 10%–100%–10% criterion. What EO engineers have decided upon is a $1/e^2 \, I_{max}$ point, which corresponds to $1/e$ of the maximum field amplitude. Thus, we shall call 'beam diameter' the value $2r$ for which I is $1/e^2$, i.e. 13.5% of I_{max}. For historical reasons EO engineers call the corresponding radius w and not r. It is useful to know that the light flux (power) contained within the diameter $2w$ (the circular integral of I within the $-w$, $+w$ limits) represents 86.5% of the total beam power P_{TOT}, 99% of which will be contained within a circle of radius $1.5w$. Regarding numerical values, a

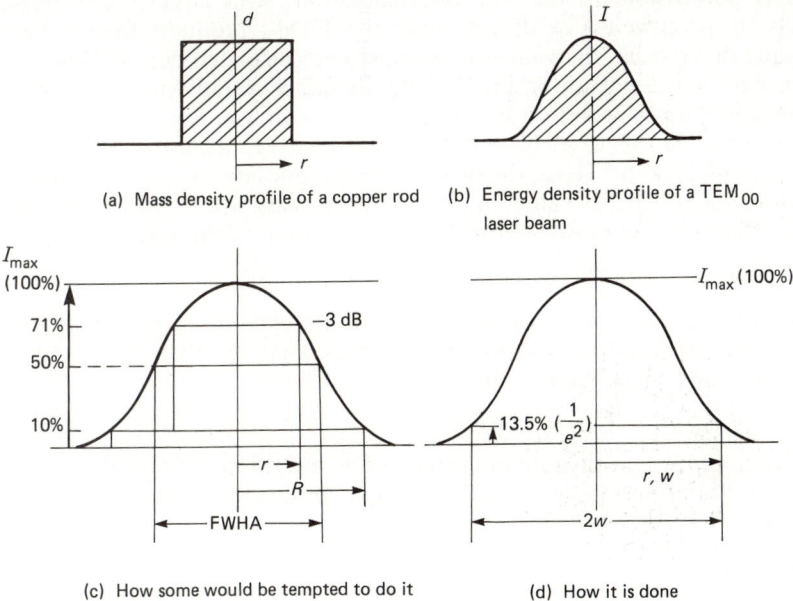

(a) Mass density profile of a copper rod

(b) Energy density profile of a TEM_{00} laser beam

(c) How some would be tempted to do it

(d) How it is done

Figure 10.11 How is the diameter of a cross-section which lacks sharp boundaries defined?

typical low power (5 mW) HeNe laser will have a beam diameter of 0.8 mm near its exit.

(b) Beam divergence

Just how constant is this diameter as we move away from the laser exit along the propagation axis z? We mentioned in Section 8.3.1 on free space optical communications that point source light can be shaped into parallel beams (collimated). Even though TEM_{00} beams behave as if they were originated by a point source and were well collimated, their light still spreads a little laterally, i.e. the beam exhibits a slight divergence. This is caused by *diffraction* (see Section 10.3.2(b)) and can be neither avoided nor *completely* eliminated, though, as we shall see in the following section,

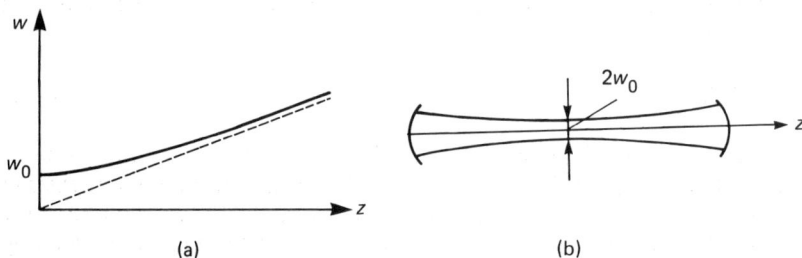

(a)   (b)

Figure 10.12 An example of diameter variations of the beam inside a laser

its effects upon the beam spread can be minimised by means of added optics. (Diffraction was disregarded in the earlier chapters of this work. This is permissible in the first approximation, with large TXAs using LEDs. It nevertheless exists.) At least the TEM_{00} produces the smallest possible divergence, the value of which will be calculated below. Near the laser exit and, indeed, inside it, the diameter variations are a little complicated and depend on the types of mirror in use. Figure 10.12 gives an example. In this case, both mirrors are concave, spherical and have a common focal point. Here, the beam diameter goes through a minimum in this focal plane and its value $2w_0$ is called the *beam waist*. Away from the waist, the radius w grows accordingly to a relation of the form:

$$w_z = w_0 \sqrt{[1 + (kz)^2]} \tag{10.6}$$

Whatever the value of k, which depends on the type of optical cavity and can vary from author to author, for a distance z greater than a few metres $(kz)^2$ becomes much greater than unity and Equation 10.6 reduces to:

$$w_z = w_0 kz \tag{10.7}$$

This is of interest, as the rate of *increase* of w_z *with distance*, namely w_z/z, is no other than the already familiar half-angle of divergence $\beta/2 = kw_0$ (see Section 8.3.1(a)):

$$\frac{\beta}{2} = kw_0 \tag{10.8}$$

Knowledge of k is necessary for the calculation of $\beta/2$. It works out that

$$k = \frac{\lambda}{\pi} \frac{1}{w_0^2}$$

[3(b), p. 326; 181, p. 281] which, for HeNe lasers, gives a practical formula:

$$\frac{\beta}{2} = \frac{0.2}{w_0} \qquad (10.9)$$

with the waist w_0 taken in millimetres and the divergence half-angle in milliradians. For a practical HeNe laser ($w_0 = 0.4$ mm) this gives $\beta/2 = 0.5$ mrad. It is worth noticing that this is only half the value of the diffraction-caused divergence of a plane, monochromatic *uniform* intensity wave of the same wavelength, 0.6328 μm, exiting a circular aperture of the same diameter as the laser waist, $D = 2w_0$, calculated from the textbook equation [e.g. 1, p. 304]:

$$\frac{\beta}{2} = 1.22 \frac{\lambda}{D} \qquad (10.10)$$

in which D is defined for very nearly the same power contents (84% of P_{TOT}) as that of $2w_0$ (86.5% of P_{TOT}). As already mentioned, the Gaussian beam has, in fact, the smallest possible divergence. We shall now see how this already small divergence of a 'naked' laser can be further reduced by means of external optics.

10.3 Beam treatment

10.3.1 Beam expansion

A divergence of 0.5 mrad may seem little, until we realise that it represents a beam 'fattening' rate of 1 m/km. Of course, observed on a laboratory bench, a beam may appear almost parallel, as its initial diameter of 0.8 mm swells to, say, $0.8\sqrt{[1 + (1 \times 1.25)^2]} = 1.28$ mm over a 1 m path. A 10 km communication link using the same laser without any external optics would be most inefficient, with its 10 m beam diameter at the RX terminal, while an Earth–Moon or even an Earth–satellite link would be utterly useless. (The Earth–Moon distance is 386 882 km and the beam diameter on the moon would be 387 km.) A beam *expander*, paradoxically, brings a remedy to this situation. Paradoxically, because it *increases* the beam diameter at the TXA exit. Here is a typical case of making things worse before they can get better, as, thanks to this device, the far-field beam diameter is much reduced. The expander diminishes the beam divergence. The ray tracing of Figure 10.13 shows how. Things are really quite simple: two lenses of different size but identical f number placed on-axis in a tube, so that their focal points coincide, do it all. The smaller lens is the device entrance, the larger the exit. The case of an idealised, parallel, on-axis incident beam is shown in Figure 10.13(a). The entrance lens L focuses the incident rays in its focal point P. As P is also the focal point of the exit lens, L_2, the latter collimates them back into a parallel bundle. This emerging bundle, however, is wider than the incident one, the numerical value of the

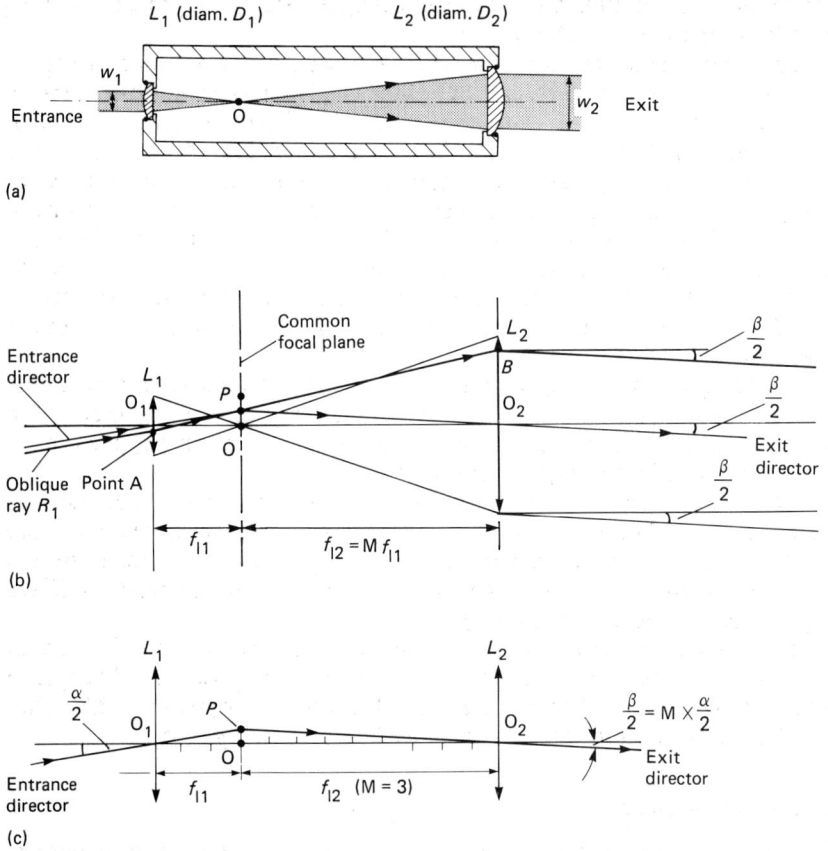

Figure 10.13 Beam expander. (a) Basic structure. Beam diameter is increased. ((b) The beam expander reduces oblique ray inclination and therefore beam divergence. (c) Director rays as ancillaries of oblique ray tracing

expansion factor w_2/w_1 being simply equal to the ratio of the lens diameters D_2/D_1 or their focal lengths F_{L2}/F_{L1}. So much for beam expansion. Let us see now how the divergence reduction takes place. What happens to an *oblique* entrance ray can be seen, at a glance, in Figure 10.13(b). Entering the expander with an inclination $\alpha/2$ (this is the divergence half-angle of the incident beam), this oblique ray R_1 is bent (upwards) by the entrance lens L_2 in point A, travels straight through the inter-lens space traversing the *common* focal plane, is re-bent (downwards) by the exit lens L_2 in point B and emerges from the expander with an inclination $\beta/2$. The bending angles at points A and B have been determined by what is a standard ray tracing technique: an ancillary entrance *director ray*, parallel to R_1 but going through the centre O_1 of lens L_1 gives the point P in the focal plane where it *must* be met by R_1. Similarly, an exit director ray PO_2 determines the degree of bending of the oblique ray at point B. (The exit director transverses L_2 unbent. As P lies in the focal plane of L_2, this lens performs a collimation and thus the oblique exit ray R_2 must leave L_2 parallel to the exit director.)

For clarity's sake, the director rays have been retraced alone in Figure 10.13(c), for an expansion factor $M = 3$. Here, it can be seen from triangles O_1PO and O_2PO that the exit inclination $\beta/2$ is F_{L2}/F_{L1} times smaller than the entrance inclination $\alpha/2$. The beam expander is, in fact, a telescope used back to front, giving a divergence reduction (called the *expansion ratio*) equal to its magnifying power $M = D_2/D_1$. Commercially available expanders have, typically, M values of 5, 10, 20 and 50. In fact, there are cases in laser use where the beam diameter needs *reducing*, e.g. in order to increase the frequency bandwidth of an external modulator (see Section 10.3.2(b)). The 'expander' can then be used in reverse, i.e. like a telescope.

The oblique ray tracing technique has been introduced above for its general interest to the reader. This simple technique is of great help whenever one is faced with the puzzling question of what happens to an off-axis ray incident obliquely upon a lens.

The expander described above is known under three names: Keplerian, astronomical and co-focal. Another, more compact, type of telescope, the Galilean, can also be used for beam expansion/reduction, but it suffers from the sometimes serious disadvantage of making the use of *spatial filters* impossible (see Section 10.3.4(b)).

Prior to closing this section, it ought to be mentioned that beam expanders can be used purely as *expanders* as opposed to divergence reducers. This application is of great value when illumination with coherent light of a sizeable object is required, e.g. in holography.

10.3.2 Deflection

Photons have no electric charge and therefore, unlike electron beams, laser beams cannot be deflected by electric or magnetic fields. Other methods have thus to be used in applications in which the pointing direction of a beam needs controlling, as, for example, in target location or scanning. These rely on mechanical, electro-optic or electro-acoustic phenomena, or on holography.

(a) Mechanical deflectors

Conceptually, mechanical deflectors are simplicity itself: the laser beam is made to fall onto an orientable or a continuously moving mirror which directs it to the wanted location or 'address'. However, when it comes to designing and manufacturing deflectors so that they fulfil the requirements of speed, resolution, repeatability, low maintenance, long life, etc., imposed by the system they are destined for, things are far from simple and sophistications such as angle linearisation, servomechanisms or noise reducing air bearings creep in. So much so that entire companies make a living from their expertise in this field.

Some deflectors are based on the very old mirror galvanometer. Moving coil or moving iron mechanisms respond to the deflection controlling current and torsion returns them to the zero position. Obviously, a mirror rotation α_R produces a beam deflection $\alpha_D = 2\alpha_R$. Here, as in all

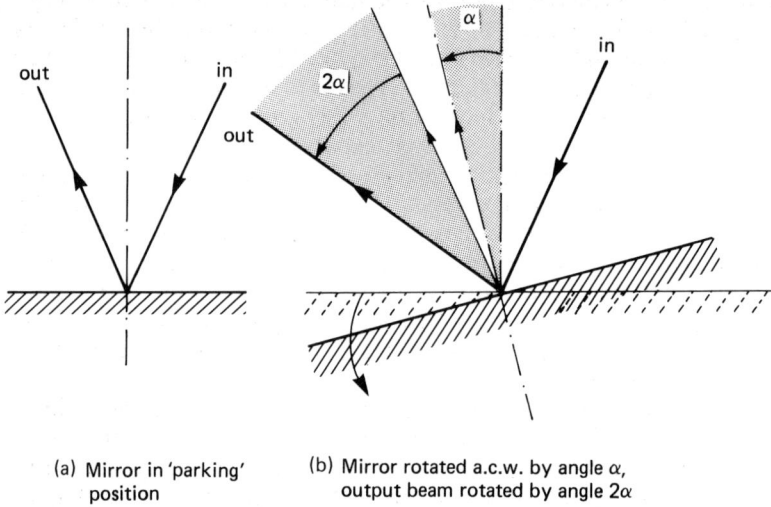

(a) Mirror in 'parking'   (b) Mirror rotated a.c.w. by angle α,
    position   output beam rotated by angle 2α

Figure 10.14 The incident laser beam is made to fall on to a rotating first surface mirror M which directs it to the wanted address by deflecting it by the angle 2α. Note the angle doubling

reflective deflectors, first surface mirrors are used to avoid glass refraction (Figure 10.14).

Some galvo deflectors work in a 'random access' way, i.e. the quantized deflection commands are aperiodic, sending the beam to the required address, X, only when asked to and leaving them there for the required duration, t. These are the 'orientable' mirrors. Some others are harmonic – the oscillating mirrors referred to above as 'continuously moving'. These are very often operated at a fixed frequency. In such cases, the mechanical

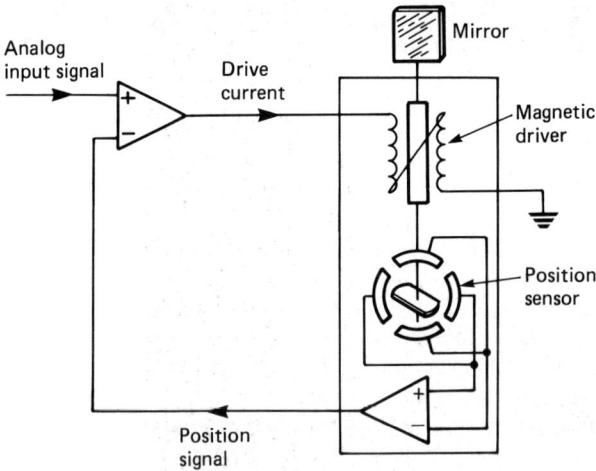

Figure 10.15 Galvo scanner with pick-up position sensor for feedback (Courtesy of General Scanning Inc)

self-resonance of the galvo can be taken advantage of. The deflection angle obeys a sinusoidal law, so that the spot writing speed is larger in the vicinity of the central position than towards the ends of scan excursion. When constant spot dwell times are required, a linearisation servo, based on a capacitive pick-up within the moving mechanism, can be used (Figure 10.15).

In resonant galvos linearisation is obviously restricted. Here, variable dwell times can be compensated for, as far as exposure is concerned, by controlled beam intensity variations (e.g. for photographic processes). When equispaced beam writing or reading across a scan line is required, a self-timing information input/output may be used. A periodic galvo deflector can resolve a few thousand discrete positions per scan and operate at scanning rates of up to 2000 per second, while the resonant types can be some 2 to 5 times faster [195].

Rotating mirror deflectors can be divided into two categories: those in which the beam deflection angle is twice the angle of rotation ($\alpha_D = 2\alpha_R$, as in galvos) and those in which there is no such angle doubling ($\alpha_D = \alpha_R$). The first category uses multifacet (polygonal) mirrors (Figure 10.16). Here, as in a galvo, the incident beam is perpendicular to the axis of rotation. The second category uses either a truncated shaft reflector (Figure 10.17(a)) or a pyramidal multifacet mirror (Figure 10.17(b)). In the first case, the incident beam is collinear with the axis of rotation, and the deflected beam describes a plane perpendicular to it. In the second case, the incident beam is parallel to the axis of rotation and the deflected beam remains within a plane perpendicular to it.

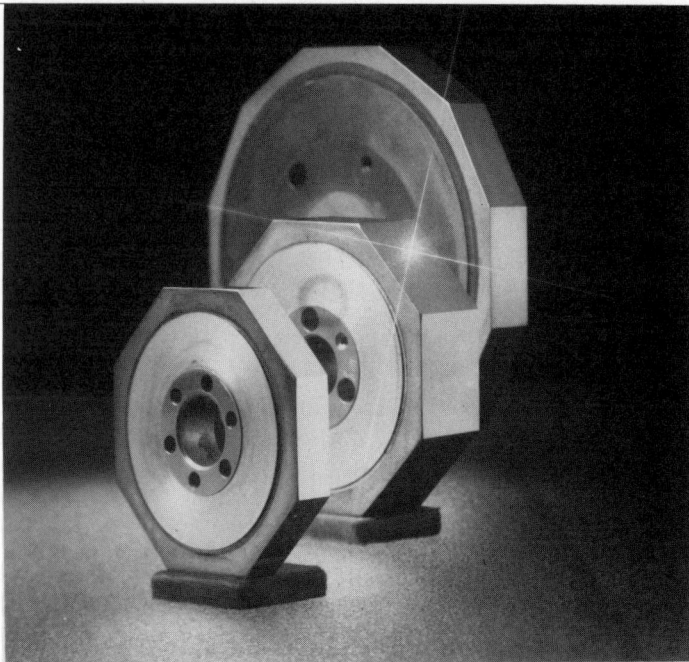

Figure 10.16 Multifacet polygonal rotating deflector (Courtesy of Lincoln Laser Co Inc)

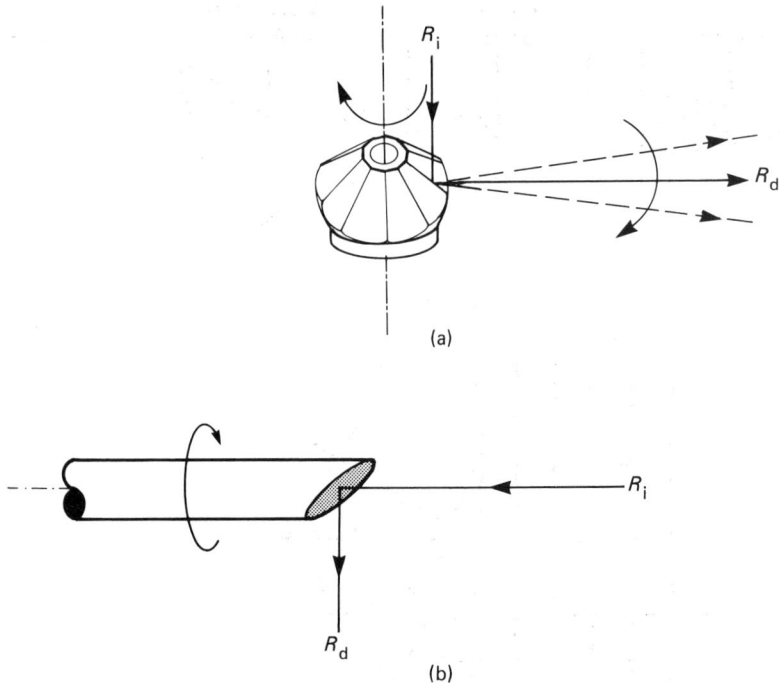

Figure 10.17 (a) Truncated shaft deflector and (b) pyramidal deflector

The most severe problem associated with rotating multifacet reflectors is that of facet alignment with respect to one another. Imperfect alignment manifests itself in imperfect superposition of successive scan lines (each produced by a facet) which results in irregular line spacing of raster (Section 10.3.2(d)) scans for which these devices are used.

Electronics engineers have a dislike of moving-part mechanisms. Laser beam deflectors exist which have none. To date, they belong to two categories: the acousto-optic and the electro-optic deflectors. I would venture the opinion that, conceptually, there is more elegance in them than in rotating mirrors.

(b) Acousto-optic deflectors

Acousto-optic (AO) deflectors are based on the *diffraction* of light. Diffraction plays an important part in electro-optics and its understanding is a prerequisite to the understanding not only of AO deflectors and modulators, but also of spatial filtering, optical signal processing and holography. The subject forms such an important part of all optics that most textbooks devote over 100 pages to it. A brief refresher follows.

In fibre optic communications, bending the fibre does not suppress the light flow; reflection is responsible for this. Diffraction is yet another phenomenon that makes light go round corners. The term stems from the latin word *diffractus*, which pertains to breaking up or scattering. Whenever a wavefront hits an abrupt discontinuity of the propagating medium,

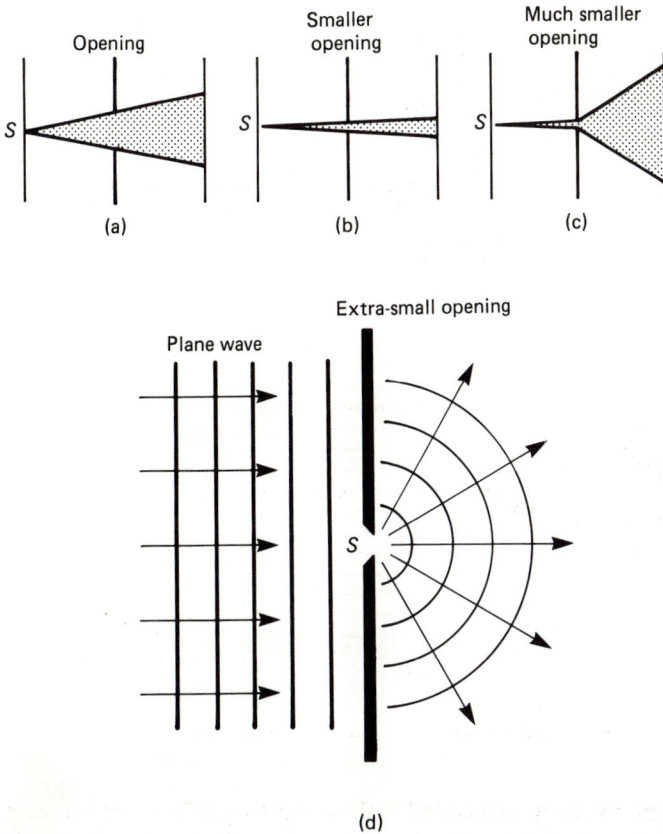

Figure 10.18 Whenever a wavefront hits an obstruction or opening, diffraction takes place

e.g. an obstruction or an opening, new waves are generated at the hit point – diffraction takes place (Figure 10.18(d)). This can be witnessed in everyday life when mechanical waves in water meet a breaker, a rock or just a twig. Figure 10.18(c) shows clearly that, with light, no geometrical shadow is formed at a distance by a *very small* hole, justifying the statement that light goes around corners.

Figure 10.19, adapted from *Lasers and Light* [7] for its convincing clarity, illustrates with great beauty what happens when the obstruction has *regular, repetitive* features. The incoming radiation is plane, monochromatic. Merging together and interfering with each other, the spherical (or cylindrical) secondary waves form a series of plane waves, propagating in a variety of directions. (We shall see below how a variant of this phenomenon has been put to good use in AO beam deflectors.) In order to make diffraction more tangible, let us give two more examples of how it can be experienced in everyday life, together with an ensuing all-important comment. Hearing the hooting of a car from around the corner demonstrates the diffraction of *sound* waves. For light, most beautiful effects of

184

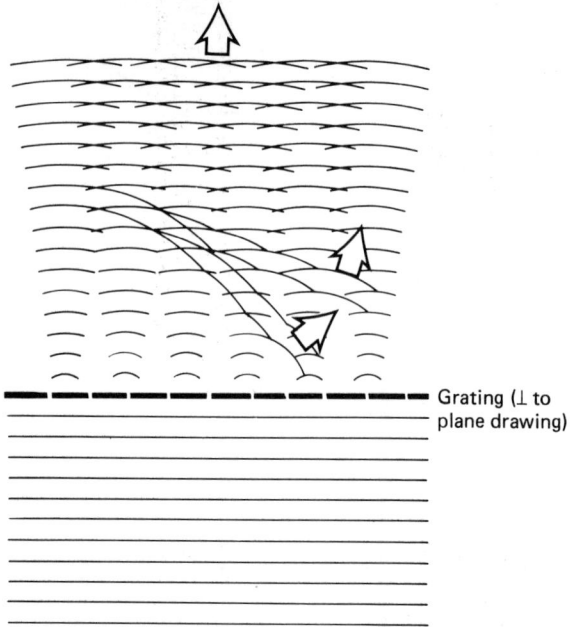

(a) Multislit diffraction grating

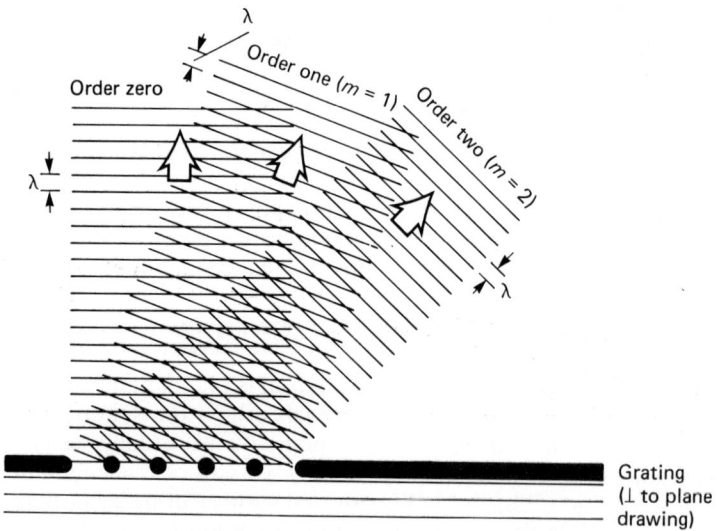

(b) Multirod diffraction grating

Figure 10.19 Two examples of diffraction gratings

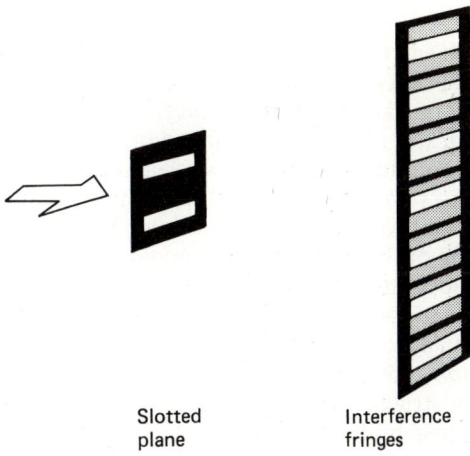

Slotted
plane

Interference
fringes

(a)

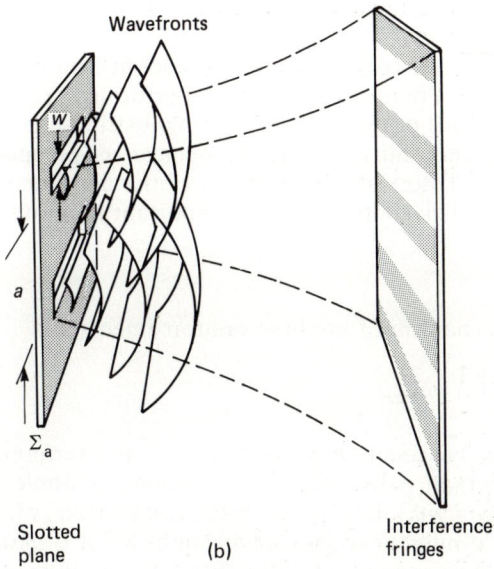

Wavefronts

w

a

Σ_a

Slotted
plane

(b)

Interference
fringes

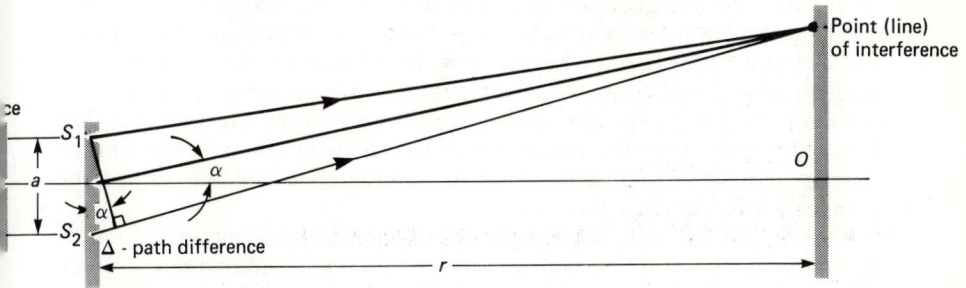

Point (line)
of interference

ce

S_1

a

α

O

S_2

α

Δ - path difference

r

(c)

Figure 10.20 (a, b) Interference fringes formed when a monochromatic coherent plane wave strikes a thin opaque plane with two neighbouring narrow slits in it (the Young Experiment). (c) The common source S_o ensures coherence

multiple aperture diffraction can be observed watching remote car headlights, at night, through a net curtain. The orthogonal threads of the woven fabric mesh produce these two-dimensional cross-like or star-like diffraction patterns made up of dark and bright *interference* patches. The most important conclusion drawn from the quoted and other observations is that diffraction is more pronounced when the wavelength is *greater* and the diffracting discontinuity *smaller*. In Figure 10.18, the smaller the hole the more diffraction and the broader the cone of light. With larger openings, the straight propagation of the original wave appears to swamp diffraction, still present, nevertheless, at the edges. In the case of the net curtain, the finer the netting the more broadly sprawled the diffraction pattern.

Fundamentally relevant to the study of electro-optics in general, and to that of AO deflectors and modulators in particular, are the iterative (multislit or multirod) structures exemplified by Figure 10.19, known as diffraction gratings. Their mathematical analysis is rather arduous and so we shall look at what happens when a monochromatic coherent plane wave strikes a thin opaque plane with just two neighbouring narrow slits in it, as in Figure 10.20, hoping that, from it, the action of the iterative grating will be perceived intuitively.

The arrangement was conceived and tried out by Thomas Young in 1806 and is known as the Young Experiment. If performed in a dark room with visible light, it yields not two but several bright stripes, separated by dark ones, on the screen. Slits S_1 and S_2 behave as secondary, line-like, cylindrical wave radiators and the light rays emerging from them interfere with one another. The simple yet most important relation:

$$\sin \alpha = m \frac{\lambda}{a} \tag{10.11}$$

expresses the mechanism of in-phase reinforcements and:

$$\sin \alpha = \frac{2m + 1}{2} \frac{\lambda}{a} \tag{10.12}$$

that of anti-phase cancellation resulting from this interference. When the optical source–screen path difference $\triangle$ equals a whole number m of wavelengths λ ($\triangle = m\lambda$) the interference is *constructive*, while for $\triangle$ equal to an integral odd number of half-wavelengths $\lambda/2$ it is *destructive*. These are the basics of the geometry of diffraction. The resulting patterns (Figure 10.21) are called *interference fringes*. (The gradual reduction of irradiance and contrast towards the edges is probably caused by the growth of the inequality of amplitudes of the interfering waves accompanied by their common decrease with the increasing distance from the source.) The Young Experiment will succeed only if the slit width w and their mutual distance a is of the same order of magnitude as λ and the slit–screen distance r much greater than λ.

If we now increase the number of slits of Figure 10.20 from two to many, we obtain a repetitive structure known as a *diffraction grating*. When illuminated from the back by a monochromatic, coherent plane wave source such as a laser, the grating will produce interference effects resulting in several beams, each leaving the structure at a different angle, creating a situation resembling that of Figure 10.22. The *diffracted* plane

Figure 10.21 Interference fringes

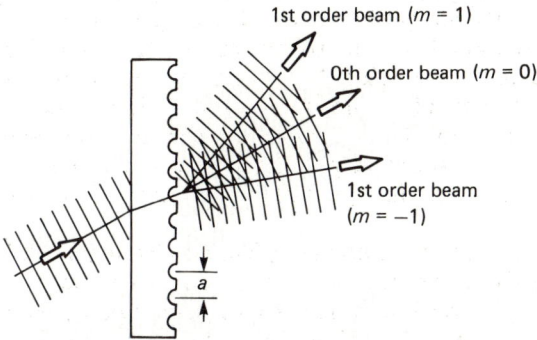

Figure 10.22 A diffraction grating produces several beams, each leaving the structure at a different angle

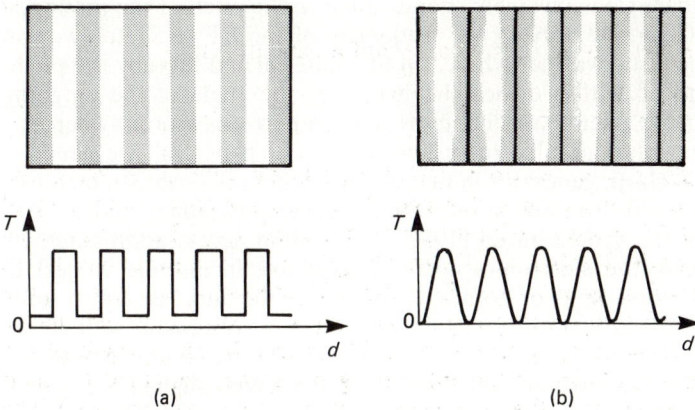

Figure 10.23 Transmittance versus distance of (a) multislit and (b) sinusoidal gratings

waves are identified by *orders*, related to the values of m of the diffraction equation: zero order for $m = 0$, first order for $m = \pm 1$, second order for $m = 2$, etc. When a few design conditions of a diffraction grating and of the geometry of its application are satisfied, the zero and first order only will be obtained. The principal condition is that the spatial distribution of the variable transmission parameter of the diffracting medium be sinusoidal, as opposed to rectangular, so far considered (Figure 10.23). This will seem

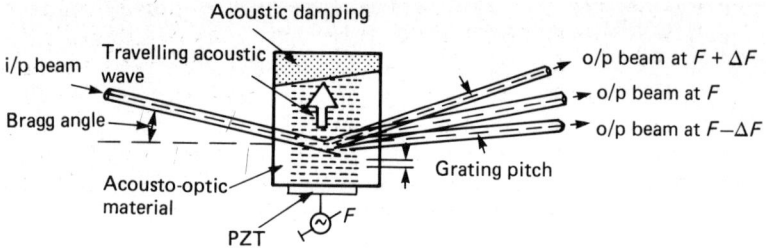

Figure 10.24 The Bragg cell. Incoming radiation must be directed at the diffraction plate at the Bragg angle. Zero-order beam not shown

quite acceptable to readers with an electronics background who will instinctively associate the sine function with a single frequency and the rectangular one with a fundamental + harmonics series. (Section 13.2 on Fourier Transforms in optics will make this even clearer.)

The second condition is that the incoming radiation be directed at the diffraction plate at a certain precisely defined angle, called the Bragg angle (Figure 10.24).

In a real AO beam deflector, the grating is not a thin plate but a crystal of a substantial thickness, while the spatially variable optical parameter is the refractive index, n. The sinusoidal variation of n is obtained by establishing an acoustic wave in this crystal. This is achieved by means of a piezoelecric transducer, also a crystal, driven by an RF oscillator. Together, they form what is known as a Bragg cell. We know already that the diffraction angle is inversely proportional to the pitch of the grating, a (Equation 10.11). By varying the frequency of the RF oscillator we can control the acoustic wavelength, λ_a, in the material and thereby the pitch. Thus, we end up with a device that will vary the angle of the outgoing beam, i.e. an AO laser beam deflector! If we require a sawtooth sweep, say for scanning a photographic recording on film or paper, all we need, in principle, is a 'chirp' generator in lieu of the fixed F oscillator. In fact, two deflectors are used for this application, an AO prescanner and an AO travelling lenslet, synchronised [195]. For random access addressing, on the other hand, the electronics makes F available in discrete values, to correspond to the x or y, or both, coordinates of the address. Up to some 2000 individual addresses in each coordinate can be obtained, with access time around a few microseconds. (The influence of the beam diameter on the response time is covered in Section 10.3.3(b).) Well aligned Bragg cells deflect about 85% of the beam intensity into the first order. This figure is termed the *diffraction efficiency*.

There is, unfortunately, a dark side to this rosy scenario of the wonderful device: the angles obtainable from a single basic deflector for random access are very small, only a degree or two. (The exceptionally high figure of 5.4° claimed by Inrad Inc [199] is still very small in comparison to CRT deflection angles.) Bulk optics can, of course, be used to multiply α by up to a factor of $\simeq 10$ but this calls for extra space, creates problems with dust, alignment, etc., and causes additional expense. AO deflectors are, nevertheless, in current use.

In practical devices the piezoelectric transducer is usually made of lithium niobate (LiNbO$_3$) and the grating typically from tellurium dioxide (TeO) or lead molybdate (PbMoO$_4$). Although the term 'crystal' has been used so far in the text for the interactive medium, high-purity glasses (e.g. tellurite glass or fused quartz) can also be used.

The supporting electronics of AO deflectors uses conveniently modest voltages for both signals and power supplies (e.g. 1 V$_{pk}$, TTL, ±15 V), has a familiar input impedance (50 Ω), and requires drive powers of 0.5–4 W. The acoustic frequency lies in the 70–300 MHz range, typically.

Before turning to the second type of deflector with no moving parts, let us note the importance of realising that diffraction is with us in all optical instruments, whether we like it or not, as there is always a lens edge or a stop limit acting as discontinuities of the propagating medium. When all the aberrations of a lens have been eliminated or compensated for, diffraction remains. The lens is then said to be *diffraction limited*. This is as far as perfecting it will go. The light from an idealised point source collimated by such a lens is still slightly divergent (see Section 8.3.1(a) on optical communications in free space). Similarly, the image of such a source formed by such a lens will always be slightly blurred, owing to diffraction, and hence no longer a point image. Of great interest is the special case of a monochromatic coherent point source, amply treated in optics textbooks [2, section 11.10].

(c) Other types of deflector
Remember that diffraction gratings can deflect a beam of parallel coherent light. With reference to Figure 10.22, if the grating is rotated by 90° so that its 'slits and rods' structure becomes parallel to the page instead of being perpendicular to it, the deflected beam will turn round and leave the plane of the drawing. Thus we see that the angle at which the incident beam is deflected, in space, is related to the relative grating/beam orientation. This behaviour is the underlying principle of *holographic deflectors*.

The above explanation is no more than a concept-oriented simplification for readers who are unfamiliar with holography (which is not introduced until Chapter 13). A holographic scanner uses a rotating disc fitted with a number of circularly distributed holograms (Figure 10.25). In addition to producing deflections of the same nature as those of a diffraction grating, holograms give us the bonus of focusing the beam to a bright, clean spot. (The object used for their making is a point source.) Bright, because diffraction efficiencies of 80% can be achieved. There is one basic snag, however, with this simple arrangement: the scan line traced by the moving spot is not a straight, but a bowed line. This difficulty can be lifted by using a prebowed (curved) recording sheet, flattened later [211].

Holographic scanning is still a young technique, but already in use in supermarket points of sale for reading bar code labels (see Section 12.6). The scanners, while at present less accurate and noisier than the previously treated types, are already much cheaper. Rotating 'hologons' present far fewer problems than spinning polygons (alignment!) and the potential for further price drops is enormous, owing to the growing ease of replication of holographic surfaces.

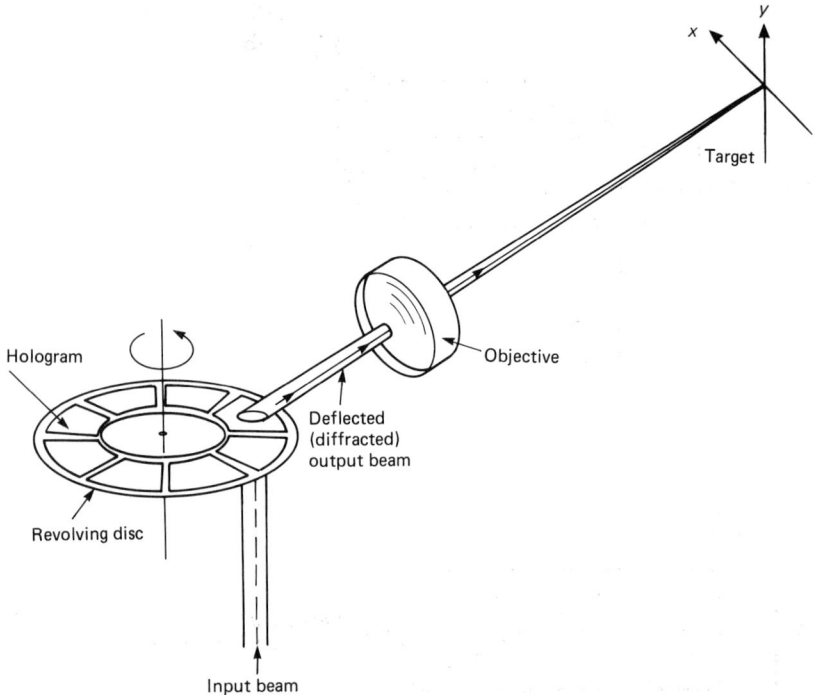

Figure 10.25 A holographic scanner uses a rotating disc containing a number of holograms

The linear electro-optic effect – the linear change of the refractive index of a material under the influence of an electric field – has found more gratifying applications in beam modulation than in beam deflection, and for this reason its treatment has been relegated to Section 10.3.3(c). The basic principle of EO deflectors is illustrated in Figure 10.26(a). Two prisms of a suitably selected electro-optic material, e.g. crystalline KDP (potassium dihydrogen phosphate, KH_2PO_4), are stuck together (in optical contact) to form a flat parallelepiped. They are so oriented that their EO effects are, as seen by the incoming light beam, 'in series opposition'. The diameter of this beam (on the left of Figure 10.26(b)) approaches the value of the crystal height and its polarisation (see Section 10.3.3(c)) lies in the plane of the drawing, oriented as shown. An electrical field, applied perpendicular to the beam, modulates the refractive indices of both prisms but, owing to their mutual orientation, when the speed of light *increases* in the upper prism it *reduces* in the lower one and vice versa. As a result, each wavefront undergoes a gradually increasing tilt (Figure 10.26(b)) as it progresses through the crystal, only to emerge from it with an inclination proportional to the value (and sign) of the applied field. The net result of this is, of course, a beam deflection proportional to the controlling voltage V:

$$\alpha = kV \tag{10.13}$$

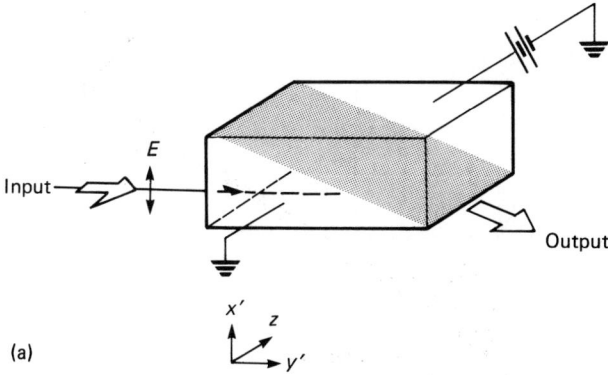

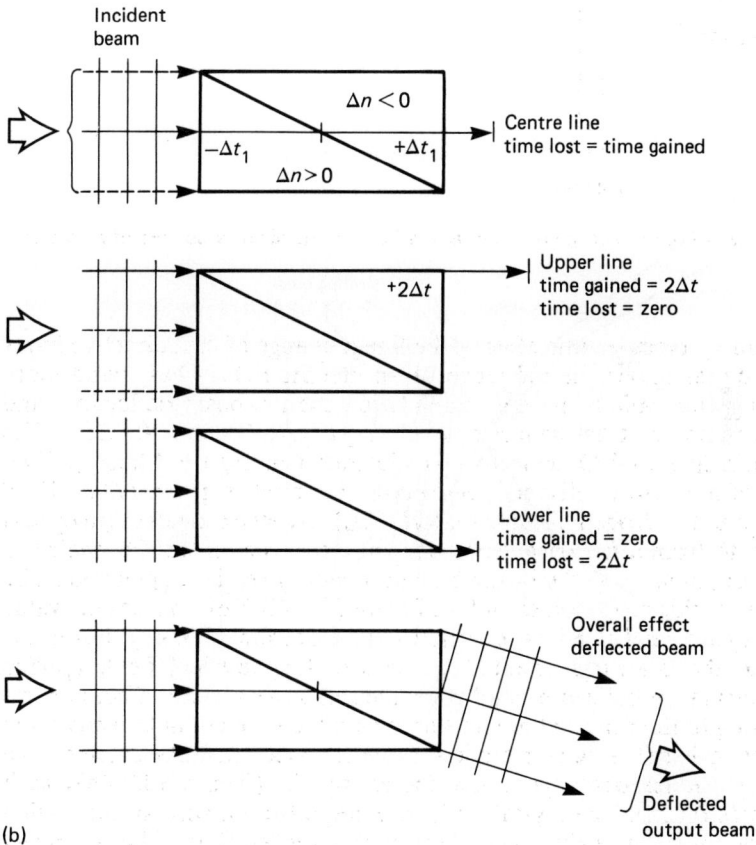

Figure 10.26 (a) Basic electro-optic beam deflector. (b) Beam deflection is obtained by tilting the wavefronts by electrically controlled differential retardation

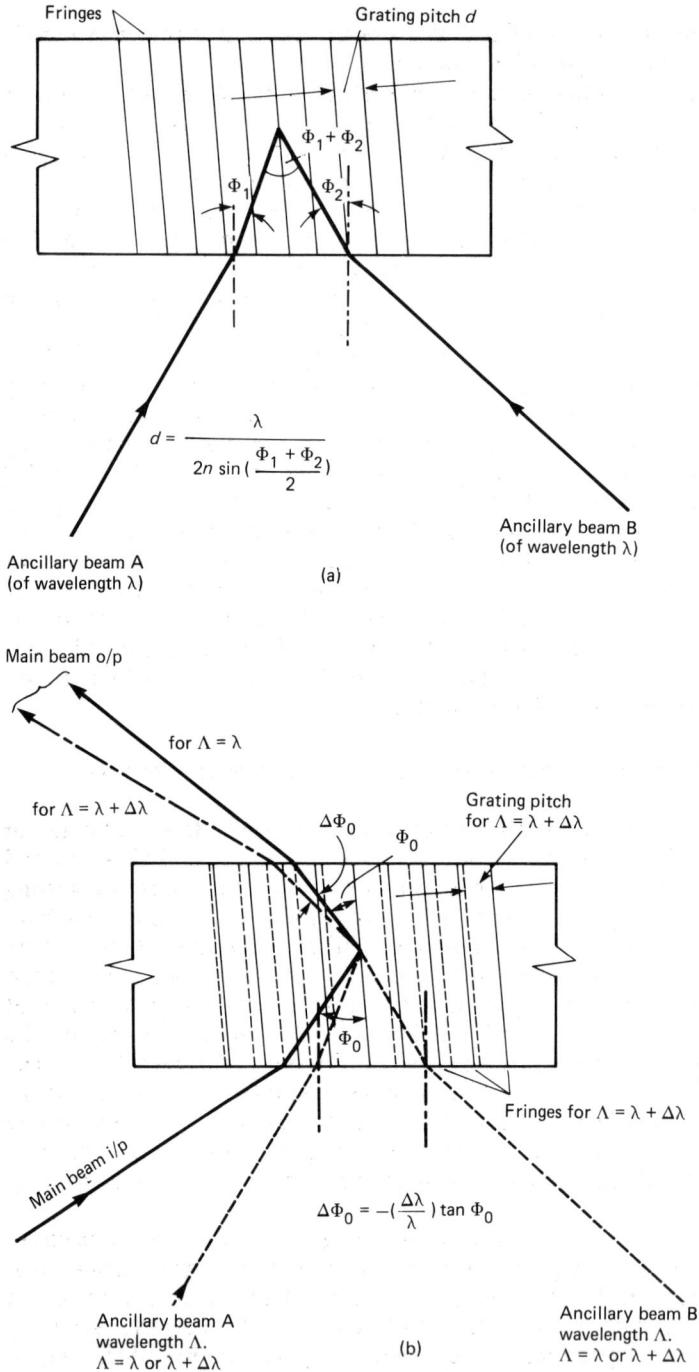

Figure 10.27 Principle of the opto-optical deflector. (a) Dynamic volume grating produced by the interference of two coherent beams in a photorefractive crystal (fringes bisect the angle formed by the two 'rays' in the crystal. Pitch $d = \lambda/2n \sin(\phi_1 + \phi_2)/2$. (b) How the wavelength of the ancillary beams (dotted lines) controls the deflection of the main beam

Note that here, unlike in a CRT, it is *not* the beam that the electric field influences but the propagating medium.

The young EO deflector has found little practical use so far and is mentioned here chiefly for its intellectual interest. The technique of *opto-optical* light deflection (Figure 10.27) is younger still. The concept is so interesting that it alone earns it a place in this book, in disregard of the fact that it has not been taken up by industry yet. Some optically non-linear materials such as barium titanate ($BaTiO_3$) or lithium niobate ($LiNbO_3$) exhibit a highly localised change of refractive index under the influence of two interfering coherent light beams. This makes it possible to create in them a transient *volume diffraction grating* [216, 212] similar to that of an acousto-optic deflector, simply by shining onto them two ancillary coherent beams at the required angles of incidence. The rest of the story reads like the explanation of an AO deflector, except that the deflection angle of the useful beam will be controlled by changing the frequency (wavelength) of the ancillary light beams. Sincerbox and Rosen [212] report that they have evolved means for automatically maintaining the Bragg condition necessary for a good diffraction efficiency, for variable wavelengths of these beams, and that their experimental device was capable of producing deflections of a HeNe laser beam of up to 11.8° – all without any moving parts! Our enthusiasm for opto-optical deflectors must, nevertheless, be tempered by the fact that their diffraction efficiency and response time are still too low and the ancillary equipment too complex for their practical application.

(d)   Field flattening, raster scanning and random access spot positioning

The deflected beam is, in most systems, focused into a fine spot made to 'write' onto a flat surface: photographic paper or film, metal plate, ground glass screen, etc. This creates two problems: firstly, the locus of a writing spot is not a straight line but an arc of circle; and secondly, a two-dimensional (x, y) spot positioning mechanism is required. Optical *field flattening* lenses (cumbersome and expensive) are sometimes used to cope with spot line curvature. (As an alternative to solid lenses, coherent bundles of optical fibres, fused together and then cut and ground, can be used to form such field flatteners (see Sections 2.2.8 and 12.4). The resolution of such FO assemblies may be insufficient in some applications.) Sometimes 'flat bed' writing is given up altogether in favour of *concave* bed arrangements in which the substrate (e.g. paper or film) is guided by a supporting cylindrical structure, concentric with the shaft of a spinning mirror.

With regard to two-dimensional positioning, the second, say y, displacement of an x, y addressing system can be obtained either mechanically (e.g. by moving the paper) or optically, for example by means of a combination of two galvo mirrors or two EA or EO deflectors. While raster scanning can be obtained by either arrangement, random access addressing of sufficient speed requires EA or EO solutions. At least one manufacturer (General Scanning Inc, USA) offers a three-dimensional (x, y, z) spot positioning system (Figure 10.28) in which the x and y deflectors are of the galvo type while the z 'excursion' (depth, for non-flat substrates) is

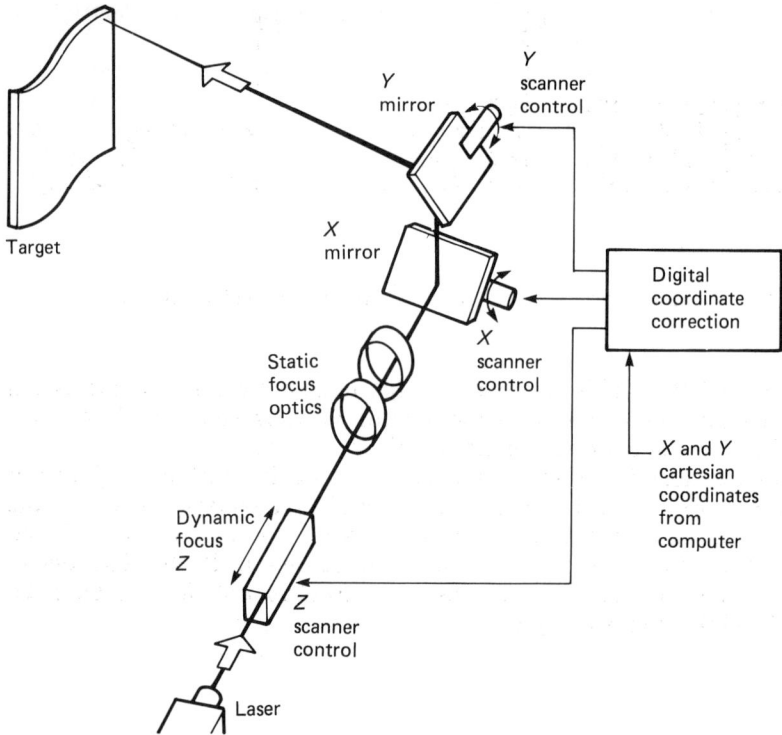

Target

Y
mirror

Y
scanner
control

X
mirror

Digital
coordinate
correction

Static
focus
optics

X
scanner
control

X and Y
cartesian
coordinates
from
computer

Dynamic
focus
Z

Z
scanner
control

Laser

Figure 10.28 Three-dimensional control of writing spot

obtained dynamically, by a computer-controlled high-speed (15 ms for up to 20 mm of displacement is claimed) optical linear translator [196].

10.3.3 Modulation

The expansion, constriction or focusing of a laser beam have all to do with its *spatial* shaping. The various earlier described deflection schemes are aimed at directing the beam to the right place at the right time. Here, both *space* and *time* are of concern to the control unit. The term *modulation* will be used, in this work, in connection with *temporal* beam shaping, i.e. with the controlled variation of its intensity in time.

In most cases, this modulation will be of the on/off type (Figure 10.29(a)). In displaying, electrophotography, phototypesetting, engraving, cutting, drilling or welding, the beam will often have to be suppressed (blanked) between two of its successive addresses, not unlike the electron beam of a CRT operated in the graphics mode. In communications, however, we may want to modulate the beam intensity in a more subtle way, for instance, sinusoidally (Figure 10.29(b)) – in systems using an RF subcarrier, itself destined to be AM, FM or on/off modulated – or analogically – in systems using direct amplitude modulation (AM) of the optical carrier by speech, music, video or process control measurands

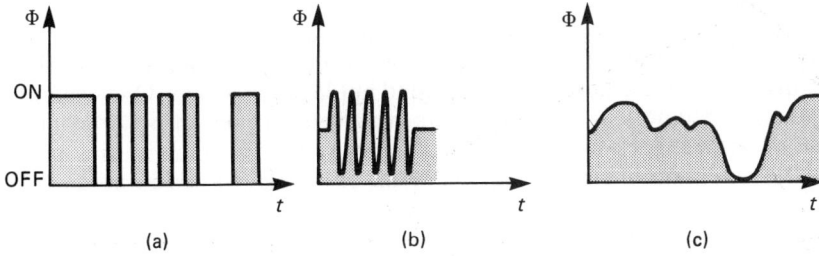

Figure 10.29 Laser modulation is the controlled variation of beam intensity with time

(Figure 10.29(b)). This long enumeration must not create the impression that analogue intensity modulation of laser beams is widely used. It is not: digital and other pulse techniques dominate the scene.

With the exception of the semiconductor laser, in which the variation of the drive current can be used, within certain limitations, to control the intensity of the radiant flux, beam intensity modulation must be effected externally, whether analogue, pulse or digital (Figure 10.30). The following text gives illustrative examples of external modulations of both the on/off and the analogue types.

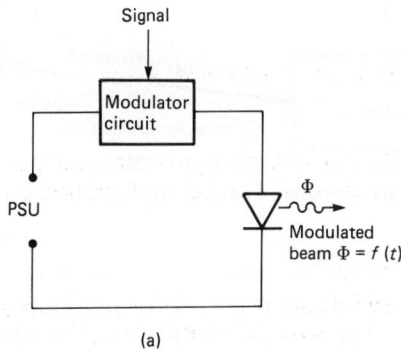

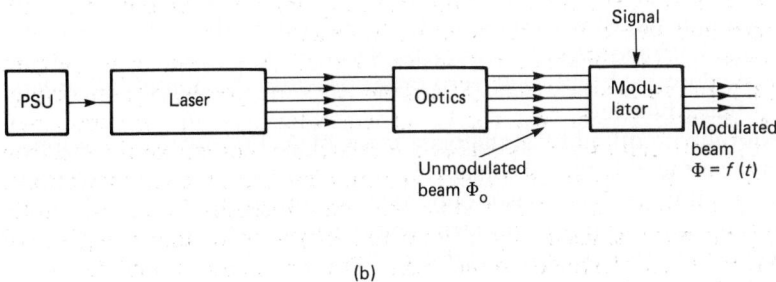

Figure 10.30 (a) Internal and (b) external laser beam modulation

(a) Mechanical modulators

Under this description come mechanical beam shutters, choppers and reflectors. Whether of the rotating wheel or of the tuning fork type, they are usually periodic and hence of limited applications scope. One example of their usefulness is as modulators in low-level measuring or control systems, where lock-in amplifier techniques require a synchronising pulse for the detection of signals that are weak or, worse, buried in noise.

(b) Acousto-optic modulators

Add a blackened knife edge to an AO deflector (Figures 10.24 and 10.31) and you have an on/off modulator. Indeed, in a Bragg cell with a very high diffraction efficiency, very nearly 100% of the beam energy can be diffracted out of the zero order. Blocking off the first order (and the residue – if any – of the higher ones) we are left with an on/off ≈ 100% zero order beam controlled by the keying of the RF oscillator (see Section 10.3.2(b), especially Figure 10.24). This can be done quite conveniently with the help of a digital chip, e.g. a 5 V TTL gate.

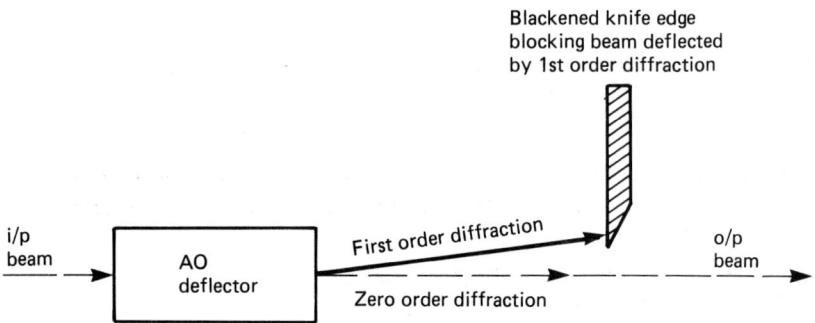

Blackened knife edge
blocking beam deflected
by 1st order diffraction

i/p beam   AO deflector   First order diffraction   Zero order diffraction   o/p beam

Figure 10.31 Add a blackened knife edge to an AO deflector and you get an AO modulator

Analogue modulation can, not only in principle but also in practice, be obtained with a Bragg cell of the above variety (through the amplitude modulation of the RF signal, responsible for the power sharing between the zero and higher orders), but is more conveniently achieved with an AO crystal of a slightly different geometry. Here (Figure 10.32), the light wavefronts are genuinely perpendicular to the acoustic wavefronts as there is no need to respect the Bragg condition (see Section 10.3.2(b), especially Figure 10.24, 'Bragg angle'). The crystal is usually much thinner than in the previously described type. The zero order provides the output, all orders $m \geqslant |1|$ being blocked.

In both the on/off and the analogue types of AO modulator, the upper limit of the attainable rate of change of the outgoing radiant flux is imposed by the duration of the acousto-optic interaction taking place within the volume, the flat projection of which is marked on Figure 10.32 as ABCD. Indeed, before the synthetic grating with its pitch, p, can be wholly replaced by a new one, with a pitch p_1, the old one must be completely evacuated. Clearly, the time this 'purging' takes is directly proportional to

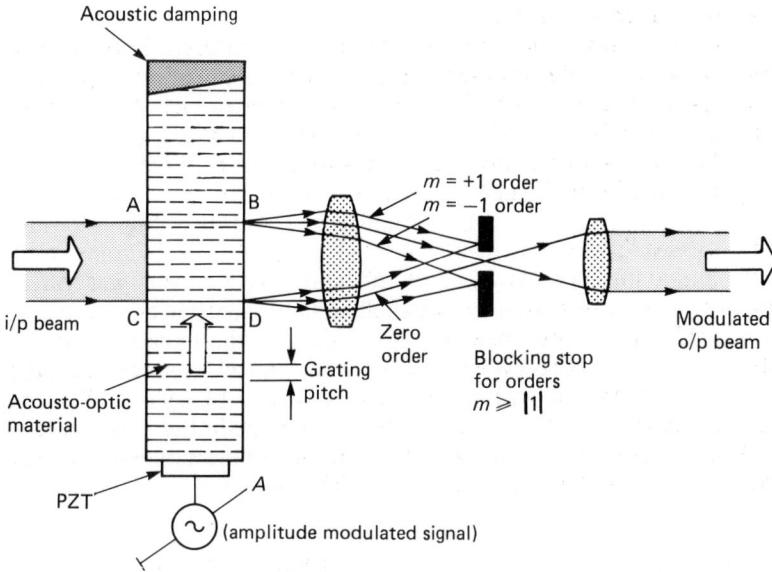

Figure 10.32 Analogue modulation with an AO crystal. Beam intensity depends on the amplitude of the modulating signal

the beamwidth, w, and inversely proportional to the velocity of the acoustic wave, v. When a high modulation bandwidth is required the designer will thus:

1. Choose a material with a high sound velocity.
2. Reduce the beamwidth, w, to a minimum.

Now, would you ever want to use a telescope back to front? The EO designer who wishes to expand w does just that: turn the expander back to front, i.e. restore its normal telescope position, to get a *beam reducer* (Figure 10.33). Placed immediately before an AO modulator the reducer increases its high-frequency response.

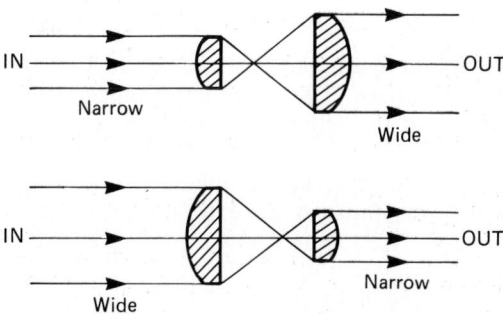

Figure 10.33 A beam expander turned back to front becomes a beam reducer. Placed before the AO modulator the reducer increases its high-frequency response

Before leaving the AO modulator, let us mention that the Bragg cell has another captivating property: it changes the actual wavelength of the radiation it modulates (or deflects). More about this phenomenon and its uses is given in Section 12.2.4.

(c)   The ordinary and extraordinary rays and their use in electro-optic modulators

More than three centuries ago a Danish man of learning, Erasmus Bartholinus, was astounded by the optical properties of 'a transparent crystal, recently brought to us from Iceland'. The crystal was *calcite* (chemically $CaCO_3$), still sometimes called Iceland spar, and the wondrous property that struck him most was that of *doubling* the images of objects looked at through it. Today, physicists and optical engineers exploit *birefringence* in many ways, laser beam modulation being one of them. Birefringence (or double refraction) has to do with the velocity and polarisation of light. In common with radiowaves, the polarisation of a luminous wave is characterised by the plane in which its electrical field lies. When **E**, the vector representing this field, and z, the direction of propagation, lie constantly in the same plane, the wave is said to be *plane* (or linearly) *polarised* (Figure 10.34). Natural light is not. In it, **E** changes

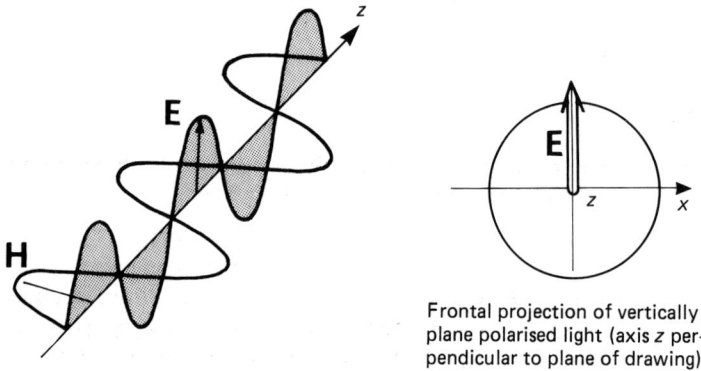

Frontal projection of vertically plane polarised light (axis z perpendicular to plane of drawing)

Figure 10.34 Plane polarised light

direction all the time, very rapidly and in a haphazard manner. It is said to be *random polarised* or simply *unpolarised*. Two more important types of polarisation, namely, the circular and the elliptical, exist (see Appendix 12).

The beams of most lasers destined for external modulation either leave the laser readily plane polarised, or undergo an external polarisation immediately upon leaving it. Figure 10.35 shows how a *polariser* works; that incident light which is already polarised in the *P* direction, traverses the device almost unattenuated, while light polarised in a direction **D** forming an angle ψ with *P* has its intensity reduced to

$$I_\psi = I_0 \cos^2\psi \tag{10.14}$$

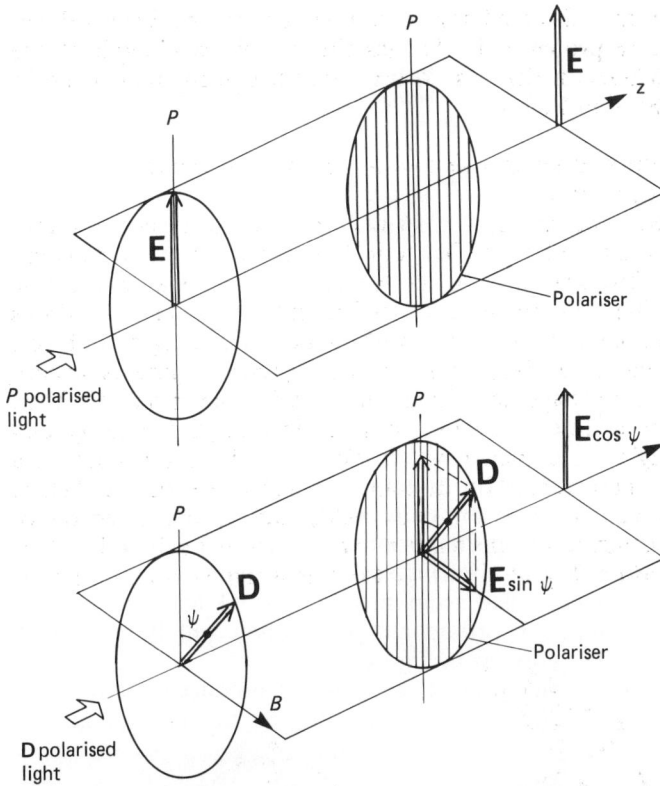

Figure 10.35 How a polariser works. The intensity of polarised light can be attenuated in a controlled way by rotating the polariser through which the light passes

The squaring of cos ψ stems from the fact that the power is proportional to the square of the amplitude. Thus the polariser 'extracts' and passes the P component. The intensity of polarised light can thus be attenuated in a controlled way by rotating a polariser through which it passes. Two 'crossed' polarisers block all light. This can be easily checked by means of two pairs of polarising sunglasses. (Instead of crystalline plates, such sunglasses use inexpensive synthetic 'Polaroid' sheet material. Molecule alignment in plastic is responsible for the effect.)

In cleaved calcite (Figure 10.36) birefringence splits a ray of perpendiculary incident light into two rays having mutually orthogonal polarisation planes. While ray 1 behaves in a very orthodox way, obeying Snell's law and other laws of geometrical optics, ray 2 breaks them all. It begins its journey in the crystal at a non-zero angle, it traverses it with a velocity different from c/n_0 and it emerges from it with a polarisation differing from that of the incident ray. In addition, it reflects from the air–crystal interface in a plane *not* containing the normal to the crystal. It *is* a truly *extraordinary ray* and such is indeed its textbook name. By contrast, beam 1 is called the *ordinary ray*.

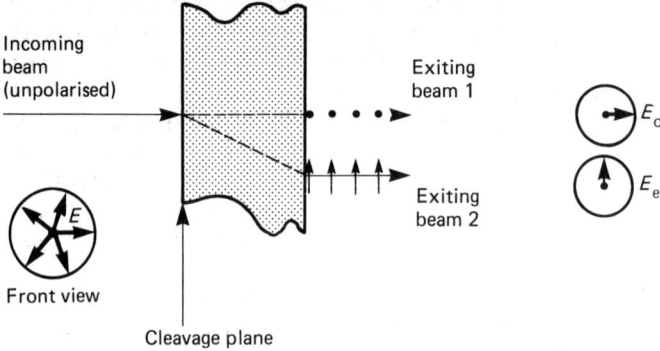

Figure 10.36 The ordinary and the extraordinary beams. Here, the incidence beam is normal to the cleavage plane of the crystal

For some chosen crystal/beam orientations both exiting rays emerge through the same point. Even then, their polarisation planes are perpendicular to each other and their speeds (prior to emergence) may differ. This is the most relevant aspect of birefringence in this context.

Rays 1 and 2 will be called from now on the O and the E rays, respectively. Their mutually orthogonal polarisation will be shown by arrows and arrowheads (dots) as in Figure 10.36.

It is helpful to view the O and E rays as the product of a vectorial decomposition of the incident ray upon entering the crystal, and as

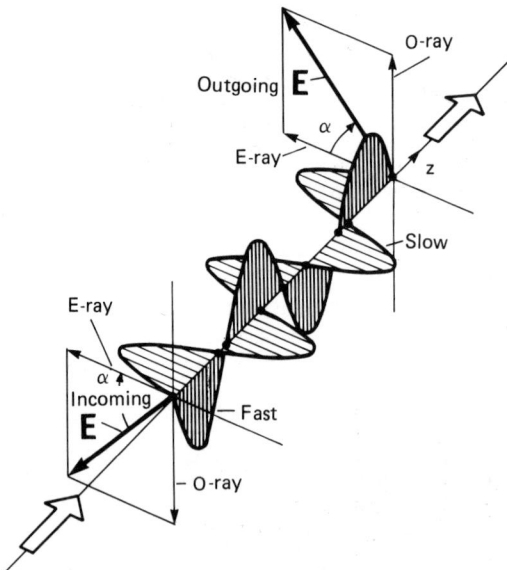

Figure 10.37 Optical retardation plate. When the thickness of a plate is such that the E-ray lags behind the O-ray by half a period, a half-wave plate is obtained. A half-wave plate rotates the plane of polarisation by twice the angle that the polarisation plane of the incoming radiation makes with the **E** plane. The diagram depicts the situation inside such a plate, except for the fact that a real half-wave retarder is some 100 wavelengths thick

vectorial components of the outgoing ray. An important consequence of the difference between n_o and n_e is that the slower E ray experiences an ever growing *phase lag* in relation to O, within the crystal: the thicker the crystal the greater the phase lag (Figure 10.37). (This is so in so-called 'positively birefringent' materials, such as quartz. In 'negatively birefringent' substances, such as calcite, it is the O ray that is the slowest.) The speed difference between the E and O rays results in *optical retardation*. Some of the fascinating uses of optical retardation (quarter-wave and half-wave plates) are described in Appendix 12, with the use of electrical analogy.

So far, the relevance of birefringence to laser beam modulators may not be obvious until one learns that in some materials birefringence can be controlled, or even induced, by an electrical field. This is the *electro-optic effect*. If the refractive index n_e of a material could be affected differently from n_o by such a field, *the optical retardation could become electrically controllable*. This is, in fact, the case of the KDP (see Section 10.3.2(c)), in which, for a certain beam/crystal/field orientation, the $n_o - n_e$ difference grows with V_{signal}. Figure 10.38 shows how the *phase* (optical) of a plane polarised laser beam could be modulated, and Figure 10.39 demonstrates how a single step – the addition of a polarising element (called here the analyser) – converts this phase modulation into an amplitude modulation, in accordance with the equation in the figure.

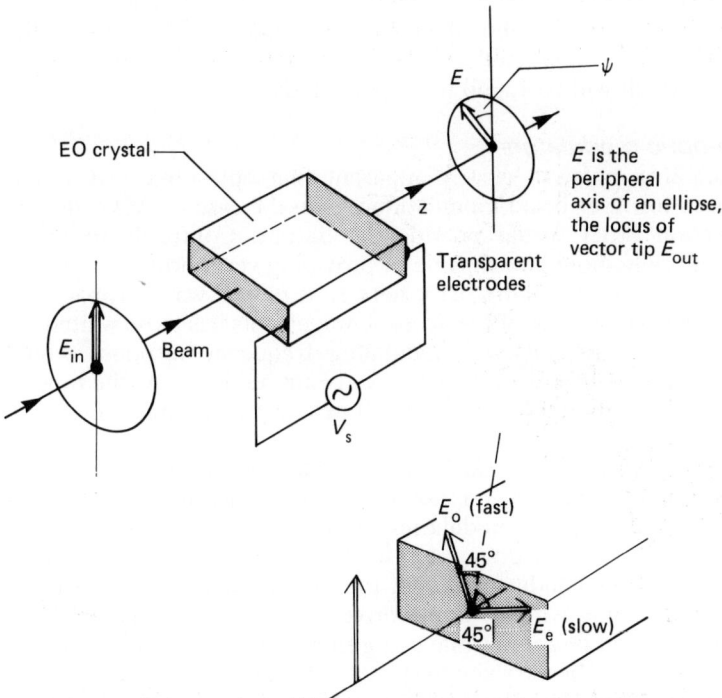

Figure 10.38 Phase modulation of a plane polarized beam in an EO crystal. Beam, crystal and field must be adequately orientated. Note that the phase shift ϕ (retardation angle) is proportional to the controlling voltage, V

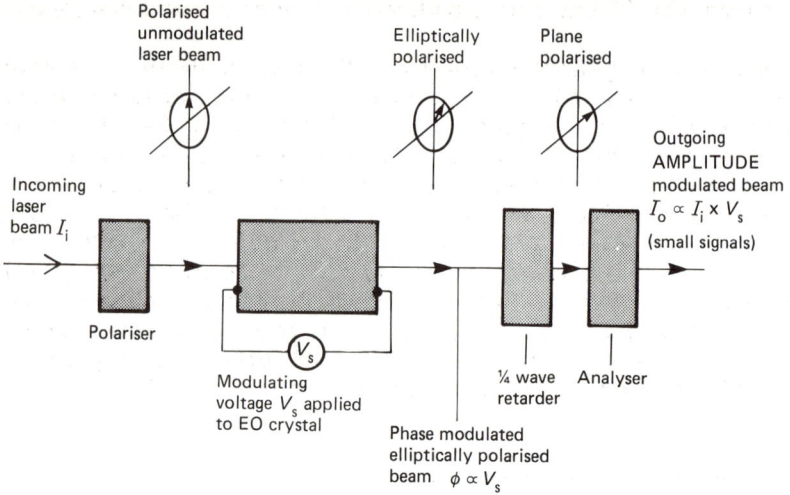

Figure 10.39 An AO modulator of the longitudinal type

A possible arrangement of a real EO modulator is shown in Figure 10.39. In it, the orientation of all the components in relation to each other is of crucial importance for the achievement of successful operation, e.g. a good modulation linearity. (For details see, for example, Ref. 192.) The presence of ellipses and circles in the last two figures calls for an explanation, which will be found in Appendix 12.

(d)   Opto-optic modulators
Section 10.3.2(c) makes it clearly apparent that opto-optic (OO) cells could be used for laser beam modulation. As in the case of AO cells, the zero diffraction order would provide the output. Contrarily to them, however, OO cells should be capable of providing very short light pulses, as *all* wave propagation taking place here is at *optical* wave propagation speeds and no slow purging of the acoustic wavefronts has to be waited for [212]. Aperture constraints on modulating frequency response would thus be lifted, and, with them, the need for beam reducers. At the time of writing, however, OO modulators are still of scientific interest only.

(e)   Internal modulation of GaAlAs and other semiconductor lasers
One of the main attractions, if not *the* main attraction, of the semiconductor laser is its ease of modulation. Simply stated: you modulate the drive current and the laser does the rest. As for modulating the current, being basically a semiconductor diode, the device makes the electronics engineer feel at home with it. Some driver circuits have been given in Section 9.4.2. As can be seen, while being simple, things are not quite the same as with LEDs. The very simplest of on/off modulation schemes only can use circuits identical or very similar to those employed with identically modulated high-power LEDs (Sections 5.5 and 5.6). Even here, some design modifications (e.g. the addition of reactance assisted energy storage) and rescaling will often be necessary, as drive currents can be much

higher than with LEDs. This is particularly true with single heterojunctions.

To ensure the constancy of light output the laser temperature will have
to be regulated by a Peltier cooler (Figure 10.40) (an electrically driven
heat pump with no moving parts, based on thermoelectric P–N junctions)
and the I_F value controlled by opto-electronic feedback. Advisable with
on/off modulation, the method is mandatory with analogue signal handling. Indeed, a quick glance at Figure 9.14(a) will show the drastic
dependence of the threshold current I_{th} on temperature, as well as the
almost dramatic steepness of the ϕ versus I_F characteristic of a typical
GaAlAs laser. The OE feedback helps with the latter. It also corrects the
ageing effects of the laser. The power supply of LD modulators should
have no switch on or off transients and no interference-caused spikes, as
these could cause irreversible laser damage.

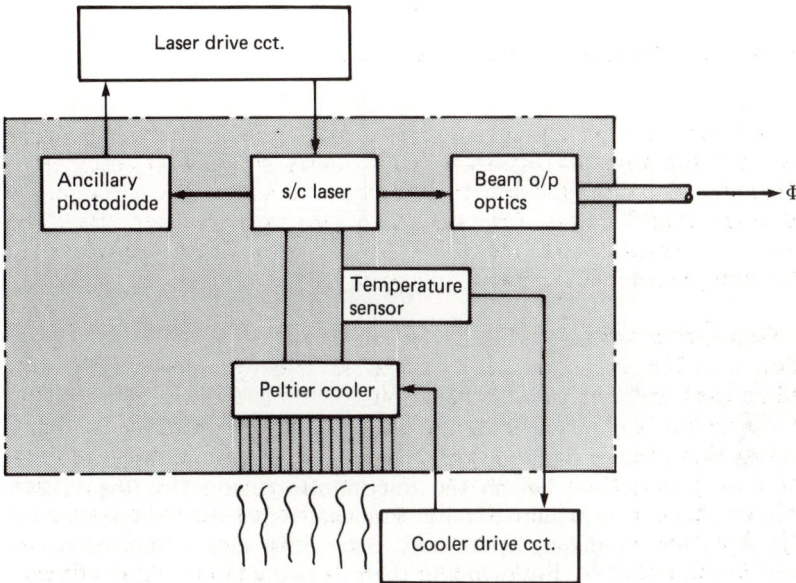

Figure 10.40 Schematic of a ϕ stabilisation arrangement

10.3.4 Various beam treatment techniques

Five other important processes must be added to the techniques of beam
expansion, constriction, deflection and modulation so far covered. They
are: spot forming, spatial filtering, fine collimation (for semiconductor
lasers), Q-switching and mode locking. Their common aim is that of
producing an intense and neatly delimited irradiance on the selected area
of the workpiece in hand, be it an eye retina, a steel plate or a roll of
phototypesetting film.

(a) Spot forming

This is probably the most fundamental technique of the four. We know, of course, that using a lens we can focus the usually neatly collimated beam into a small spot. The question arises: how small? The theoretical absolute limit is set by diffraction. A *perfect* lens of an active diameter D and a focal length f_1 *overfilled* by a laser beam of a diameter and modal distribution such that it irradiates the lens *uniformly* (Figure 10.41) would produce a spot of a diameter:

$$(d_s)_{th.1} = 2.44 \frac{\lambda}{D} f_1 \tag{10.15}$$

d_s being delineated by the circumference of the Airy disc (the first zero of the diffraction pattern, see Section 10.3.2(b)). Such a spot would contain 83.8% of the energy leaving the lens. We note that $(d_s)_{th.1}$ is proportional to f_1/D, the $f_\#$ of the lens. We also note the proportionality to λ implying that a HeNe (red) laser will produce a larger spot (smaller irradiance) than

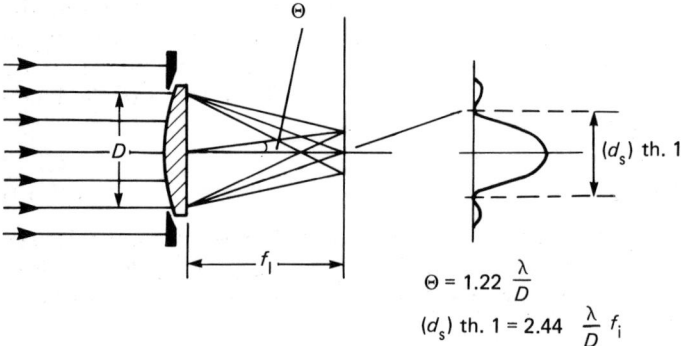

$$\Theta = 1.22 \frac{\lambda}{D}$$

$$(d_s) \text{ th. } 1 = 2.44 \frac{\lambda}{D} f_i$$

Figure 10.41 A perfect lens of effective diameter D overfilled by uniform radiation produces a spot with diameter $2.44(\lambda/D)f_L$

a HeCd (blue) one of equal power. In terms of orders of magnitude, this theoretically limited d_s lies in the micrometre region (for the visible) assuming lenses with $f_\#$ above unity. Practically speaking, single-element spherical lenses producing spots only some 35% larger than $(d_s)_{th}$ are commercially available, but owing to their $f_\#$ being far in excess of unity, the spot diameters lie now in the 10–50 μm region. For smaller spots, a microscope objective with a large NA (large magnification, small $f_\#$), centred and focused carefully, will succeed, despite the fact that it hasn't been designed for the job. For still smaller spots, specially designed multi-element lenses for infinite conjugate work are required.

With Gaussian beams (single-mode TEM$_{00}$ lasers) not overfilling the lens, another formula sets the absolute limit to the spot diameter, namely:

$$(d_s)_{th.2} = \frac{4}{\pi} \frac{\lambda}{D} f_1 \tag{10.16a}$$

giving:

$$(d_s)_{th.2} = 1.27 \frac{\lambda}{D} f_1 \tag{10.16b}$$

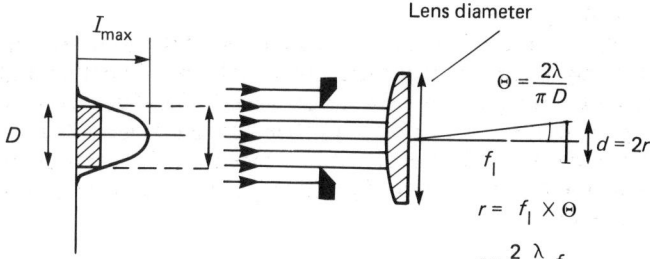

Lens diameter

$$\Theta = \frac{2\lambda}{\pi D}$$

$d = 2r$

$r = f_1 \times \Theta$

$$r = \frac{2}{\pi}\frac{\lambda}{D}f_1$$

$$d = \frac{4}{\pi}\frac{\lambda}{D}f_1 = 1.27\frac{\lambda}{D}f_1$$

Here D is NOT the lens diameter but the beam width for which $I \geqslant I_{max}\ (\frac{1}{e^2})$

Figure 10.42 A Gaussian beam underfilling the lens produces a smaller spot – the diameter is now 1.27 $(\lambda/D)f_L$, but D has a different meaning from that in Figure 10.41

When comparing $(d_s)_{th.1}$ and $(d_s)_{th.2}$ care must be taken in choosing the right D. In Figure 10.42, lens underfilling takes place with D expressing the beam – not the lens – diameter, with D expressing the beam bounds within which the intensity I is greater than $I_{max}(1/e^2)$. This portion of the beam contains 86.5% of its total power.

In the useful numerical relationship:

$$d_s \simeq 0.8f_\# \quad \text{(in microns)} \tag{10.17}$$

for HeNe lasers, remember that $f_\#$ is *not* the f number of the whole lens but only that part of it which is irradiated by $I < I_{max}\ (1/e^2)$, characterised by D. It is important that practical, real lens diameters be 1.5–3 times greater than D.

(b)   Beam purification

In the case of a TEM$_{00}$ beam from a medium-power HeNe laser having undergone a few reflections and refractions in the optical train of a piece of equipment, careful analysis of the beam intensity profile or the beam's projection on a good clean screen will reveal a certain departure from the nice smooth bell-shaped Gaussian expected. Small blemishes, bright or dark, if not minor interference fringes and rugged edges, will be noticed. Dust on mirrors and small lens imperfections have caused them. A process of theoretical elegance and great simplicity of implementation can get rid of them. It is called *spatial filtering*. In this case, spatial filtering is implemented by placing a thin piece of blackened metal, pierced in the centre, inside the beam expander. This pinhole is not an ordinary diaphragm, however. (The pinhole acts as a 'stop', too, in an ancillary way, by blocking spurious, inter-element reflections of the optical train. Spatial filters can be used with astronomical (Keplerian) expanders only as there is no *real* image present inside the Galilean ones.)

How does a pinhole restore the blemished beam profile to its neat, textbook Gaussian shape? Recall from diffraction (Section 10.3.2(b) and Figures 10.18–10.22) that the finer the netting, the more badly sprawled the diffraction pattern. From this, we can intuitively deduct that smaller

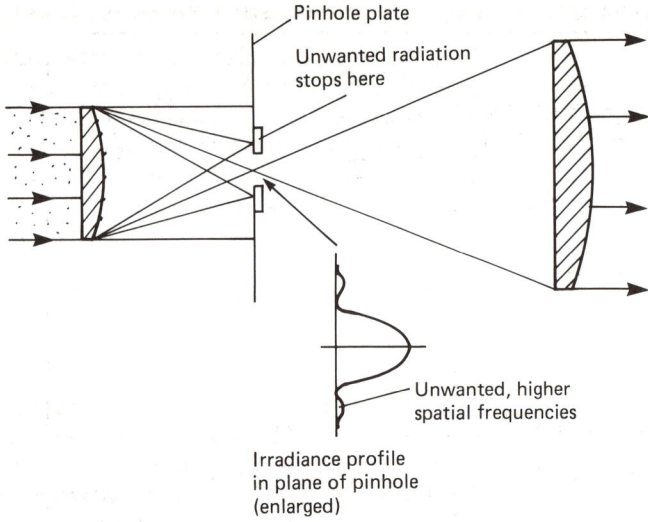

Pinhole plate

Unwanted radiation stops here

Unwanted, higher spatial frequencies

Irradiance profile in plane of pinhole (enlarged)

Figure 10.43 Spatial filtering by means of a pinhole. The pinhole stops radiation diffracted by dust particles, lens scratches, etc. This is only possible because smaller obstacles produce larger diffraction angles

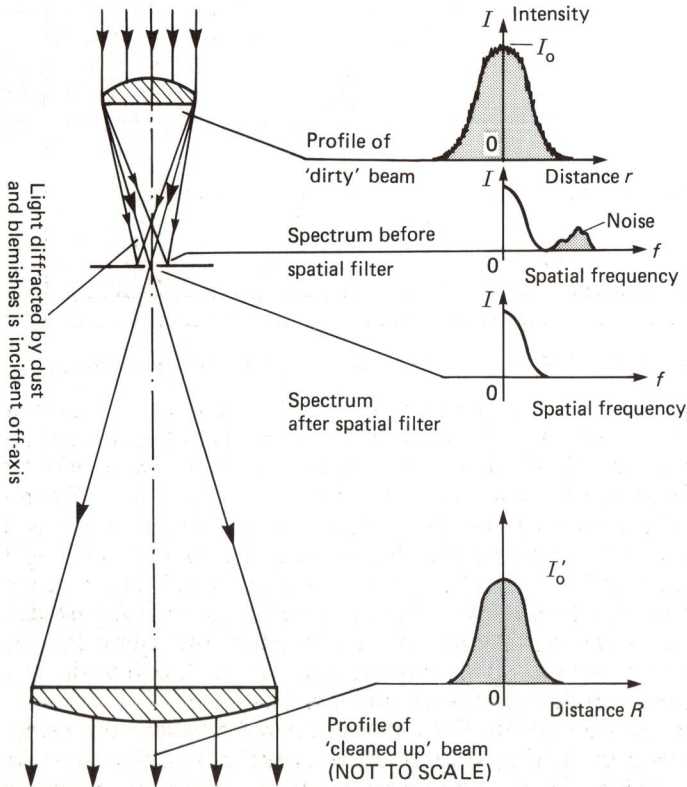

I Intensity

I_0

Profile of 'dirty' beam

0

Distance r

I

Spectrum before spatial filter

Noise

0

f

Spatial frequency

I

Spectrum after spatial filter

0

f

Spatial frequency

Light diffracted by dust and blemishes is incident off-axis

I_0'

0

Distance R

Profile of 'cleaned up' beam (NOT TO SCALE)

Figure 10.44 Thanks to the pinhole, the expander produces a cleaned-up Gaussian beam

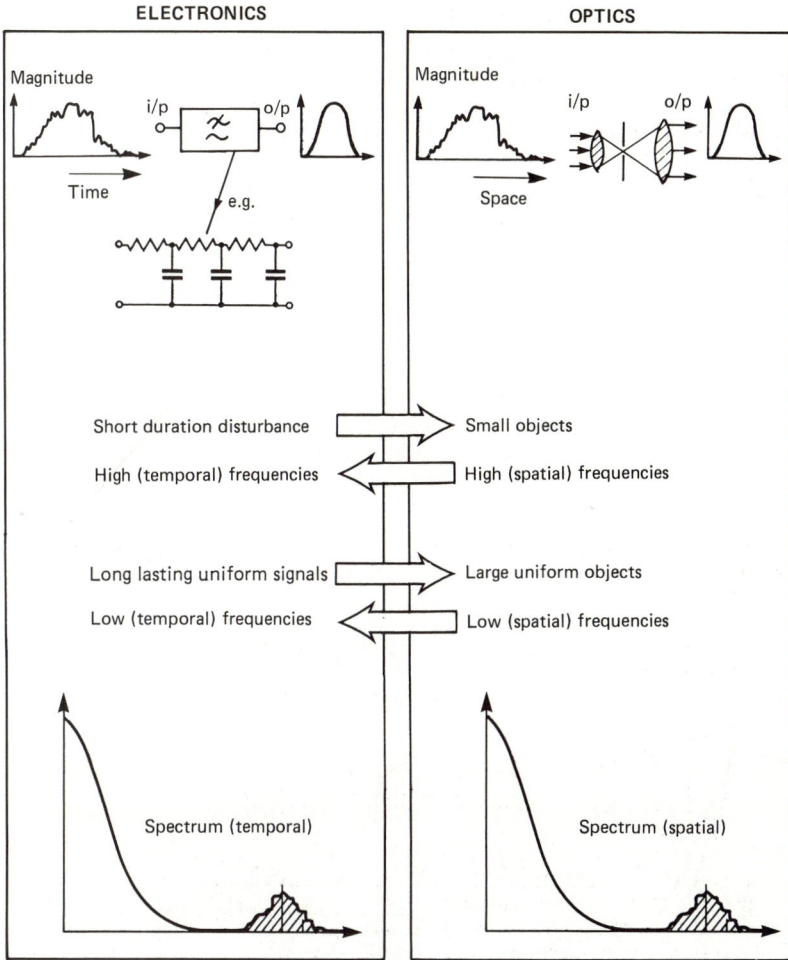

Figure 10.45 The analogy between electronic and optical low-pass filtering is striking

objects impeding the beam produce wider diffraction angles. This is supported by Equation 10.11, in which d can be thought of as representative of the object size. This brings us to *spatial frequencies: the smaller the obstruction the higher the associated spatial frequency* ($f \propto 1/a$). Figures 10.43 and 10.44 (and see also Figures 13.13–13.16) show that, in the focal plane of the first lens of the expander, the irradiation produced by the light diffracted by obstructions lies off centre of the main lobe. The higher the spatial frequency, the further away from the centre the resulting irradiation. By accurately placing a pinhole of the right size in the right place, the unwanted light diffracted by dust and scratches can be blocked off. The expander will then produce a cleaned-up Gaussian beam.

The analogy between the filtering out of unwanted high spatial frequencies and the low-pass filtering of temporal frequencies used in electronics engineering is striking (Figure 10.45): an RC ladder network clearing up HF interference polluted pulses is an electronic pinhole!

Spatial filtering rates among the most exciting topics of modern optics, connected with optical computing and signal processing. As such, it will be given some coverage in Chapter 12.

(c) Collimation of semiconductor laser beams

The divergence of a semiconductor laser beam is not only markedly larger than that of beams produced by other types of laser (degrees rather than milliradians) but suffers also from strong asymmetry: it is much smaller in the plane of the junction than orthogonal to it. The effect stems from the confinement sculpted geometry of the active region (Section 9.4.1). Figure 10.46 shows typical x and y beam spreads. The ensuing far-field cross-section is elliptical. Collimating such a beam is further complicated by the fact that there is not one but two planes from which the radiation seems to originate (virtual object planes), tens of microns apart, one for each of the above-mentioned axes. Special attention must thus be paid to the design of the appending optics for applications calling for good collimation or fine focusing, as, for example, in the case of optical information storage on

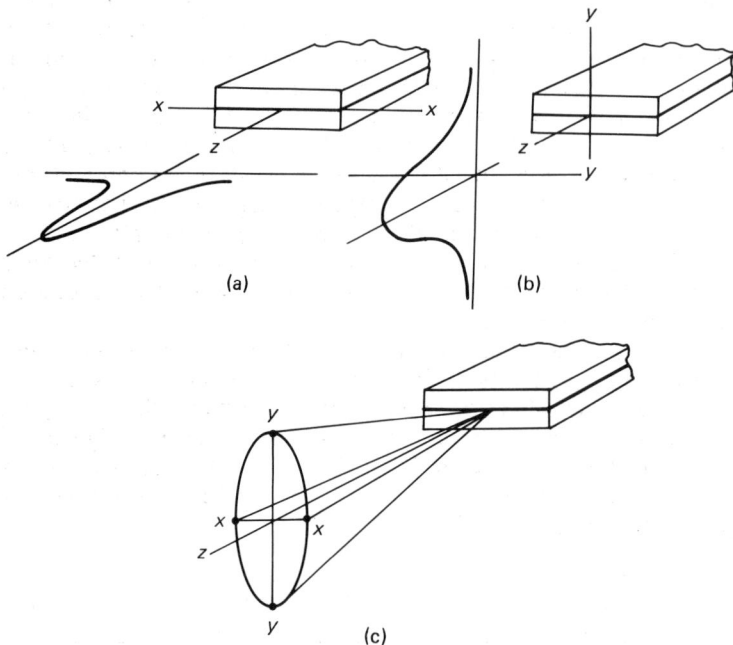

Figure 10.46 Far-field angular intensity in the (a) x–x and (b) y –y planes. (c) The resulting elliptical beam cross-section

disc, where very small circular spots are required (Section 12.5). At least one manufacturer has put on the market miniature precollimated lasers packaged in small tightly toleranced tubes. (For example, devices of the CQL 10 series, by Mullards Ltd. The 34 mm long tubular package has an 11 mm diameter toleranced to + 0–11 µm! See Ref. 219 or Mullards Technical Publication M82-0120, and CQL 13 and 14 Data Sheets.) The inside houses a combination of a precision three-element lens and a single-element cylindrical lens, the latter individually matched to each particular laser. This optics converts an astigmatic elliptical beam with divergences:

$$\left(\frac{\beta}{2}\right)_x \simeq 25° \quad \text{and} \quad \left(\frac{\beta}{2}\right)_y \simeq 15°$$

into a narrow (5.4 mm diameter) circular beam with an almost uniform intensity profile and a divergence of 'better than 0.3 mrad', meaning:

$$\left(\frac{\beta}{2}\right)_x = \left(\frac{\beta}{2}\right)_y \leqslant 0.15 \text{ mrad}$$

(d)   Q switching

Q switching is a frequently used technique for obtaining very short but very powerful output pulses from lasers. Remember that for laser action to take place, both the population inversion and the high selective positive feedback conditions must be satisfied (Section 9.9). Giant intensity pulses can be obtained from lasers by holding back the positive feedback until the build-up of the population inversion has reached a very high degree. By so repressing the continuous degradation (through stimulated emission) of the population inversion of an ordinarily operating laser, very high energies are accrued. Then, at the critical moment in time, when the pumping has gone far enough, the indispensable positive feedback for laser action is unleashed and the stored energies released. The intense stimulated emission quickly depopulates the upper energy level, exhausts the inversion and stops the laser action. The controlled Q becomes low and the next cycle is about to begin. The whole process is reminiscent of electronic pulse generators based on the gradual build-up, followed by a sudden release of reactively stored energies.

The actual damping and undamping of the laser cavity can be either externally or internally controlled. External, also called *active*, Q switching is either by means of an AO or an EO modulator built into the cavity, but controlled externally by an electrical pulse train or by means of high-speed rotation of one of the end mirrors. With the latter, the feedback loop is closed only for a brief moment, during which M_1 and M_2 are sufficiently close to parallelism (Figure 10.47).

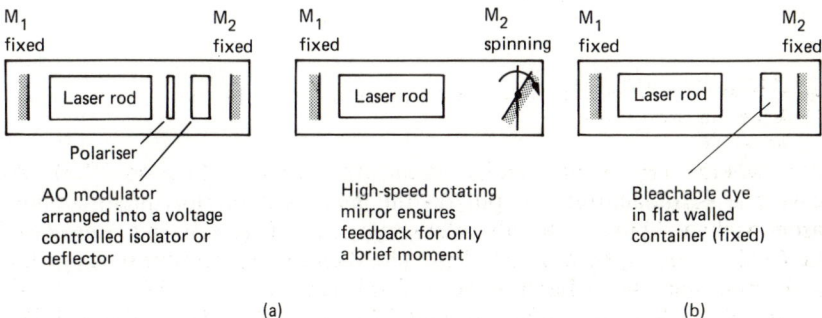

Figure 10.47 (a) Two types of active Q-switching. (b) Passive Q-switching

Internal, also called *passive*, Q switching has the advantage of simplicity. Here, the 'shutter' is a liquid chemical, a dye, which is more transparent for intense than for feeble irradiances (think of it as the optical analogue of a non-linear resistance *decreasing* with increasing voltage). At the beginning of a cycle the almost opaque dye absorbs heavily the still weak incident radiation. (The working principle of the dye shutter rests with the match between its absorption band and the working wavelength of the laser.) Then, as the irradiance grows, the dye absorbs more and more of it, becoming more and more transparent, until it can absorb no more. In this saturated state the 'bleached' dye is quasi-transparent. This allows the pent-up energy to be unleashed in a huge, but brief, pulse. The dye is then unbleached and the next cycle is about to begin. Note the absence of external stimuli.

The effectiveness of the Q switching technique is aptly illustrated by the performance of a kilowatt Nd-YAG laser which, when Q switched, will deliver mighty megawatt pulses of a few to a few tens of nanoseconds duration.

For further reading, see Refs 181 and 192.

(e) Mode locking

Mode locking is another technique for producing powerful brief laser pulses. It relies on an organised cooperation between the many longitudinal modes (see Section 10.2.2(a)) that exist in many types of laser. Normally these modes coexist quite independently and whatever temporary phase concurrence may exist between two or more modes is fortuitous, with no guaranteed repeat cycle. When an N-mode phase concurrence takes place, the field amplitude is NE_o, but the beam intensity (power) is $N^2E_o^2$. Without mode concurrence, however, the combined power of N modes will be only NE_o^2. This is the crux of the argument. Organised periodic phase concurrence between N modes increases the pulse peak power by a factor N. Even when the amplitudes of the various modes differ, as they usually do, the gain is considerable, particularly with solid-state lasers, where N is very large (see Figure 10.48).

An interesting calculation (see Ref. 192, p.245) shows that the regular interval between two consecutive pulses of a mode locked laser (period T) depends solely on the physical length, L, of the laser:

$$T = \frac{2L}{c} \qquad (10.18)$$

The pulse duration, t, is:

$$t = \frac{1}{N}\, T \qquad (10.19)$$

In T we recognise the round trip duration between mirrors (Equation 10.4) and from t we deduce the interest of the large number of longitudinal modes in a solid-state laser, already acknowledged by the expression for peak power, $P_{pk} = N^2E_o^2$, for this type of work.

The results of mode locking are truly spectacular: mighty gigawatt pulses of picosecond duration with gigahertz repetition rates. Such can be

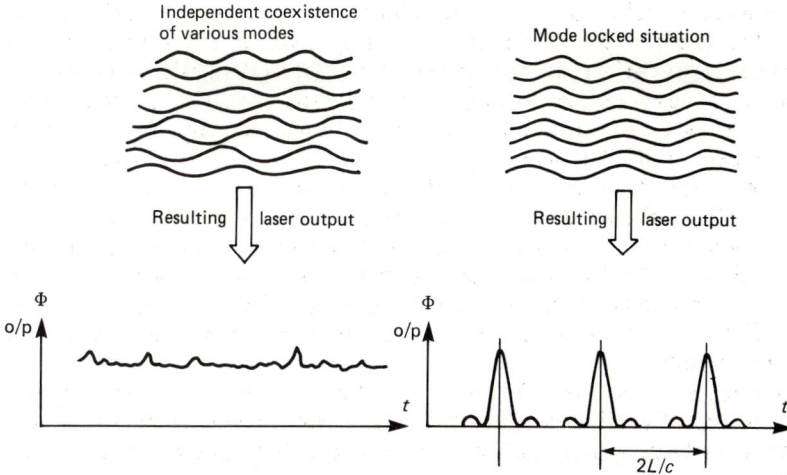

Figure 10.48 Mode locking explained graphically

their brevity that the length of the resulting wavetrain in free space $l = ct$ could be a mere fifth of a millimetre (8 thou) containing just under 300 individual sine waves!

Means of achieving mode locking fall, as in the case of Q switching, into two categories: *active* (external) and *passive* (internal). In active mode locking, an AO or EO modulator is used as a gate controlled open by a pulse generator with a repetition rate $1/T = c/2L$ and a pulse duration $\simeq t$. Passive mode locking can be achieved with bleachable dyes as discussed in Section 10.3.4(d). Their non-linear behaviour repeatedly favours the stronger radiations (initially, those resulting from fortuitous phase concurrences of a few modes), while attenuating the weak (non-concurrent) ones. The former grow, at the expense of the latter, and soon strong, brief, repetitive pulses become firmly established. With the saturable dyes method, both Q switching and phase locking can take place in the same laser, resulting in brief trains of extra-brief pulses.

For further reading, see Refs 181 and 192.

Chapter 11

Especially for electronics engineers

11.1 Non-electrical transmission of electrical signals by means of insulated signal couplers

The long-felt need of in-circuit signal transmission without metallic continuity or radio can, today, be easily fulfilled by means of conveniently applicable optical coupling. The Insulated Signal Coupler (ISC, often called an opto-coupler or opto-isolator) represents a rare, perhaps unique, case of an opto-electronic device that an electronics engineer can treat like just another purely electronic component. It is included in this text, firstly, because it is seldom covered in textbooks on electronics engineering; secondly because a fuller understanding of what goes on inside this photon-coupled device may help its better and wider use; and thirdly, and most importantly, because its operation relies so strongly on the interaction of light and electricity.

The versatility of this coupling element is amazing; we sometimes look at a circuit and wish we could do things to a part of it without affecting the rest. Effectively, this is the kind of freedom the ISC gives us. The ISC's contactless signal coupling makes it possible to operate freely almost any kind of loading, offloading, cross-connecting and d.c. level shifting at the output, without its having the slightest repercussion at the input. We could even short-circuit the output or put the mains across it, without the input taking any notice! It is an ideal buffer, unlike a transformer. Unlike a transformer, too, *it will transmit* – still unidirectionally – the *d.c. component* of the *input waveform*. This is very useful in the transmission of pulse trains of variable pulse duration, repetition rate or both, and of other complex waveforms. When used in lieu of an a.c. transformer, the ISC can have very wide bandwidths – from the lowest conceivable frequencies to hundreds of megahertz. As a relay, it will switch loads without the millisecond long delays of an electromagnetic relay, without bouncing, sparking, contact dirtying or wear. In addition, the ISC is usually smaller and lighter than its wound counterpart. The above-mentioned uses involve in-circuit applications. As an inter-circuit element (e.g. between the analogue and digital portions of a complex piece of circuitry, or a computer–peripheral link), the ISC will eliminate earth loops, exhibit a high degree of noise immunity (more about this later) and show an appreciable degree of immunity to electromagnetic interference.

The possibilities the ISC opens to a circuit designer are such that I did not hesitate to hail it as 'A wish come true' in an early article on its applications [93].

11.2   Basic structure and operation

The basic structure of an ISC is illustrated in Figure 11.1. On the input side, a light emitting diode (LED) converts the electrical signal into infrared light. The signal travels then as light, impinges onto the output element – a photodiode or a phototransistor – is reconverted and re-emerges in its original, electrical form! Thus, within the ISC the information transfer is effected by non-electrical means, allowing the electrical insulation to be as high as the maker cares to design it. In addition, the information flow is strictly unidirectional.

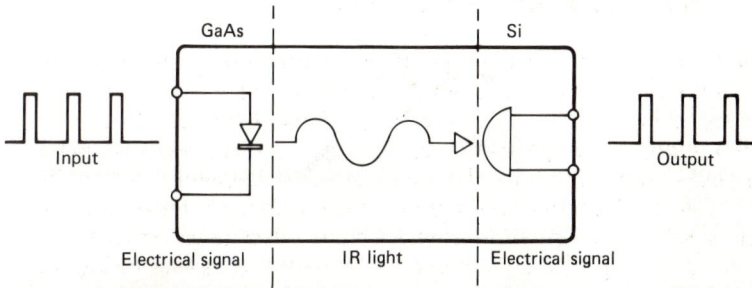

Figure 11.1 Basic structure of an ISC

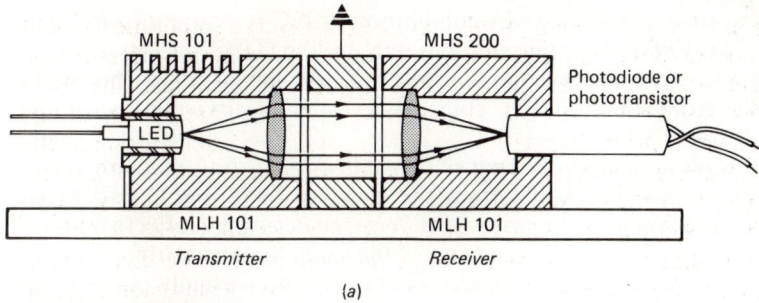

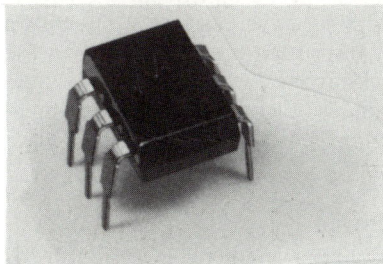

Figure 11.2 ISCs old and new. (a) An early developmental model of the ISC (MCP Electronics Ltd, 1966). (b) A more recent model, the MOC1000 (Courtesy of Motorola Opto)

In fact, the ISC is a self-contained, fully enclosed, miniature optical one-way communication link, complete with its transmitter (TX), receiver (RX) and intermediate light-propagating medium. This aspect of ISCs is perhaps best illustrated by an early developmental model (1966) complete with its collimating and decollimating lenses (Figure 11.2).

11.3   Variations on a theme

The three constituent parts of commercially available ISCs need not necessarily be those quoted in the above description of the basic device (though very often they are). Variations on a theme are easily imagined, with VLEDs, IREDs, neons, even miniature incandescent lamps as the TX element; air, glass (flat or lensed), acrylic, transparent Teflon (FEP), epoxy or optical fibres as the medium; and photodiodes, phototransistors, fotofets, photo-SCR, or light sensitive resistors as the RX elements. In addition, amplifiers, Schmitt triggers, logic gates and other bits of circuitry can be squeezed into the ISC enclosure (and often are), usually (but not exclusively) on the RX side, either in the hybrid (thick film) or in the monolithic (IC) form. These permutations give the designer a wide choice, dictated by applications and economics. The categorisation of what is available is best made from the viewpoint of device behaviour, with regard to parameters such as speed, sensitivity, output power, current transfer ratio (CTR), input/output insulation voltage, linearity (hence analogue and digital types) and phase reversal. Some 'multiples' also exist, with up to five ISCs in a single package. All being said, however, the IRLED/epoxy/phototransistor combination represents the bulk of what there is on the market.

The external appearance of a contemporary ISC is a far cry from Figure 11.2(a). Many 'singles' come in 6-pin dual-in-line (DIL) packages (Figure 11.2(b)) and most multiples in 16-pin DILs. As such, they do not differ, externally, from standard TTL chips. A few species survive in their older, tubular form.

In some photon-coupled devices the light propagating medium (air) is deliberately made accessible from the outside. Such is the case of *gap detectors* (interruptor switches) and *reflectance detectors* (reflective switches). While these devices are made up of the same three constituent parts as ISCs, I shall not treat them as such because, here, usually no electrical signals are fed to the input and hence there can be no question of signal coupling. The input receives, instead, a *fixed* drive, usually d.c., with a zero information content. Modulation is imprinted mechanically, via the coupling medium rather like in an optical barrier (see Section 12.2).

Viewing ISCs as electrical components leads to a better understanding of their categorisation, which follows.

11.3.1   ISCs with and without gain

The most important parameter of this circuit element is probably its current transfer ratio (CTR). This is the ratio of output current, I_o, to input current, I_i, usually expressed as a percentage. It will easily be seen that the

CTR of a basic LED/bonding/photodiode arrangement cannot exceed the value of a small fraction of a per cent. Indeed, the conversion efficiency η_{TX} of a GaAs LED is usually less than 1%, the TX/RX light coupling cannot exceed 50% and the conversion efficiency of a silicon photodiode η_{RX} lies in the region of, say, 7%. The CTR, being the product of $\eta_{TX} \times \eta_\phi \times \eta_{RX}$, would thus be, at most, $0.01 \times 0.5 \times 0.07 = 0.035\%$. This would mean, in practical terms, an output current of no more than 35 μA for a 100 mA input: not a very useful circuit element. When the RX photodiode is replaced by a phototransistor (the photon impact playing the role of the absent base connection) with a gain of 100–300, the situation is much improved. Some of the speed of response is then sacrificed. On the other hand, the photodiode sensor could be followed by a fast transistor, or an IC amplifier, inside the ISC, preserving most of the speed response. Sometimes extra components are added upstream of the device, i.e. on the TX side, to increase its overall sensitivity, as well as its CTR. In practical ISCs the values of CTR range from 0.1 to 10, and thus we have ISC *without* and *with* current gain.

11.3.2 Analogue and digital

Linearity of the transfer function $f = I_o/I_i$ (or the constancy of CTR) will usually be aimed at, but seldom achieved (Figure 11.3). There is, however, a group of ISCs in which non-linearity is sought, for digital applications. Hence the *analogue* and *digital* ISCs. In ISCs especially designed for digital work, the CTR term is replaced by the fan-out factor, familiar to TTL users. This may reach values of 5 or even 10.

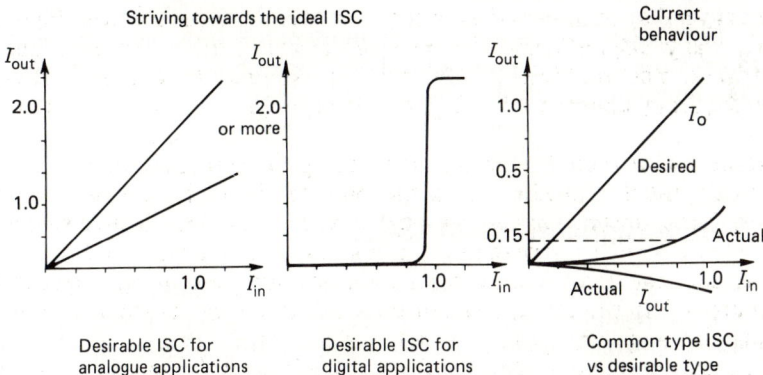

Figure 11.3 Striving towards the ideal transfer function of an ISC

11.3.3 Audio and RF – slow and fast

The speed of response is another parameter giving rise to the following categories of ISCs: audio, RF, slow or fast digital, depending on the –3 dB point, frequency bandwidth or the rise or fall time, or pulse propagation delay. Hundreds of megahertz and tens of nanoseconds are achievable.

11.3.4 Low and high voltage

The input/output insulation (electrical isolation voltage) of an ISC can be as low as 1000 V or as high as 15 kV. Hence, we have low and high voltage devices, with 1500 V usually as the demarcation point.

11.3.5 High sensitivity and high power

The value of the required input current at the working point is another categorising criterion. 10 mA is a typical value, but some line receivers use 0.5 mA and they will be termed *high-sensitivity* couplers. Output power levels (often the current sinking capability) will also depend on the application, providing yet another division of ISC into *low* and *high power* categories. (One could classify solid-state relays, capable of switching 10 A at 235 V a.c. among high-power ISCs.)

11.3.6 High CMR units

Worth mentioning among 'specials' are couplers with an electrostatic screen, a fine wire mesh, providing a high degree of input/output capacitive decoupling, while leaving the light coupling almost undiminished. (The screen will be connected to a fixed potential, usually 0 V.) Common mode rejection (CMR) can be down to −120 dB.

11.3.7 Singles and multiples

Some of the above categories of couplers, more particularly the digital and line terminal varieties, are made not only in singles but also in multiples. Among the multiples, quads are of greatest interest.

11.4 Some applications of ISCs

Some examples of circuits using ISCs (Figure 11.4) will now be given to ensure that the reader leaving this section has more than a theoretical, perhaps even abstract, notion of this versatile component. However, detailed circuit design will be avoided as this is not part of electro-optics proper, and excellent application notes for the ISC can be found in designers' manuals, application reports and product data sheets published by various manufacturers of opto-electronic devices. Many have the advantage of being periodically updated, an important factor in this fast moving technology. (See particularly Ref. 106, also Refs 10, 109–111 and, to a lesser extent, Refs 8, 96, 97, 100, 101, 105, 107, 113–117, 119 and 123. Ref. 112 (pp. 142, 147, 148, 151 and 152), though not an opto manual, contains some interesting circuits. An extensive analytical treatment of the ISC will be found in pages 3.1–3.61 of Ref. 10.)

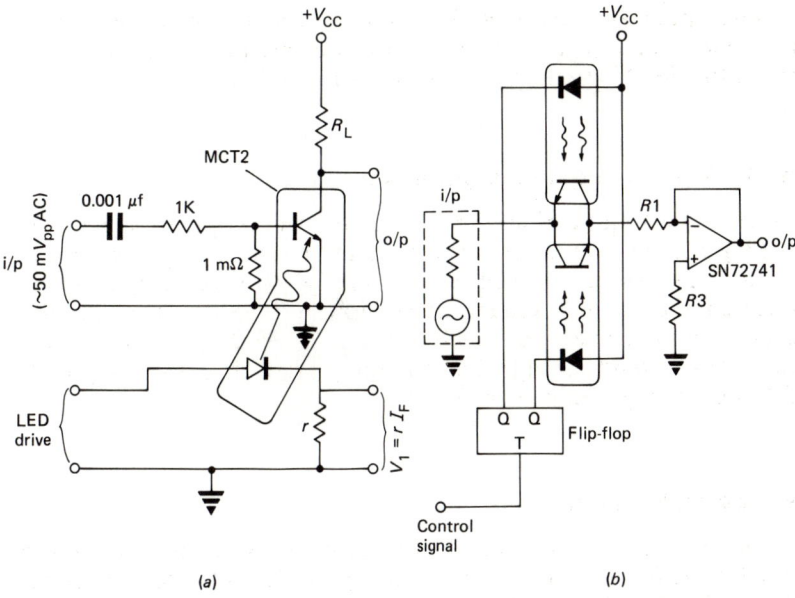

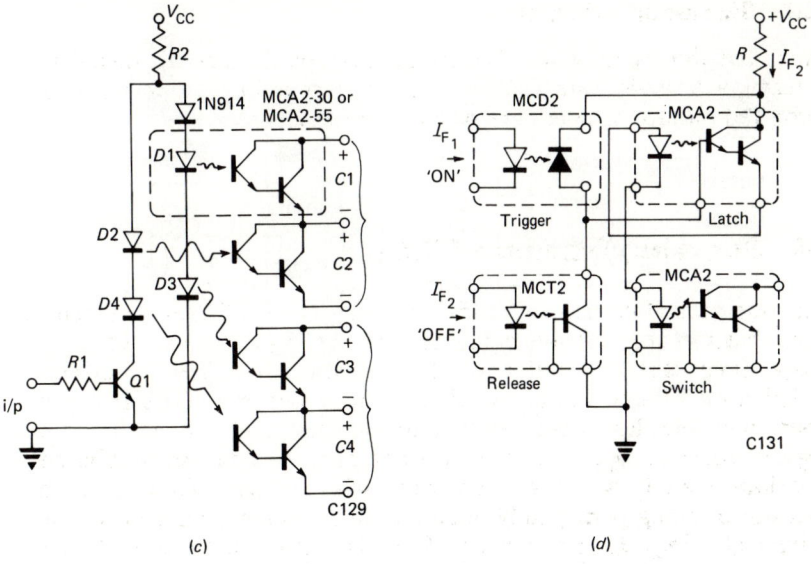

Figure 11.4 Some applications of ISCs: (a) simple gated amplifier; (b) simple chopper circuit; (c) DPDT relay; (d) latching relay

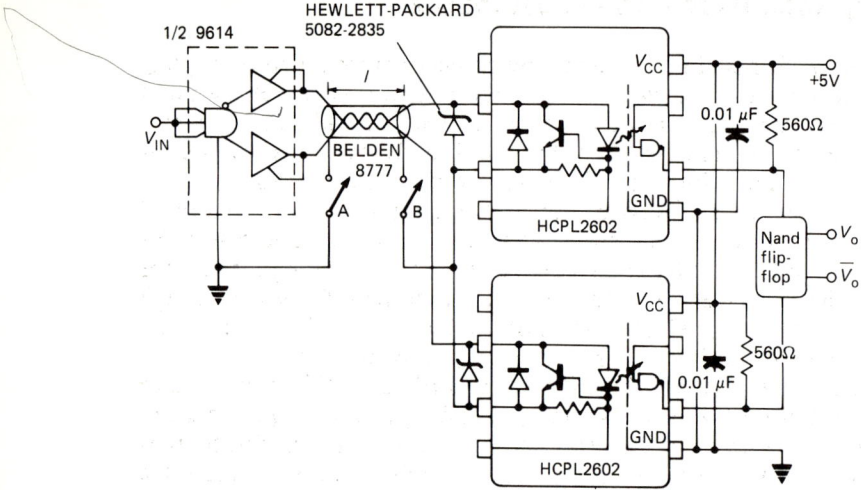

	Without Schottky diodes			With Schottky diodes			Switch A	Switch B	Units
l	<1	30	90	<1	30	90	—	—	m
	100	165	340	55	75	215	Open	Open	ns
t_p	45	125	310	45	70	185	Open	Closed	ns
	45	60	125	45	60	125	Closed	Closed	ns

(e)

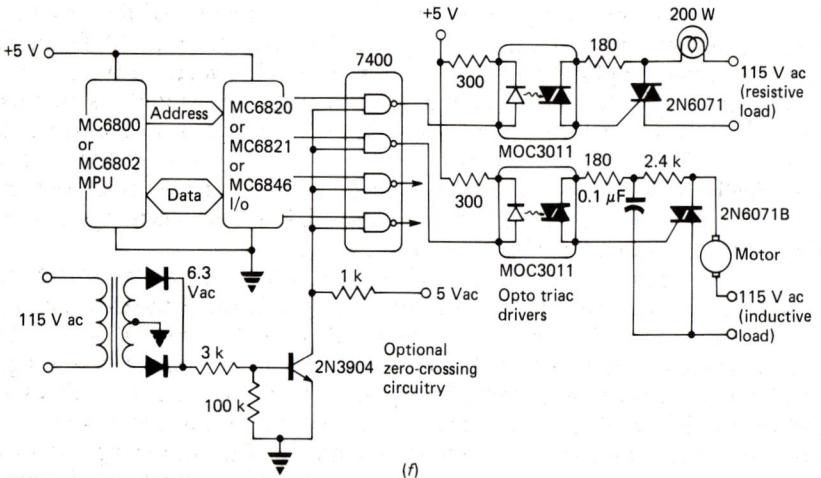

(f)

Figure 11.4 (e) polarity reversing split phase line receiver (note the current clamps on I_F); (f) interfacing microcomputers to mains-fed peripherals. The insert shows TTL to mains via triac coupling

11.5   Optofollower – the supercoupler

I coined this neologism in response to my enthusiasm for an opto-coupling arrangement that will produce at its output an *exact replica* of the input signal. None of the variations described in Section 11.3 will do it. The term opto-follower is a direct descendant from emitter follower, itself a successor to the old cathode follower of the days of valve electronics.

It has been mentioned (Section 11.3.2) that the input/output linearity of an analogue ISC is usually aimed at but seldom achieved. The reason for variations of the CTR with input current is twofold: (a) the light emitter is not a truly linear current-to-light converter, especially for low and high values of I_F, and (b) neither the phototransistor nor the photodiode/transistor combination is a linear light-to-current converter.

Apart from the lack of linearity of an ordinary ISC, temperature variations of both conversion factors η_{TX} and η_{RX} can be quite severe. While it is true to say that the drop of η_{TX} with rising temperatures is partly offset by the rise of η_{RX}, a residual overall variation of the CTR persists [10, p. 332]. In addition to these short-term variations, there is a long-term decay of the CTR, caused by the time degradation of LEDs. Thus, all in all, in applications requiring signals to be transmitted faithfully over octaves of amplitudes, for a broad range of temperatures and for long periods of time, as in bioresearch, medicine and process control instrumentation, the transformer-like action of the straight ISC is unsatisfactory, despite its splendid isolation and unidirectionality.

The optofollower puts an end to these imperfections. In it, the skills of electronic circuit design come to the rescue of optoelectronics. Here, as in so many other situations, the principle of negative feedback is used to linearise, as well as stabilise, the overall performance of a combination of imperfectly linear and imperfectly stable components (Figure 11.5). The trick consists of constantly comparing the output current of the arrangement to its input current, and feeding back the error function (difference) into the output amplifier to achieve a close match between them. In itself, this tracking operation is trivial enough: gain stabilised amplifiers all over the world use it. The designer's wiles, however, come from the use of a second, ancillary optocoupler in the scheme, in order to keep the output electrically isolated from the input [106]. The tracking is organised by causing the output current I_o to drive the LED of the ancillary coupler ISC_2, as well as being available for external use.

The I_o/I_i comparison is effected via the ISCs, by judging the effects of both LEDs on the two photodiodes Ph.1 and Ph.2. Light fluxes are being compared instead of electrical currents and, provided the CTRs of both couplers match closely within the whole of the working temperature range and have similar ageing characteristics, the effect is the same. Note that the individual ISCs need not be linear: all that is required of them is to be a matched pair. Note also that the scheme lends itself for achieving a ratio n between I_o and I_i other than unity, the equivalent of a current transformer, $n = I_o/I_i$ remaining independent of current amplitude, temperature and time. With a little ingenuity, n becomes programmable! At the time of writing I know of no single package optofollower that is commercially available, so this elegant and useful circuit arrangement has to be a DIY

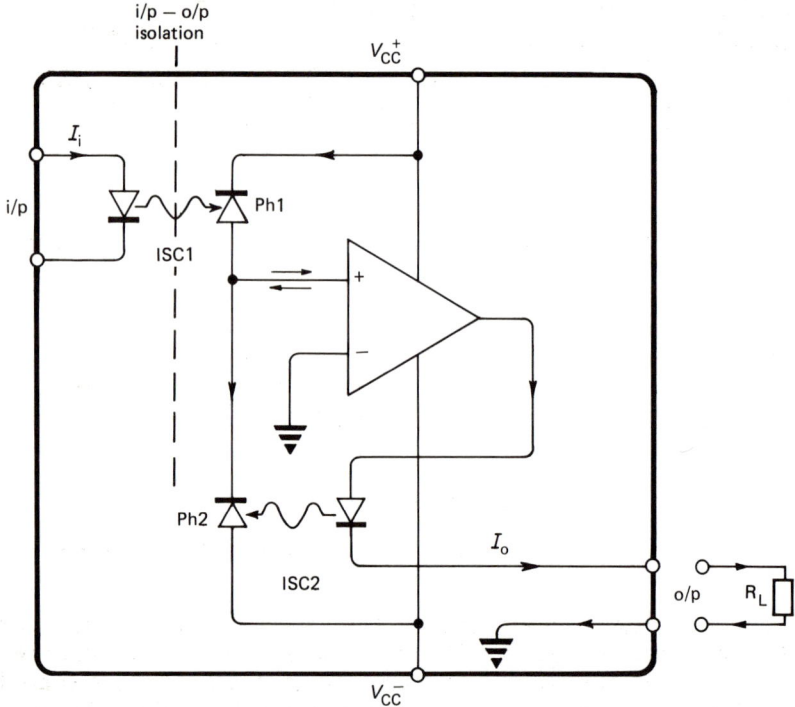

Figure 11.5 Negative feedback used to linearise as well as stabilise the performance of a combination of imperfectly linear and imperfectly stable components

job. Some manufacturers, however, offer twin analogue ISCs suitable for this application [106, 107]. What is commercially available is a high-performance optically coupled isolation amplifier, in which linearisation has been achieved in a similar, while not identical, way to the above. It uses 1½ couplers, namely one LED and two photodiodes. The device is claimed to have a very high input impedance (10^{11} Ω), a common mode rejection of 120 dB, an a.c. isolation of 2000 V and an output capability of ± 10 V and ± 5 mA. The gain linearity (as well as the 1000 h stability) is claimed to be better than 0.05% [97].

11.6   Digital transmission of analogue signals

Lateral thinking has led to yet another solution to the problem of optical coupling of analogue signals in a way that escapes the prejudicial effects of LED/photodetector non-linearity, thermal variation or ageing. It does more than that: it provides a solution which can also cope with transmittance variations of the light coupling medium. *Digitisation* is the keyword of this solution. An A/D converter on the TX and D/A converter on the RX side are used, so that on/off digital signals propagate in the medium [10, p. 336]. Sufficient gain is designed into the circuit to guarantee a secure saturation of the RX switching element for the whole of the dynamic range

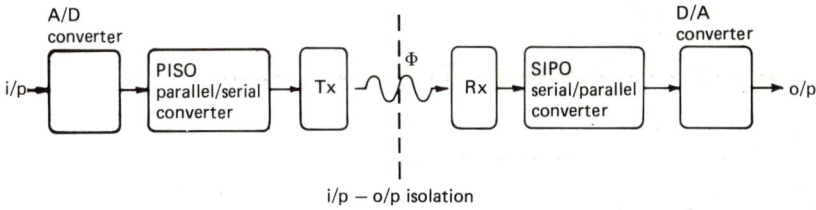

Figure 11.6 Digital transmission of analogue signals through use of A/D and D/A converters

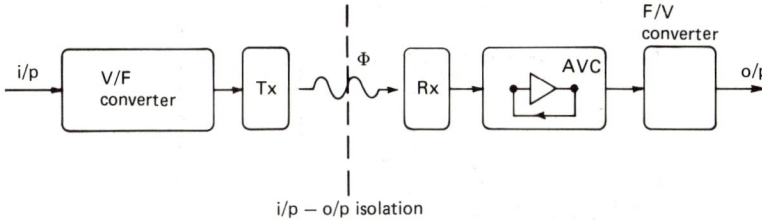

Figure 11.7 Digital transmission of analogue signals through V/F and F/V converters

of the signal strength. Even simpler, but still in the same vein, is the method of voltage to frequency (V/F) conversion on the TX end, combined with an F/V signal recovery on the RX side. Automatic volume control copes with any amplitude loss likely to occur, within the designed in limits. (See Figures 11.6 and 11.7.) Pulse width modulation can also be used [10, p. 338].

Despite commercial availability of D/A, A/D, V/F and F/V chips, the digitisation and V-to-F conversion methods are more complex and more expensive than the optofollower and are mostly of interest in genuine optical linkages (as opposed to in-circuit isolating couplings) in which the insertion losses of the medium, such as air or water, are likely to vary as a result of distance, temperature, pressure, pollution or other alteration. As for a third possibility, the PWM method, the residual non-linearity of its transfer function is at least one or two orders of magnitude higher than that of the optofollower.

Chapter 12

Red and not-so-red rays for engineering

Chapters 7–9 showed some of the capabilities of electro-optics in communications, medicine and manufacturing. We return to these areas now that the principal techniques of lightwave technology have been exposed, respecting the same order as that of the above mentioned chapters.

12.1 The sensing fibre

Chapter 7 showed us how superbly optical fibres are suited to the task of transmitting information. Fibres can do more, however, than to act just as a transmission medium: they can convert mechanical, electrical thermal and other parameters of the physical world into variations of intensity, phase, speed, modal distribution and other characteristics of the light wave, and thus act as *transducers*. This is not only fascinating but also very valuable: in its optical guise, transducer-generated information can travel great distances, unharmed by electrical interferences, while generating none itself and remaining intrinsically safe to the environment through which it travels. This is certainly more than can be said of electric transducers. Thus, in the years to come, Optic Fibre Sensors (OFSs) are likely to invade process control plants in which a small spark can cause a great conflagration, places where chemical agents threaten metallic leads with corrosion, and hospitals where the absence of all galvanic patient–equipment connection offers a guarantee of intrinsic safety.

Some sensors are *intrinsic*, i.e. sensors in which the parameter to be measured causes a change of one or more characteristics of the fibre itself, such as the refractive index or the modal distribution (see Section 7.2.3). Others are *extrinsic*. In these, the fibre acts as a highly convenient light transmission medium used for carrying the information generated just outside it, such as the change of colour, under the effect of a temperature variation, of the emission of a fluorescent pellet. Because of their intimate relation with the change producing element and of the analogue as opposed to digital character of the information they carry, such fibres belong, generally, to OFS and not FOC technology.

As some sensors rely for their operation on coherent light while others do not, we have a further categorisation of OFS into *coherent* and *non-coherent* types. Intensive research and development goes on in the

222

field of both the intrinsic/extrinsic and the coherent/non-coherent OFS, witnessed by well attended international colloquia and a vast literature already in existence (see bibliographies in Refs 200–209, especially Ref. 202). Here are a few illustrative examples of FOSs, some of which are still under development.

12.1.1 Pressure sensor

The straightforward non-compensated pressure sensor is probably the simplest of all OFSs. In it, the pressure to be measured is applied to the rearside of an elastic membrane which converts it into a displacement (Figure 12.1). The front side of the membrane has a mirror-like finish.

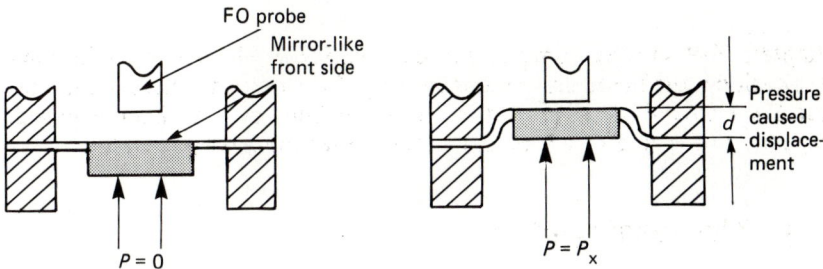

Figure 12.1 The pressure to be measured is applied to the rear of a membrane, the front of which has a mirror-like finish

Light is shone upon it by one or more fibres and is reflected back onto one or more fibres. The effectiveness of the fibre–mirror–fibre throughput is representative of the membrane displacement, d, and thus of the pressure, P. In other words, the transfer factor $T = \phi_{out}/\phi_{in}$ is a known function of P:

$$T = f(P) \tag{12.1}$$

In a commercially available fibre optic pressure sensor (FOPS), the transmitting and receiving fibres are intermingled in a semiflexible bundle up to 8 m long. The transfer function is shown in Figure 12.2. That

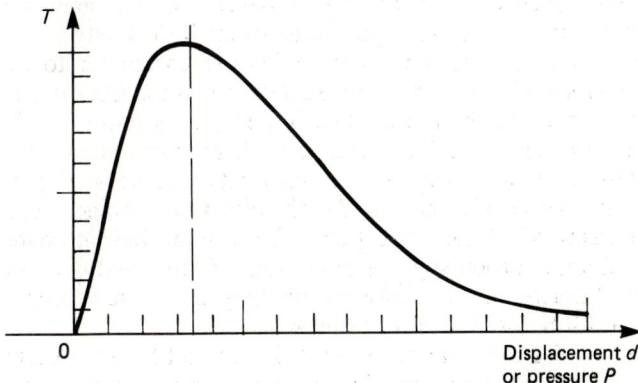

Figure 12.2 Transfer function of a FOPS using a fibre bundle (arbitrary units)

T should be zero for $d = 0$ results from the fact that, with this geometry, both the light-emitting and light-collecting fibres are obstructed by the membrane when it touches the bundle tip. The range of measurable displacements is $\simeq 40\ \mu m$ and the instrument's resolution can be as much as $0.05\ \mu m$. This corresponds to a differential displacement of only 0.125%. The light source is an incandescent lamp.

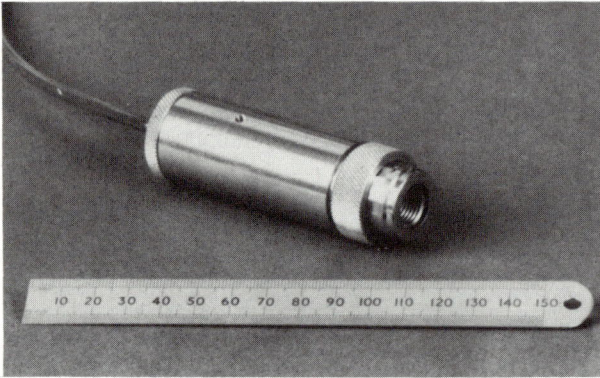

(a)

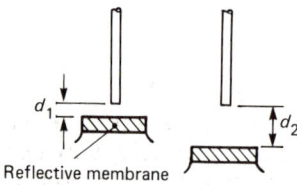

Reflective membrane

The proportion of recaptured power depends on the displacement d

(b)

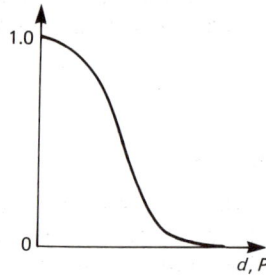

The transfer function is monotonic

(c)

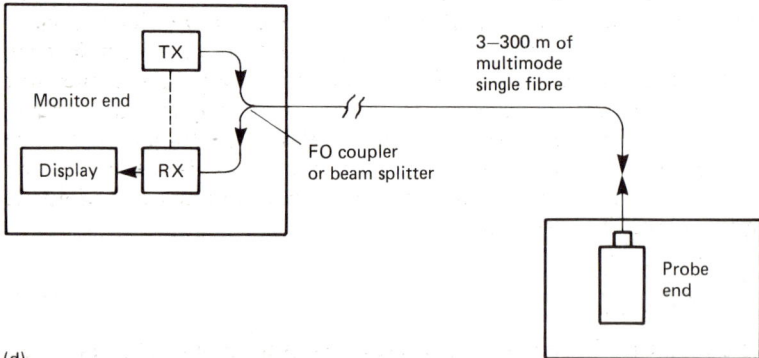

(d)

Figure 12.3 (a–c) University College London's FOPS I uses only one single fibre. (d) A beam splitter or an FO coupler separates the TX and RX functions of the single fibre (Courtesy of University College London)

A more recent version of FOPS, developed by University College London (UCL Project S28, supported by the British Technology Group), uses one single fibre and thus has a *monotonic* transfer function containing a useful linear portion, as seen in Figure 12.3(c). The fibre tip performs the double function of radiating and collecting light. To the contrary of the previously described model, there is no obscuration but maximum reflection here, for $d = 0$. A beam splitter located at the monitoring end of the equipment separates the TX and RX functions of the fibre (Figure 12.3(d)). The probe–monitor distance, L, i.e. the overall length of the connecting FO cable, is limited by the fibre attenuation only, which means that very nearly all industrial applications ($L = 3$–300 m) should be feasible. The UCL FOPS I [206] uses a 200 μm multimode all-glass fibre and a modulated LED. Its useful pressure range is 100 mbar to 20 bar (dependent on the membrane stiffness) and its absolute noise-limited displacement resolution is 50 nm.

For repeatable results non-compensated FOPSs require:

1. Stabilised light sources
2. Non-tarnishing membranes
3. Constant attenuation FO links

In an attempt to eliminate these and other constraints on OFSs in general, research workers have evolved various compensation schemes over the last few years.

12.1.2 Rotation sensors

When two beams of coherent light are launched into a *stationary* coiled length of optical fibre from both ends A and B simultaneously, they exit the coil through B and A simultaneously too, hence *in phase*. If, however, the coil is made to spin on its axis, say clockwise, exit A advances towards the beam, while exit B recedes from it, this resulting in a phase difference between the exiting beams (Figure 12.4). (The wavefront representation helps towards an intuitive representation of the situation: the rotating fibre either *runs* forward towards the advancing wavefront or *slips* from it.) Using single-mode fibre makes it quite easy to produce a clear interference between the exiting beams. The rotation-induced difference between the two optical paths of the counter-propagating beams will produce a double phase shift (one in each beam) proportional, among other things, to the angular velocity of the coil, ω. The 'other things' are the constructional constants of the instrument. It will be seen that the monitoring and interpretation of the interference pattern can yield ω. Quoting from Ref. 201 we have:

$$\omega = \Delta\theta \left(\frac{\lambda c}{8\pi\, NA} \right) \tag{12.2}$$

We thus end up by having an FO rotation sensor. Figure 12.5 gives an example of the interference pattern of an experimental sensor and its variation. Arranging three such sensors in an orthogonal *xyz* structure and adding some signal processing electronics gives a navigational gyroscope.

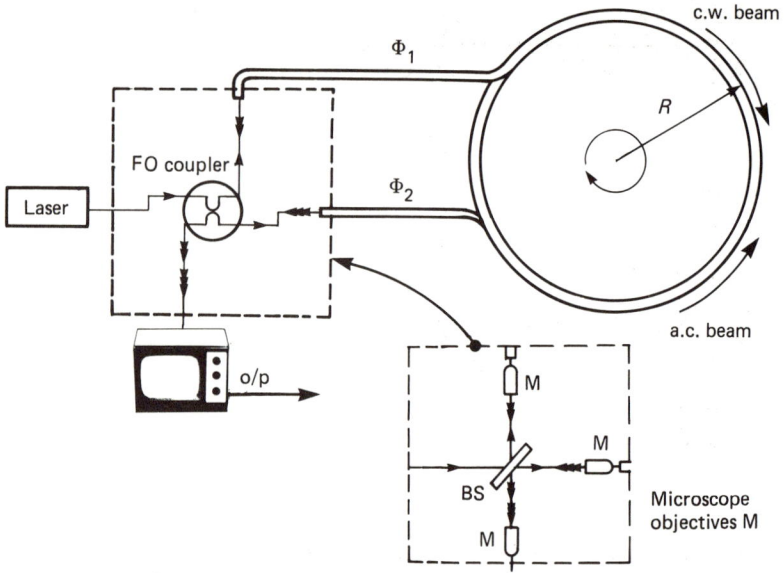

Figure 12.4 The two beams ϕ_1 and ϕ_2 are derived from the same laser by means of an FO coupler (or a beam splitter, BS) which also acts as an interference-producing beam recombiner

(a)

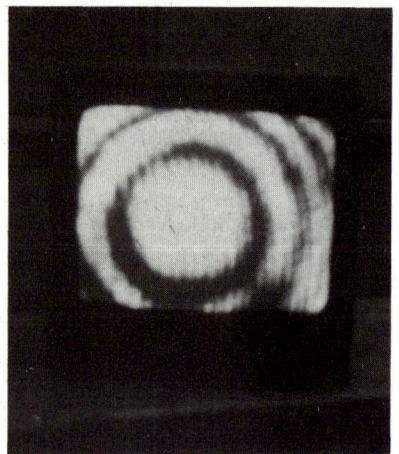

(b)

Figure 12.5 Examples of interference patterns of an experimental FO rotation sensor: (a) anticlockwise rotation; (b) clockwise rotation

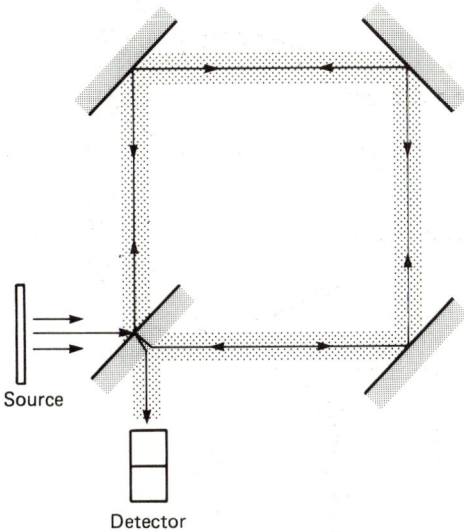

Figure 12.6 The half-century-old (1913) Sagnac interferometer

(For further reading see Refs 200–203, each of which contains a rich bibliography of its own.)

The OF gyro has great sensitivity: earth rotation (15°/h) can be detected without much difficulty, 1°/h has been quoted frequently and, at the time of writing, figures down to 0.05°/h have been reported [200] for an all-fibre instrument. The latter has also a mass production potential. The term 'all-fibre' refers to the replacement of the bulk optics cube beam splitter/combiner by a four-port fibre coupler (see Section 7.6). This kind of replacement is illustrative of a prevailing trend in OFS technology. To close, note that the basic configuration of the OF gyro corresponds to the over half a century old 1913 Sagnac interferometer (Figure 12.6).

12.1.3 Non-electric probes measure electric quantities

(a) Sensing current with optical fibres
This intriguing technique exploits elegantly an effect discovered by Faraday over a century and a half ago: the plane of polarisation of light in glass rotates under the influence of an axially applied strong magnetic field (Figure 12.7). (This does not affect optical communication lines – not only because the magnetic fields encountered there are far too weak to do so but also because the luminous *intensity* remains unaltered.) A single-mode laser-powered fibre is wound around a power cable, enabling the current-created magnetic field to exert its rotating power over a great length of it. Two photodiodes, each with its own analyser, follow a Wollaston prism [1, p. 504] and detect the amount of rotation of the beam by measuring the parallel and orthogonal components of the exiting light, thereby giving a measure of the magnetic field, H, and hence of the current, I. Electronic

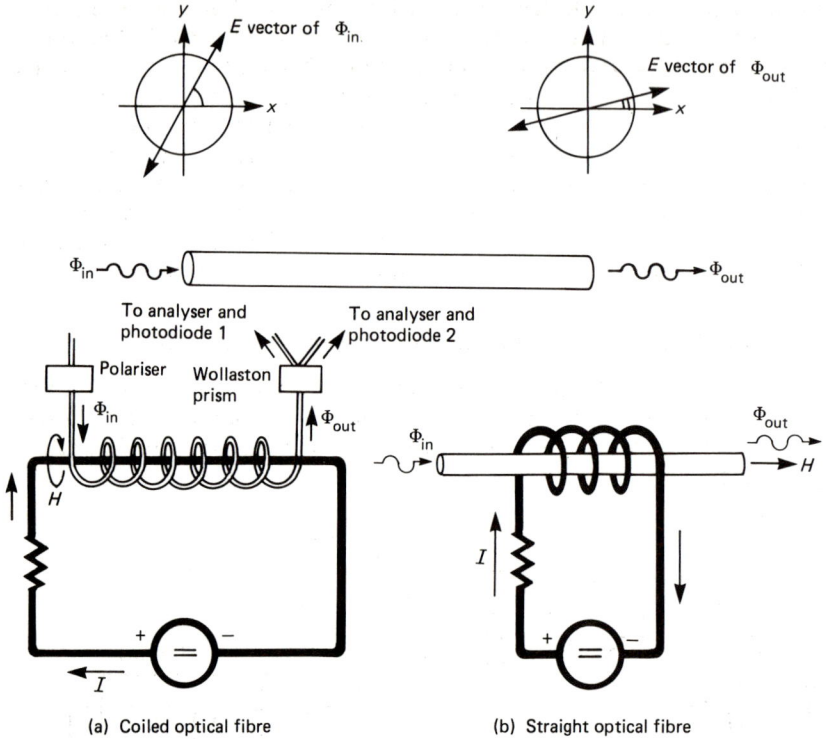

Figure 12.7 Two possible ways of generating a longitudinal field H in an optical fibre. Current sensing based on the Faraday effect. The plane of polarisation of light rotates under the influence of the axially applied magnetic field

circuitry eliminates the influence of light intensity variations. Accuracies of the order of 0.25% have been reported, in the 50–200 A range [204, pp. 11–12]. Low birefringence fibres (accidentally) give smaller stress induced errors. Note that the Faraday effect angle of rotation is independent of the direction of light propagation but does depend on that of current. Thus an a.c. current produces an a.c. output signal. Although power lines have usually a frequency of 50 or 60 Hz only, frequencies well into the megahertz region – limited by the far end electronics only – can be coped with.

The chief attraction of the Faraday sensor lies in the non-electric character of the link between the point of test and that of display which can be tens of kilovolts apart.

For further reading, Ref. 17, p. 261 gives an excellent account of the Faraday effect. FO current sensors are treated in some detail in Ref. 204, pp. 4–12. The EO Faraday isolator, analogous to the microwave one, is described in Ref. 192, p. 110.

(b) Sensing voltages with optical fibres

If the detection of a magnetic field by a fibre can be used for measuring electric current, one could logically expect the possibility of measuring voltage by sensing an electric field. In fact, not one but several electro-optic effects exist. (The reader will remember the laser beam EO modulator (Section 10.3.3(c)) exploiting one of them.) None of these effects lends itself to voltage sensing with fibres as directly as the Faraday magneto-optic effect does to current measurement, not only because the manifestation of the EO effects becomes sufficiently pronounced only in rather unusual, most often crystalline materials of which fibres cannot be made, but also because the non-linearity of the conversion processes involved would impose restrictions on the dynamic range of an instrument in which a crystal would generate and the fibre simply convey the information.

12.1.4 Heterodyning and phase-lock looping with OFSs

Some interferometric OFSs would be highly impractical, were it not for the application in them of the principle of *frequency shifting*, universally used in radio receivers, or of *phase-lock techniques* so frequently used in control engineering.

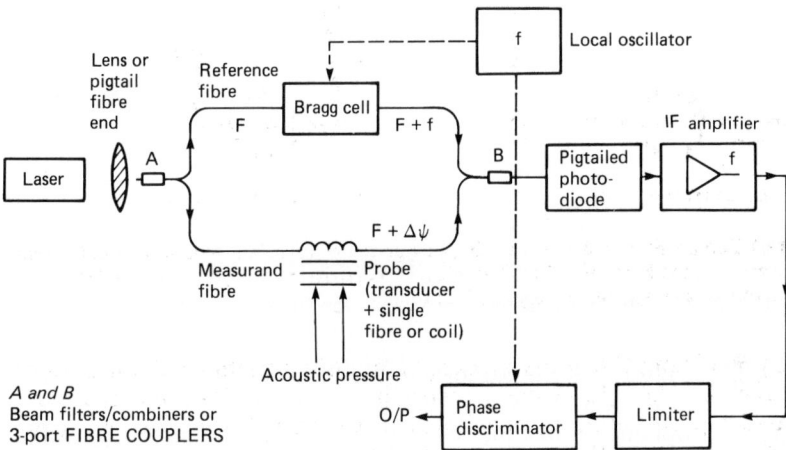

Figure 12.8 OF hydrophone in a Mach–Zehnder interferometer configuration with a Bragg cell frequency shifter in its reference arm

Figure 12.8 shows an OF hydrophone (submarine sound detector) using single-mode fibres in a Mach-Zehnder interferometer configuration with a Bragg cell frequency shifter in its reference arm. The coherent laser light, divided in two beams at point A, propagates through the measurand and reference arms, both subject to whatever spurious acoustic noise might be present. Obviously, its effects cancel out. The light of the measurand arm alone, however, undergoes phase changes $\Delta\psi$ induced by the acoustic signal. Reunited in point B the beams produce an interference which varies in harmony with the acoustic signal. $\Delta\psi$ results from the mechanical strain in the fibre affecting its length and its refractive index, n. Coiling the fibre increases greatly its length and thereby the sensitivity of the arrangement. Sound, ultrasound, pressure or temperature variations can be detected.

Looking at Figure 12.8 the electronics engineer will recognise the hetero-
dyne principle. Here, as in a superhet receiver, two frequencies F and $(F + \triangle F)$ are mixed to give an easily manageable intermediate frequency IF. The
local frequency $F + \triangle F$ is obtained by a sideband of the modulation of F
by the already described Bragg cell deflector (Section 10.3.2(b)). (The
Bragg cell was used in Section 10.3.2(b) as a *dynamic* diffraction grating. A
static (mechanical) diffraction grating, while capable of deflecting a laser
beam, is incapable of producing a frequency shift of light. The Bragg cell
produces this shift because it is 'live', i.e. because of the periodic, *temporal*
sinusoidal variation of the refractive index in any given location of the cell,
$\triangle f$ being equal to the drive frequency of the cell.) 'Electronic magic'
transposes the phase shifts from the incredibly high optical frequency (5×10^9 MHz) domain into the radio domain of tens or hundreds of megahertz
only [168].

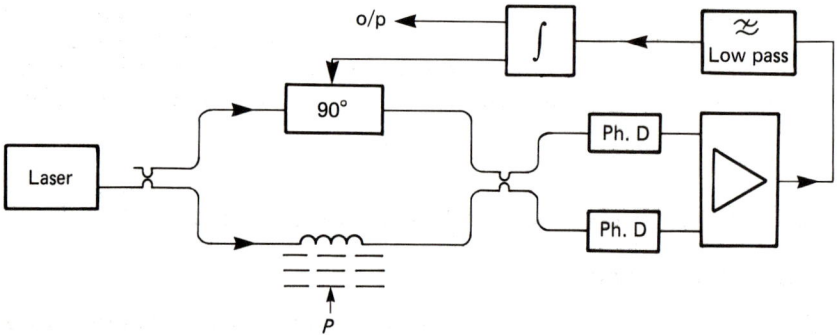

Figure 12.9 OF homodyne hydrophone in a Mach–Zehnder configuration

Figure 12.9 illustrates the use of another electronic inspired technique,
the phase-locked loop, for signal recovery in the same type of interferometric hydrophone as has just been described [200]. The Bragg cell is
replaced by a 90° (when at rest) electro-optic phase shifting component.
This could be one of the devices covered in Chapter 10 or, if bulk optics
were to be avoided, a PZT cylinder with the reference fibre wound round
it. One recognises, of course, homodyne detection. According to Ref. 200,
phase shifts as small as 0.1 µrad have been measured, at around 1 kHz,
with this sensor architecture.

   OF hydrophones are still under development. Their special attractions,
apart from those common to all OFSs mentioned at the beginning of
Section 12.2, are: superior sensitivity (10 dB plus, over and above
piezoelectric models) and ease of shaping for application-dictated
directivity.

12.1.5   Concluding remarks on OFSs

Optical fibres can do more than simply transmit information. When used in
OFSs they can convert very aptly strain, pressure, flow, temperature,

displacement, rotation and velocity, as well as electric, magnetic and many other physical parameters, into variations of intensity, speed, phase polarisation, modal distribution and possibly other characteristics of the light they transmit. They can further convey with intrinsic safety this or other transducer-generated information in hazardous atmospheres of the process control industry and in applications implying invasive biomedical *in vivo* monitoring and testing.

Elegant and effective schemes have been evolved for the compensation of such factors likely to affect the repeatability of OFS performance as fibre ageing or reconnection, the ageing of light sources, the tarnishing of reflective surfaces, and the variation of ambient temperature (see, for example, Refs 206 and 209).

In many of their applications OFSs outdo older techniques on sensitivity, dynamic range, freedom from interference and, potentially, cost.

While it would have been wrong to burden this book with detailed description of a swarm of realisations belonging to a still predominantly developing, and hence very fluid, technology, it would have been equally wrong to totally ignore in a work of this nature a branch of lightwave technology holding such great promise for the future. Hence this middle-of-the-road attitude of introducing OFSs by a few illustrative examples.

12.2   Optical barriers and laser 'chalk lines'

12.2.1   Optical barriers

The subject matter of this section is much simpler than that of the rest of this chapter. This simplicity does not make it any less interesting. Extremely useful optical barriers are today widely employed in industry and in everyday life. Basically, they are modern versions of the photoelectric controls of the 1930s and 1940s, the two best known of which are the automatic door opener and the mechanical escalator's start switch.

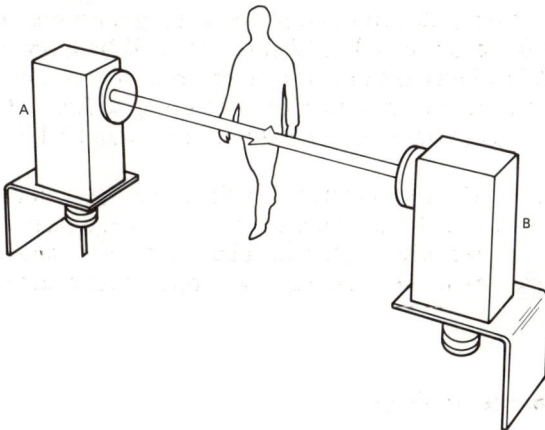

Figure 12.10 An optical barrier continually testifies the integrity of the A–B optical path

Today's optical barriers (Figure 12.10) are opto-electronic linkages operating on the *by default* principle. Here, the *raison d'etre* of the light beam is *not* to transmit information from point A to point B, but to continually testify the integrity of the A–B optical path. Once this integrity is violated, the RX at point B generates a signal which initiates some predetermined action. The applications are many, from machine guards, through conveyor batch counters, to intruder alarms. Opto-electronic barriers feature enormous advantages over their photoelectric ancestors: size, weight, drive power economy, longevity, immunity to ambient light, and an almost total imperviousness to malicious fooling. The ease of modulation of OE light sources, their collimation facilitating smallness and their monochromaticity lie at the root of their superiority. When it comes to intruder alarms, the invisibility of the infrared beam they use is their undeniably unique asset.

(a)   By transmission barriers

The basic barrier of Figure 12.11 is still often used. Such a simple arrangement is capable of no more than detecting an unsuspecting intruder crossing a straight-line boundary. To extend its use to perimeter guarding, prisms or mirrors can be used (Figure 12.12). Intruder height discrimination can be designed in by the use of two or more beams to selectively ignore (or detect) small children, animals or fence jumpers, etc. To assure

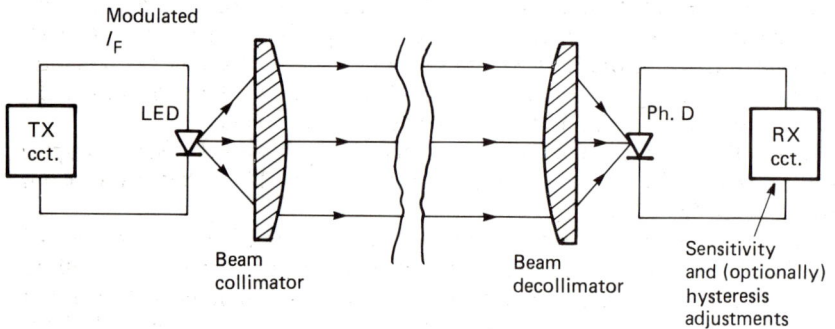

Figure 12.11 Basic EO barrier

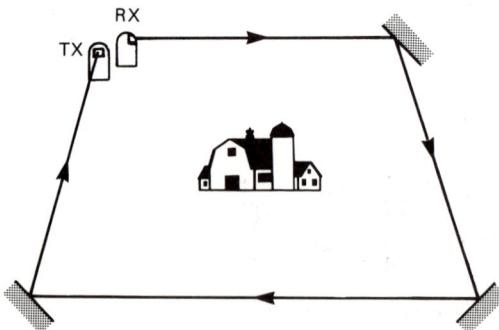

Figure 12.12 Reflectors extend the use of an EO barrier to perimeter guarding

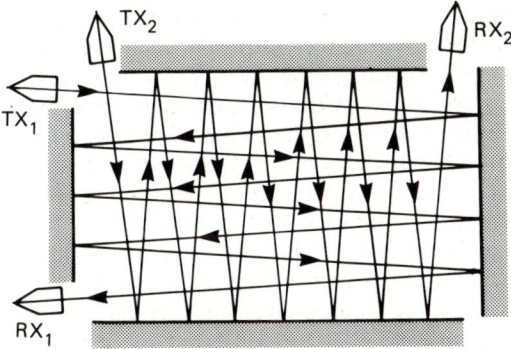

Figure 12.13 An infrared webb for the detection of small objects within a large cross-section

the safety of a press operator's hands, an OE 'curtain', also using mirrors, can be installed. The breaking of the light beam anywhere within the opening disables the press ram. When a simple detection of a small object anywhere within a large cross-section is to be assured, a fine infrared web suffices (Figure 12.13). For the detection of small objects traversing a restricted area, e.g. falling down a chute, a beam waist offers a solution (Figure 12.14). (I like referring to this arrangement as 'an optical eight', not only because of its resemblance with the figure ∞, but also because the total conjugate length usually equals $8f$.) An x,y grid of several independent beams, however, will also reveal its position (x,y coordinates). Extending the principle to three dimensions, x,y,z enclosures can be made in which, for example, infrared beams help research workers to study flight patterns and other behavioural characteristics of insects.

Most OE presence detectors are intrinsically 'failsafe'. For example, tampering with the mirrors of Figures 12.12 and 12.13 will interrupt beam propagation. So will an LED failure.

The use of time constants in the receiver circuits offers possibilities of ignoring fast moving interceptors (e.g. birds, mice or flying stones getting past an OE fence). While any type of pulse modulation makes all securing barriers impervious to fooling by torches, individualised PCM with time-variable codes offers a higher degree of sophistication when greater safeguards, e.g. against highly organised crime, are required. (So would a CW system with non-repetitive FM.)

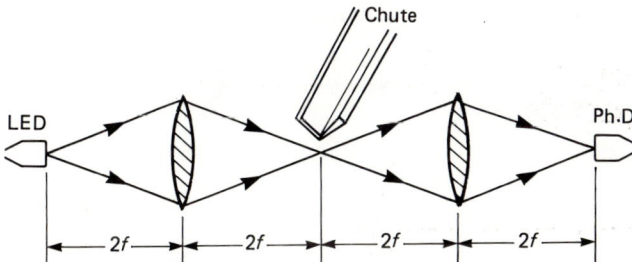

Figure 12.14 An 'optical eight' for the detection of small objects falling down a narrow chute

These are but a few examples of what optical barriers can do. Regarding intruder alarms, many more OE anti-crime measures probably exist as closely guarded trade secrets of the security industry.

All the above described barriers were of the *transmittance* type, with separate TX and RX terminals. In situations in which it is desirable to have all the electronics in the same location, *reflectance* barriers provide the answer.

(b) By reflection barriers

In preference to a mirror, a retroreflector or *retroflector* may be used (Figure 12.15). (This contraction of the word retroreflector will be used henceforth.) The enormous advantage of such a reflecting device is that it is very tolerant of misalignment. A retroflector sends back a ray of light in an antiparallel direction to its forward travel. It does so for a broad range of angles of incidence. While a tilt $\triangle\alpha$ of an ordinary mirror produces a deflection $D = 2\triangle\alpha$ of the reflected ray, a retroflector exhibits a

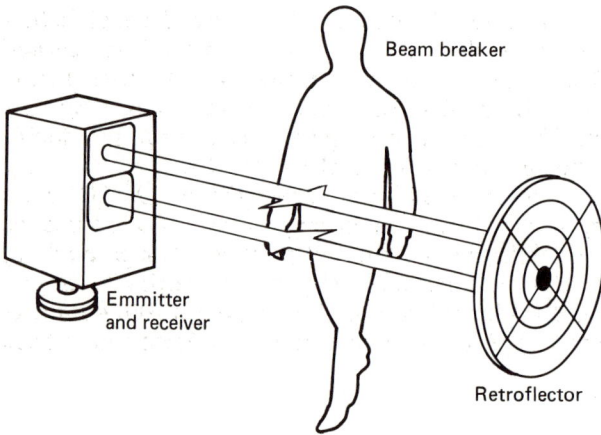

Figure 12.15 A 'by reflection' EO barrier with retroflector

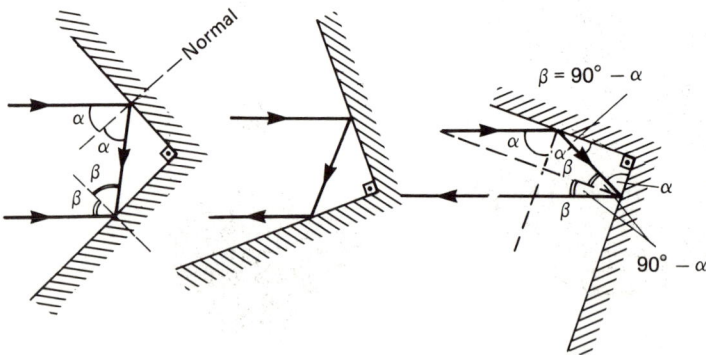

Figure 12.16 A 'cube corner' retroflector returns the beam towards its origin, regardless of the angle of incidence

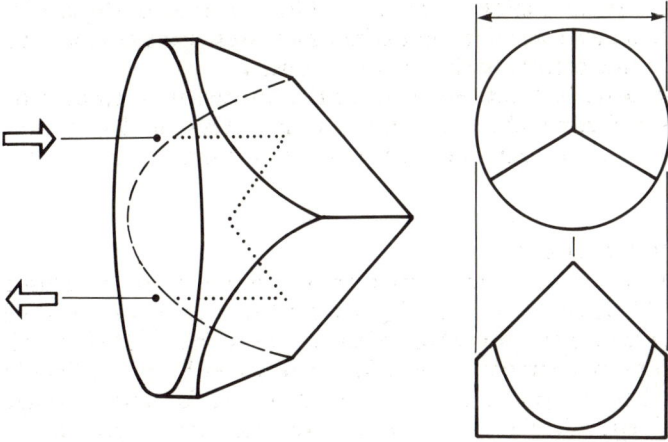

Figure 12.17 A cube corner (trihedral) retroflector

remarkable $D = 0$ for $\triangle\alpha$ values of within $\pm 10°$ or more. To explain how such a device works, we must digress for a moment to the 'cube corner' retroflector. Two reflective surfaces at right angles to each other (Figure 12.16) form a retroflector for incident rays contained in a plane perpendicular to both of them. This interesting property results from the double reflection $(2\alpha + 2\beta)$ being identically equal to $180°$ (as $\beta = 90° - \alpha$). Three mutually perpendicular reflective surfaces (Figure 12.17) form a cube corner (trihedral) retroflector, returning towards the source all incident rays within its effective aperture *regardless of their angle of incidence*. Practical cube corners are made of glass and, relying as they do on TIR, have acceptance angle limitations. Silvering the working surfaces can remove those limitations. (Hollow cube corner retroflectors have been

Figure 12.18 Retroflective targets (Courtesy of Honeywell)

brought out recently by at least one US company. Their unique advantage is the broad spectral response, from ultraviolet through the visible to infrared.) For security or industrial applications sufficiently large (or even clustered) cube corners would be too expensive. They are therefore replaced by embossed-and-filled targets (Figure 12.18) containing large numbers of small, rather rudimentary retroflective cavities flat-filled at the front, not unlike the still cheaper plastic catadiopters (sometimes called Cataphots – a trade name) for cars and bicycles. The misalignment tolerance of such retroflectors is usually $< \pm 20°$. (The type described is in general use with optical barriers. For other uses, other types exist, e.g. the 'cats eyes' for road traffic at night.)

Based on a similar principle is the selfadhesive retroreflective tape marketed under the trade name of Scotchlite 2275 Blue and 2870 Silver by the 3M Company.

(c) Range considerations

One is tempted to state that the TX–RX distance of practical transmittance barriers ranges from just under a metre (ticket gates in Underground stations, narrow conveyor belts in industry) to a few hundred metres (OE fences guarding larger properties). However, this would ignore not only the good many OE barriers in industry, working over, say, a 5–50 cm range, but also the probably even more numerous gap detectors (Figure 12.19) in which the TX–RX distance is no more than a few millimetres (sequencing wheels, limit detectors, etc.).

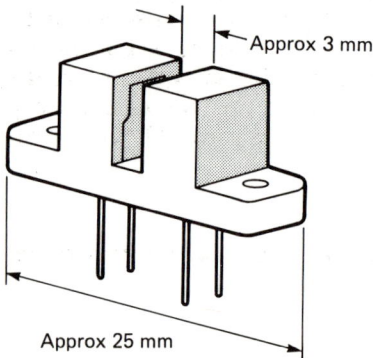

Approx 3 mm

Approx 25 mm

Figure 12.19 A GE gap detector (interrupter) with a 3.2 mm TX–RX distance

What is the maximum distance a given equipment will span? The answer will be found in Chapter 8, as the relationship between the design parameters and the range of FSOC links applies here too. Stretching the length of a security barrier by pushing the gain of the RX amplifier excessively will obviously cause a deterioration of the SNR leading, in turn, to false alarms. For a given TX, the signal strength will be proportional to $1/R^2$, in clear weather (less in fog or snow). In reflectance barriers, however, the signal decay with distance is much more severe: as the retroflector is usually smaller than the cross-section of the incident beam, the inverse square law has to be applied *twice* (once on the incident and once on the reflected beam) yielding $S \propto 1/R^4$, with atmospheric attenuation applying over a $2R$ distance. Thus, retroreflective linkages

Figure 12.20 The retroflective panel ($\sim 50 \times 50$ cm^2) of cube corners left on the Moon by astronauts Aldrin and Armstrong on 21 July 1969 can be seen on the far left in the background (Courtesy of NASA)

cover, as a rule, distances of under 5 m. One sizeable exception to this rule, however, concerns a linkage (though not a barrier) working over nearly 400 000 km. This is the Earth–Moon laser boomerang (Figure 12.20), established on 21 July 1969, when astronauts Aldrin and Armstrong sited a large panel of high-precision cube corners on the surface of our natural satellite, thereby helping to measure more precisely than ever before the Moon–Earth distance. (According to ref. 237, the centre-to-centre distance was found to be 384 404 km, within a 0.5 km error. Other sources claim even higher accuracies.)

12.2.2   Light sheets and light beams for construction, surveying and agriculture

(a)   The builder's chalk line
A builder's 'chalk line', horizontal, vertical or skewed, can be obtained from a low power HeNe laser easily and cheaply (Figure 12.21) by fitting an add-on 'line-optics' (Rofin Ltd, UK) onto the laser. A bright red and *very straight* line will be projected instantly onto the structure of interest, with nothing to erase or cover up once the job is done! The optics is basically a cylindrical lens. A combination of two lenses at 90° will generate two perpendicular sheets of light, the projection of which on a wall will form a cross. Trivial an application of lasers this may well be, but how useful!

Slightly more elaborate is the 'laser-level' contractors' alignment aid (Figure 12.22) (Spectra-Physics Ltd, UK). Here, a rotating beam produces a horizontal or vertical trace on a construction – whether plane or three-dimensional. Self-levelling (or self-plumbing) assures nearly perfect

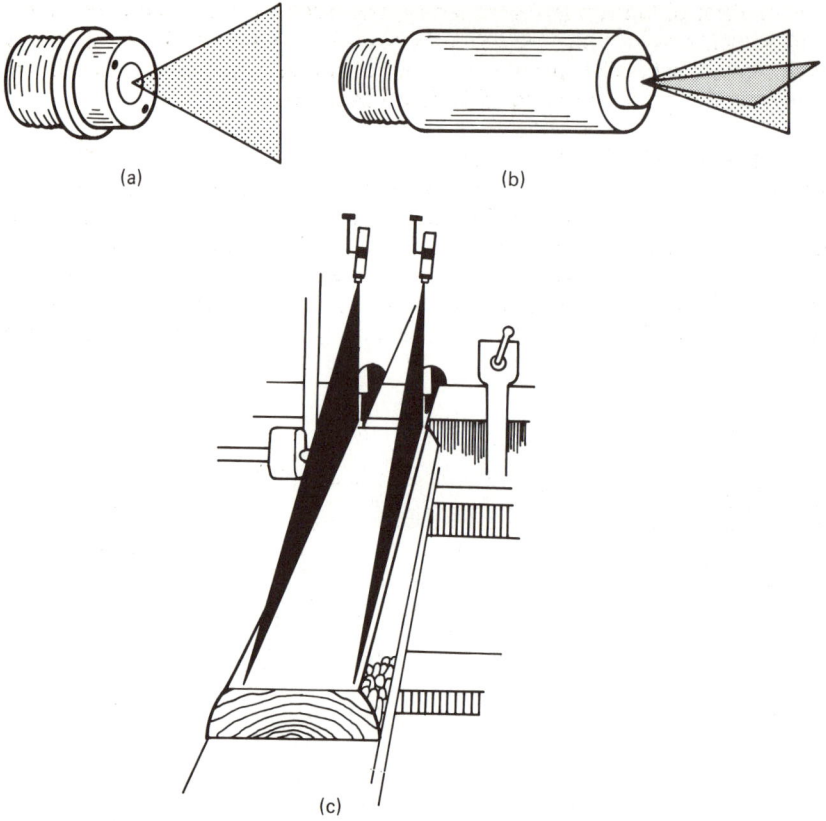

Figure 12.21 (a) Line and (b) cross generation by the addition of 'line optics' to the laser. (c) Application of two (a) devices in a saw mill

Figure 12.22 The contractor's 'laser level' showing a Spectra-Physics transmitter with dual receivers on a motor grader (Courtesy of Spectra-Physics Ltd)

level or vertical lines, e.g. for screeding crane rails or for erecting columns, steel structures or other elevation work. The 'painted' stripe is a few millimetres thick. The instrument is used in conjunction with the 'laser-Rod', a sensing and beam locking EO 'grade pole'.

(b)   Lasers and LEDs for alignment

Accurate alignment is one of the fundamental operations in engineering. A TEM_{00} laser and a *quadrant detector* (Figure 12.23) provide an excellent combination for revealing misalignment. Indeed, the Gaussian beam profile of such a laser will cause an imbalance to the inputs of the analogue circuits unless all four quadrants are equally irradiated, i.e. unless the beam is accurately central. The analogue circuits are designed to produce

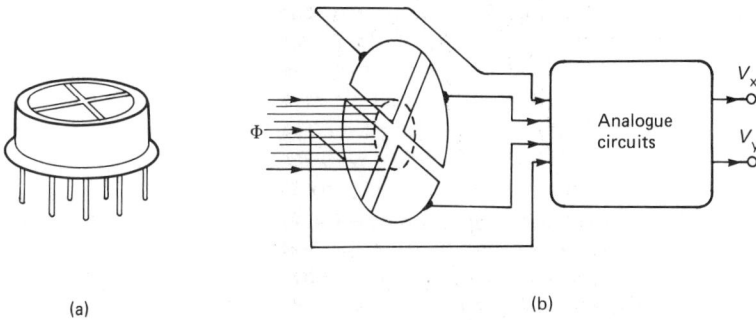

(a)

(b)

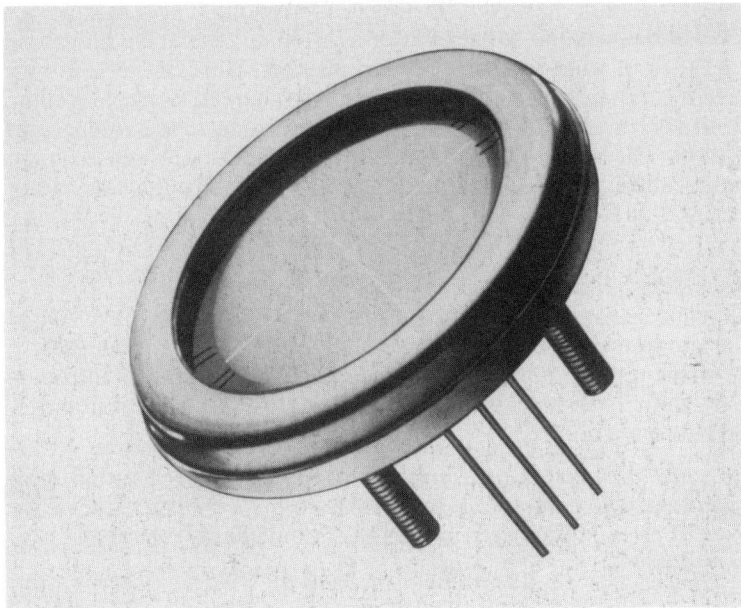

Figure 12.23 The silicon quadrant detector is a four-element photodiode. An output signal is generated when the incident light spot is not accurately centred (Photograph courtesy of Centronic Ltd)

V_x and V_y outputs proportional to the respective misalignments. An *autocollimator* uses a high-quality mirror well secured to the part to be aligned, to return the incident, well collimated light beam to its origin very accurately. This will not happen unless the mirror is perfectly perpendicular to the incident beam. Thus, perpendicularity, parallelism and flatness of various constituents of an assembly can be checked. Autocollimators were around well before either the laser or the LED was born, using incandescent bulbs. The modern EO version uses the not-so-red (infrared) radiation of an LED and photodetectors. It features many advantages over the older types. (For details, see, for example, technical information on the UDT 1000, an electronic autocollimator by United Detector Technology Inc, USA.) The older autocollimators call for two operatives (as the misalignment is read on the *fixed* part) while the modern ones and the laser/quadrant combination can manage with one. High pointing stability is a stringent requirement for laser/quadrant arrangements. The use of a beam expander improves it, by virtue of the same mechanism that reduces the divergence of a laser (see Section 10.3.1). Nevertheless, a warm-up period of the instrument may still be called for. Good instruments attain accuracies of around 10 or even 1 μrad of misalignment.

Applications of laser techniques have penetrated surveying and even such an unlikely branch of human endeavour as agriculture. Strongly concerned with the measurement of angles and distances – in addition to some levelling operations – surveying is such a highly specialised technique that it will not be covered here. Excellent books on the introduction of electronics and electro-optics into it have been published recently [280–282]. Some indications of distance measurements by laser in general will be found, however, in the next section. One example for agriculture will be included – that of laser-assisted *automatic* land levelling – to give an idea of what lasers can do for farmers. Here, a HeNe slowly rotating beam generates a plane of reference for the quasi-perfect levelling of a field, so very useful for reducing salination through the avoidance of over-irrigation. The beam guides a land scraping machine which does the rest automatically [244]. (Raising agricultural yields through the use of lasers is treated in Ref. 263.)

(c) Distance measurements

Instruments measuring the distance of a target, usually called *rangefinders*, are of interest to the military, the geodesic surveyor, the meteorologist, the space travel expert, etc. (see Section 12.3.1(c)). The introduction of lasers to these instruments has greatly improved their performance. Two principles are exploited:

1. *Elapsed time instruments* measure, by means of high-resolution electronic counters, the time a short pulse takes to make a return journey to the target. The unknown distance is thus found simply from:

$$D = \frac{ct}{2} \qquad\qquad (12.3)$$

assuming the speed of light in the medium is c. Other than in air-free space, a correction has to be introduced for $n \neq 1$ (from measured values of temperature and air pressure) for high-accuracy measurements. (n differs from unity by $\simeq 3 \times 10^{-4}$. Ref. 245 gives the formula $n - 1 = 77.6[1 + 7.52 \times 10^{-3} \times \lambda^{-2} (P/T) \times 10^{-6}]$, with P in millibars, T in kelvins and λ in micrometres, for dry air.) Obviously, the shorter the pulses, the higher the resolution but, also, the greater the bandwidth and power requirements of the system. This 'echo' technique is sometimes called lidar (Light Radar), especially in the context of atmospheric (clouds, pollutants, turbulence) measurements. Accuracies of 10^{-5} to 10^{-6} have been claimed for distances in excess of 150 km or so.

2. *Phase comparison instruments* use amplitude modulation by a radio frequency F_M of a continuous lightwave carrier. D is derived from the phase difference $\triangle \psi$ between the outgoing and the returning radiation, but only within $M \times 2\pi$ ambiguity. Using a frequency sweep makes it possible to find the integer M. It also makes possible the use of a 'null phase' method for F_M, doing away with the phase measurement as such. F_M lies in the 30 MHz region. This CW method calls for less power and bandwidth. Accuracies of 10^{-6} are not too difficult to achieve and figures of 10^{-7} have been claimed for distances in excess of 30 km [76].

Realising that a 10^{-6} accuracy corresponds to an error of one millimetre in one kilometre, or approximately 1/16 inch per mile, makes us see how respectable an instrument the laser rangefinder is.

Other radar-inspired techniques, such as the linear frequency modulation of a continuous wave (FMCW) should be translatable to laser ranging but, at the time of writing, no industrial realisation based on this principle is known to this author.

Laser rangefinders for moderate distances (say 1–3 km or so) and for work with so-called 'cooperative' targets, i.e. those fitted with retroreflectors (see Section 12.3.1.2) can use semiconductor lasers, and therefore be light and small. Those for automatic tracking of a moving target, on the other hand, have to be fitted with complex servomechanisms and will therefore be heavier and bulkier.

Finally, those for submarine-to-satellite work will call for powerful krypton or xenon lasers [176] and be bulkier still though lighter than some, for yet other tasks, which use CO_2 lasers.

12.3   Illuminating and imaging FO bundles

We return to optical fibre light guides for a brief treatment of a group of applications in which guiding *light as such* is the task assigned to them. Unlike the radiation used in optical communications or in fibre sensing, the radiation used here must, of course, always lie in the visible part of the spectrum. Also in contrast with FOCs and OFSs is the fact that illuminating or imaging fibres are used not singly but grouped in bundles (Figure 12.24) containing large (when not enormous) numbers of individual strands.

What better than a flexible light guide to illuminate in all safety the inside of a combustion engine, a narrow-bore tube or a bronchial canal, when visual inspection is called for?

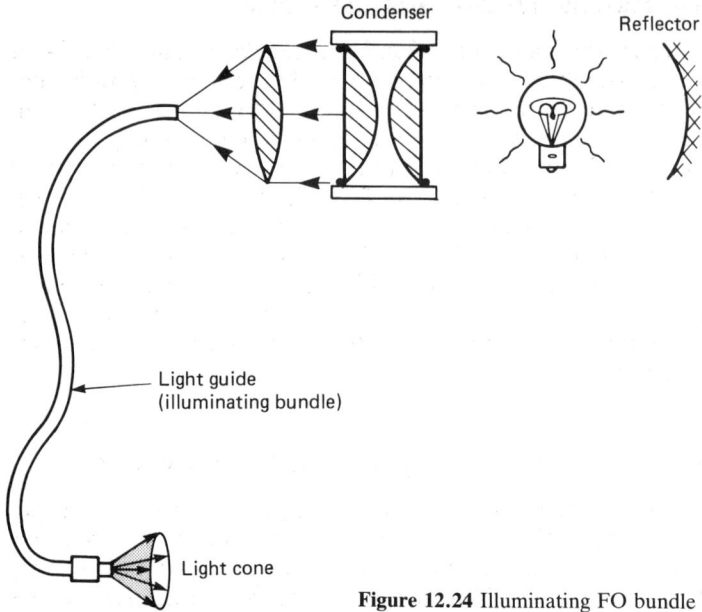

Figure 12.24 Illuminating FO bundle

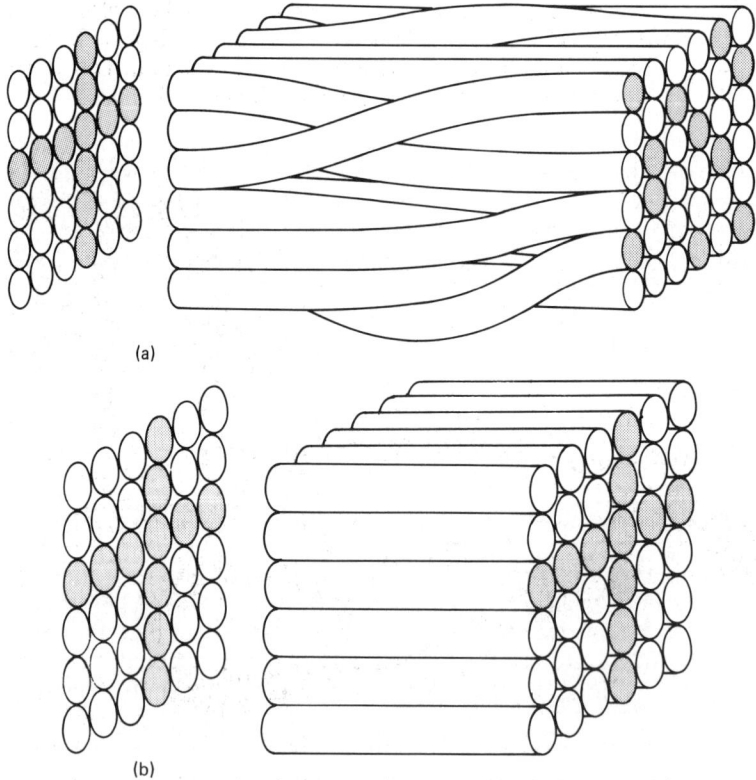

Figure 12.25 (a) Incoherent and (b) coherent light guides

12.3.1   A new meaning for the word 'coherence'

The order in which the individual strands of the *illuminating bundle* of Figure 12.24 are assembled does not matter: they can be intermingled, entwined and convoluted – the light will still come through. It is a different story, however, when it comes to the *imaging bundle*. Through it, the inaccessible object will be viewed and to avoid image scrambling a strict correspondence between in-fibre and out-fibre tips within the bundle must exist. When the strands are laid in such an orderly manner, the bundle constitutes a *coherent light guide* (Figure 12.25). A good, modern device of the kind has an active diameter of a few millimetres, is highly flexible, has a transmittance of about 50% or better for 1 m length and can resolve about 10 μm (twice the fibre core diameter per line pair). What could be more convincing than Figure 7.1?

12.3.2   Applications of illuminating and imaging assemblies

Medical and industrial instruments, using imaging light guides (endoscopes and fibrescopes (Figure 12.26), respectively) differ mostly by sheathing and accessories. Most use a combination of incoherent (for illumination) and coherent (for imaging) bundles; some are equipped with lenses, prisms and/or orienting mechanisms.

Quite different, however, is a light-guiding assembly of a face plate for a recording cathode ray tube (CRT). This assembly is a coherent bundle which grew in diameter and shrank in length to such an extent that it no longer looks like a bundle – more like a pancake (Figure 12.27).

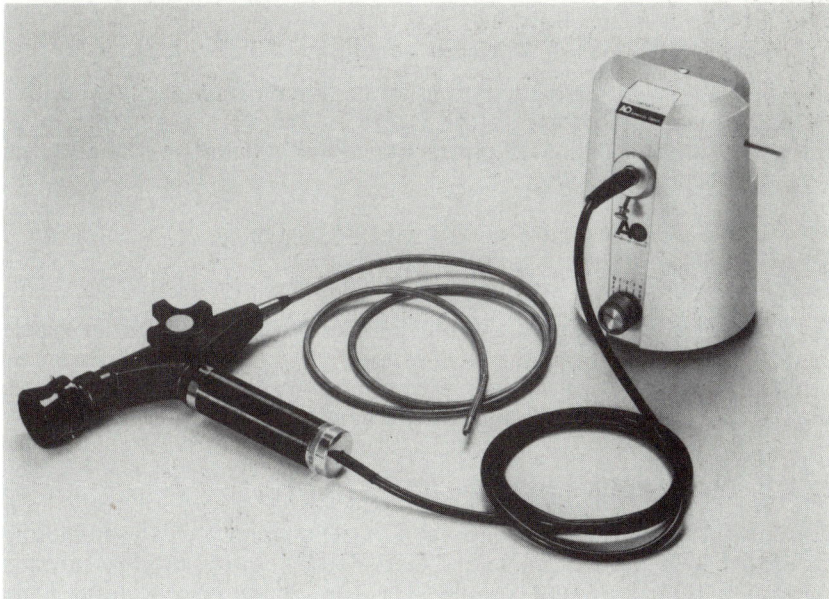

Figure 12.26 Industrial imaging light guide; FS-236A fiberscope with 1180 illuminator (Courtesy of Schott Fiber Optics Inc)

Light spot on
CRT faceplate
is 4 fibre
diameters wide.
($\sim$30 μ)

(a)

(b)

8 μm

Fibre optic
element

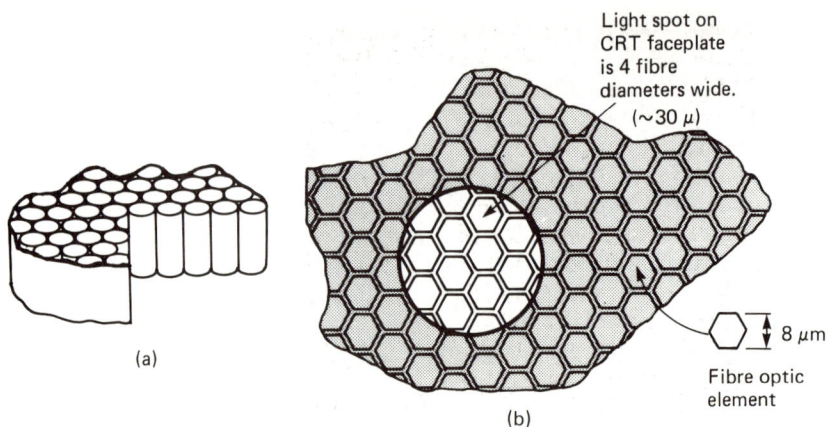

Figure 12.27 Fragment of a light guiding FO 'pancake' assembly for a recording CRT (a) before and (b) after fusing

Anyone who has tried to transfer a waveform trace from an ordinary CRT onto tracing paper by contact will have noticed the line-broadening effect. The phosphor emits light into all directions and, as the paper is separated from it by the thickness of the face plate, it is brushed not by a stylus but by the base of a cone of light. This makes it necessary to use cameras for recording CRT traces photographically. By using the FO 'pancake', not only can we skip the lenses and use contact printing instead, but we also benefit from a vastly improved light gathering efficiency. (For details of this application, see Ref. 246; and for general information on fused FO mosaics, see Refs. 2, 7 and 17.)

Two more situations in which FO bundles find applications are:

1. When 'cold' light is required, e.g. for the illumination of heat-sensitive biochemical microscopic slides.
2. In miniaturised versions of optical barriers of both the by-transmittance and by-reflectance types.

12.4 Optical data storage on disc

After the brief excursion of the previous section into the land of visible light, we return to the realm of the infrared to see how EO engineers exploit these not-so-red rays to alleviate the insatiable hunger for data storage capacity of modern information handling systems.

12.4.1 Digital optical storage

To illustrate Digital Optical Storage (DOR) systems let us examine the Laserdrive 1200 System, made and marketed by Optical Storage International Inc (OSI), a joint venture between Philips and Control Data Corporation (Figure 12.28). It uses a GaAlAs laser for writing very high density digital data on a thin metallic layer. For reading, the same laser is

Figure 12.28 Laserdrive 1200 with disc (Courtesy of Philips)

used, albeit at reduced power, in combination with a PIN photodiode. To achieve a capacity of 1 Gbyte for a 12-inch disc (equal to that of some 10–100 magnetic discs of similar size) and a user error rate of around 10^{-12}, exacting engineering problems had to be solved, embracing not only optical and electronic but also mechanical, metallurgical, logistic and control engineering disciplines. The development took several years. The result is impressive. Here are some features of the Laserdrive 1200.

(a) Recording principle
All information, whether pictorial, verbal, alphabetical or numerical, is reduced to strings of binary zeros and ones. Writing it onto the recording medium – a fine layer of a bismuth compound – is by means of ablation: a tiny circular *matt* area obtained by burning out a hole in the otherwise high-shine bismuth layer represents a one, while the unburnt and therefore highly reflective digital location will denote a zero. It can be seen that the process is irreversible, implying a read only (non-erasable) type of memory (ROM) (like most paper documents or microfilm). Each bit is checked immediately after writing (DRAW – Direct Read After Write). If an error is detected, the relevant data are rewritten – and adequately readdressed – elsewhere on the disc. (In the same way it is possible to alter a limited

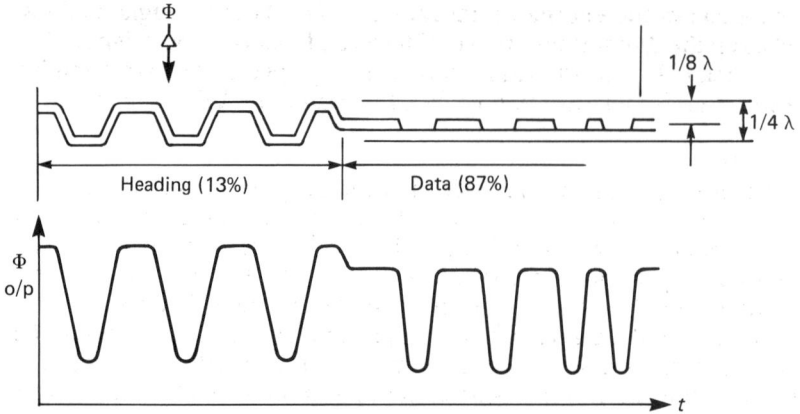

Figure 12.29 Longitudinal cross-section of track and corresponding detected signal

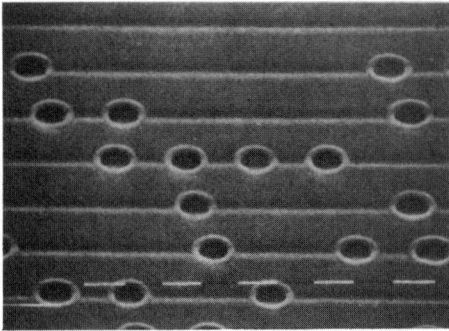

Figure 12.30 Micrograph of preaddressed and recorded tracks

amount of the recorded material by writing it elsewhere and readdressing.) The data are recorded on a continuous, tightly wound spiral, pregrooved in the polymer backing prior to the deposition of the bismuth layer (Figure 12.29), thus providing an optical 'handrail' to the write/read head. This groove contains also preformed (in the replication process) 'headings', i.e. addresses, in the shape of short sequences of binary pit and no-pit areas distinguishable from data by their reflectivity and depth (Figure 12.30).

(b) Dimension

Let us now put figures to some of the dimensions to get an appreciation of the enormity of mechanical and optical problems the designers had to tackle [240–242]. The data holes have a diameter of 1.3 μm and the between tracks distance is 1.6 μm. This corresponds to a bit density of 3×10^5 bits/mm^2 (over half a million bits per pinhead area). The pit depth is $\lambda/4$ for the headings and $\lambda/8$ for the data field ($\lambda \simeq 0.82$ μm) and the thickness of the bismuth layer is a mere 0.03 μm. Pre-grooving and pre-addressing is on a 10 μm layer of the photopolymer. Finally, two 'giant' dimensions: the thickness of the strengthening/protective glass layer is 1.1 mm and the glass lens distance, for writing and reading, is 2 mm. This last figure shows the enormous advantage of DORs over magnetic discs. Scratches on the disc surface are out of focus and, hence, mostly ignored by the opto head. No 'head crash', this haunting spectre of data handling operations that brings ruin to both head and disc, is likely to happen.

(c) Optics

The electro-optic head has four jobs to do: to keep itself on track, to keep itself in focus, to write and to read. All this out of contact, of course. It weighs a mere 40 g (~ 1½ oz) [95, 242], inclusive of laser and photodiodes. If you cannot see it in Figure 12.28, it is because the designers have, wisely, hidden it away underneath the double-sided disc. Figure 12.31 shows how the tracking error is detected. In a centred position the lens collects equal amounts of light from the left- and right-hand slopes of the groove. Otherwise, an imbalance exists, to be detected by photodiodes PD1 and PD2 which, after amplification, will feed the radial control servo. A residual error < 0.15 μm is claimed, even for a mechanical disc eccentricity of 0.1 mm! Track searching also uses optics: a fringe counter stops the linear motor controlling the radial motion of the head within 10 tracks of the wanted position, after which track number reading takes over. Both actions take, on average, 250 ms. The depth of focus is extremely small (a glance at Figure 12.29 shows that it has to be). The control of the lens–disc distance derives its error signal from the imbalance between photodiodes PD3...PDN. The result is a spot diameter ≤ 1 μm. Beam splitters 1 and 2 take care of separating the write and read operations, the latter at a much reduced laser power. Notice the isolating function of the quarter-wave plate (see Section 10.3.3(c) and Appendix 12) and the beam ellipticity correcting optics for the semiconductor laser.

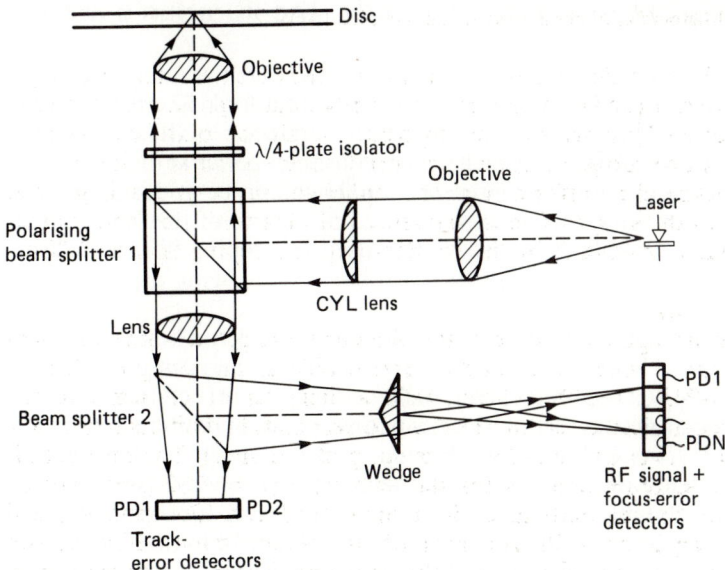

Figure 12.31 Optics of the EO head

(d)   Other aspects

The above pointers to a particular DOR system are about as much as it seems fair to include in a book on electro-optics. For details on the self-clocking scheme (which is necessary because the machine runs at a constant angular – not linear – speed) and information on totally non-optical aspects of the machine, such as modulation schemes, data format-ing, data compression (for digital picture storage), error correction, etc., the reader is referred to other sources and their particular bibliographies [240–242]. Much interesting reading can be done on optical consumer disc systems – the video VLP and the sound digital compact disc (CD) – of which this DOR is a successor.

12.4.2   Other DOR disc systems

At the time of writing, some ten firms, all American or Japanese, compete with OSI Inc (according to Ref. 242, which bases its information on a Rothschild Consultant survey and a Gerry Walter report). All offer discs of the read-only type, eminently suitable for the archival type of storage – for both text and pictures – much in demand in databases. One of them claims for their disc a capacity of 4 Gbyte and a user error rate $<10^{-13}$. Erasable media for rewritable discs are under development.

Before leaving the DOR disc, let us re-emphasize its startling storing capacity. Saying '1 Gbyte' sounds rather abstract. Saying instead '500000 A4 pages typed in single spacing' may be more meaningful: this is the amount of text contained in 2500 novels similar in size to the volume you are reading. This can be accommodated on just one disc, and 64-disc 'juke-boxes' are already on their way.

12.5   The supermarket's bar code, the librarian's magic wand

A patch of black-and-white stripes – which would look like a miniature zebra crossing, were it not for the width variations of its constituent elements – appeared on some packages of consumer goods at the beginning of the decade. By now, its presence on High Street goods is almost universal (an example can be seen on the back cover of this book). The 'bar code', as it is called, is also widely used by lending libraries. Even some publishing of computer software in bar code is being experimented with.

It takes much less time to scan the code than to read and key a price into a cash register. There is a greater accuracy, too, as the item price can be checked against the computer-held, daily updated value, countering label-ling mistakes or abuse. Automatic, paperless stock updating streamlines management. Advantages of using bar codes in other fields include increased patient safety through label checks on medicine bottles, automa-tion in programmable package and cargo routing (see, for example, Ref. 248 and its references to development work in this field in the 1970s), bus route time checks, etc. It is therefore certain that bar code identification is here to stay.

12.5.1   Code reading

Code reading has to cope with adversities such as variations of colour and texture of the substrate (not all codes are printed in genuine black and white and not always on paper), ink spread, poor contrast, presence of neighbouring text and pictures, ambient light and, above all, the demands for positional tolerance during scanning (not all packages are square, not all surfaces are flat, e.g. bottles, tins, packs of crisps, etc.).

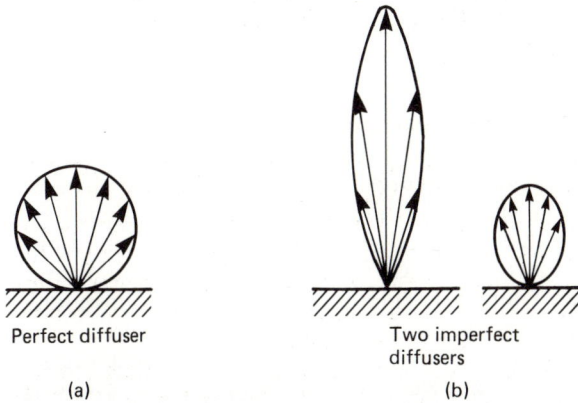

Perfect diffuser

(a)

Two imperfect diffusers

(b)

Figure 12.32 Diffuse reflection

In addition, equipment designers have to bear in mind LED and laser ageing and, most stringently, user safety. The above constraints apply to 'off-surface' scanners. Contact scanners, like the librarian's 'magic wand' have an easier life. The sensing element of the scanner is a small *reflectance detector*; an LED or laser illuminates the bar code and a photoreceiver detects the returned light, obviously stronger for the white than for the black stripes. But the term 'reflectance' does not refer, here, to the mirror-like or *specular* reflection known to us from previous chapters, rather to the backscatter of radiation, *diffuse* reflection, produced by most non-mirror-like surfaces. Figure 12.32(a) shows the polar diagrams of a perfect diffuser, behaving like a secondary Lambertian source. (See Section 4.4 and Appendix 4.) Good matt white bristol board or a chalk-covered surface comes close to this ideal. Ref. 2 quotes magnesium carbonate-covered surfaces to be 97% efficient diffusers. The Kodak card is an excellent diffusing laboratory aid. Figure 12.32(b) shows two imperfect diffusers, e.g. two grades of medium gloss paper. Scanner illumination is usually tilted to eliminate the ordinary, specular reflection, which would otherwise cause errors with glass, plastic, metal and glossy paper. Using diffuse reflection makes all item/scanner positioning non-critical.

12.5.2   Cracking the code

Several codes (five major ones) are at present in use. The most frequently encountered is the Universal Product Code (UPC) of which five versions exist. Figure 12.33 shows the Grocery A version. In it, the first two decimal

Figure 12.33 The UPC code Grocery A version. Note the presence of user-readable numerals

digits encode the country of origin (sometimes replaced by just one, the 'number system digit' used for checks, in conjunction with the check digit [81]), the next five the manufacturer, and the following five the description of the item. The last one is the all-important check digit. In the centre and at both extremities are pairs of easily recognisable guard bars. Each digit is composed of seven modules, some black, some white, forming always two black and two white bars (Figure 12.34), one to four modules wide. (For an exhaustive discussion of the UPC encoding, inclusive of the abbreviated, small size version E and of the method of calculating the check digit, see Ref. 80, part 2 and Ref. 81.) The beauty of the system lies in the method of decoding the information. After extensive processing

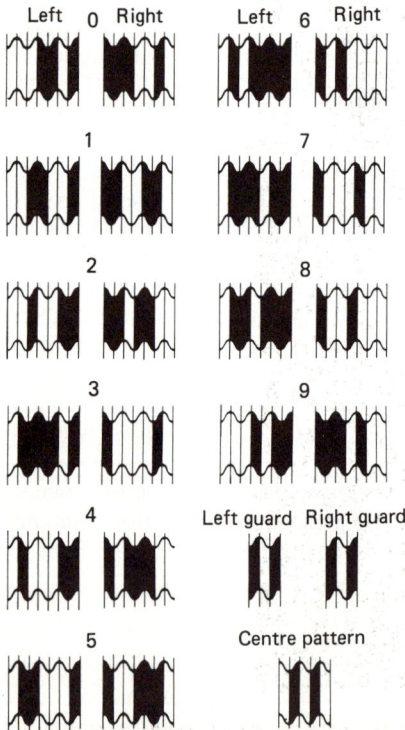

Figure 12.34 The way UPC decimal digits are formed

	UPC	
Character		*No. of width*
0		3.2.1.1
1		2.2.2.1
2		2.1.2.2
3		1.4.1.1
4		1.1.3.2
5		1.2.3.1
6		1.1.1.4
7		1.3.1.2
8		1.2.1.3
9		3.1.1.2
start/stop		1.1.1

(e.g. the signal processing used by Hewlett-Packard for their HP3000 input scanner [82]) it ends up as a series of electronic on/off signals. These are decoded by the *ratio* of the on and off durations, so that the system is tolerant not only to label size variations (1 : 2 up and 1 : 0.8 down), but also to a range of scan speeds and thereby to skewed scanning. In addition, the UCP patch can be read in both the left-to-right and right-to-left directions. It is the high degree of tolerance all round that gained the system its quasi-universal acceptance.

12.5.3   Scanning the code

Most bar code scanners fall into two categories: the *in-contact* hand-held pen-like wand (Figure 12.35) and the *out-of-contact* laser beam sweeper (Figure 12.37). The first type is suitable for light-traffic retailers, stores, libraries, computer users, etc. The second has for its major application area the mass traffic supermarket. An example of a wand reader is the HEDS 3000. Its heart, the integrated reflectance detector HEDS 1000, is of such great general usefulness that it deserves to be shown here (Figure 12.36) (see Hewlett-Packard data sheets HEDS 3000 and HEDS 1000). A red LED chip and a photodiode + transistor chip share a common substrate.

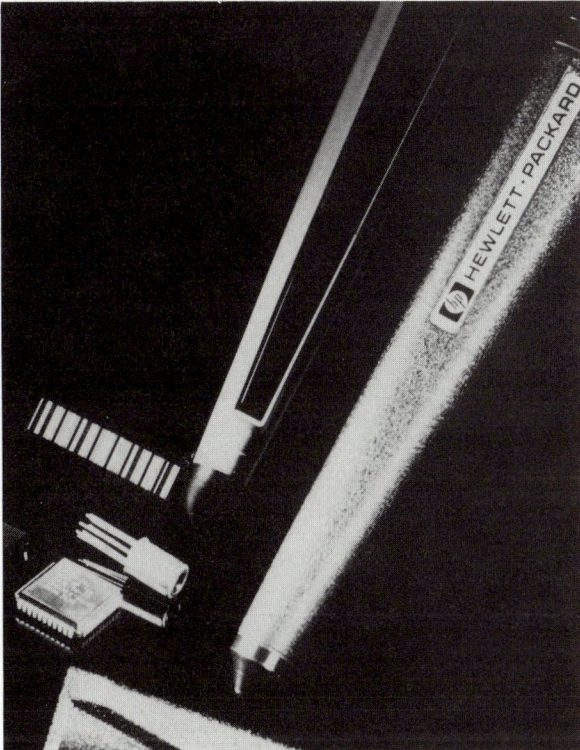

Figure 12.35 Point of sale bar code 'magic wand' reader (Courtesy of Hewlett Packard)

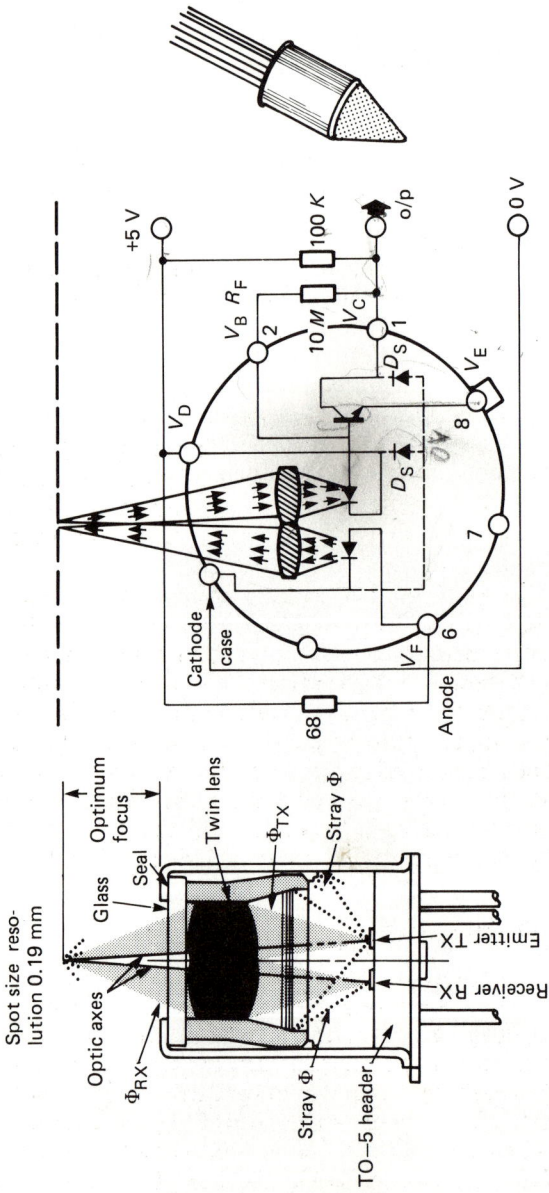

Figure 12.36 The optics, electronics and external appearance of the integrated reflectance detector HEDS 1000 (for clarity, some PIN numbers have been omitted; D_s is the substrate diode)

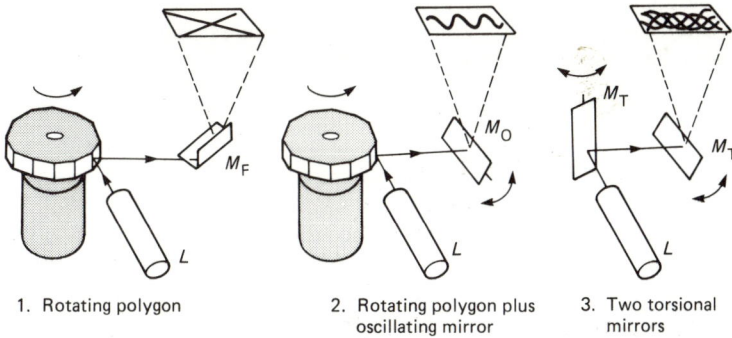

1. Rotating polygon

2. Rotating polygon plus oscillating mirror

3. Two torsional mirrors

Figure 12.37 Out-of-contact reading. The laser beam is made to trace a complex pattern on the scanned item. L, laser; M_F, folding mirrors; M_O, oscillating mirror fixture; M_T, torsional mirror fixture

A miniature twin lens combines the LED produced spot and the photodiode field of view into one area. The lens, being red tinted, doubles up as a colour filter. The whole of the HEDS 1000 fits in an elongated TO-5 can! The HEDS 3000 wand contains additional circuitry delivering a digitised TTL or CMOS compatible output. Systems receiving this output accommodate scans in either direction.

Out-of-contact reading of a bar code is effected by sweeping it swiftly with a well focused laser beam. A photodetecting system with a carefully delineated field of view registers the local *diffuse* reflectivity variations of the label as it gathers the backscatter. For secure, position-tolerant readings, several scans are necessary, a reason for which the laser beam is made to trace a complex pattern on the scanned item (Figure 12.37), generated by mechanical deflectors (see Section 10.3.2(a)). Once the correctness of the reading is checked against the check digit an audible bleep is issued. In case of heavily mutilated code bars, there is always the last resort: the human readable numerals underneath them!

12.5.4 Concluding remarks

The task of reading black-and-white stripes may appear trivial. In a point-of-sale situation it is not. The expression 'black and white' itself must sound like a near-insult to a scanner designer. The bars are *not* either totally absorbent or totally reflective. Some commercial packages are dull, others have a metallic shine. Many use colour. Effective contrast may be as low as 2 : 1. The scrutinised item can be in motion (on a conveyor belt) not always at constant speed (they may be presented by hand). Code patches vary in size in 2.5 : 1 ratio. A depth of focus of several centimetres is required. Ragged bar edges, stains and sometimes garbled printing have to be coped with.

The fact that supermarket bar code reading became possible is due both to the skill of EO designers and the capability of electronic circuit integration.

Because they have succeeded in devising equipments using no more than a humble milliwatt of laser power, these designers have made the non-dangerous laser part of our everyday life.

12.6   Laser Doppler velocimetry

'The Coloured Light of Double Stars' is the title of a paper published by Christian Doppler in 1842 [250, p. 236]. In it, Doppler stated that the pitch of sound or the colour of light changes when its source moves towards or away from the observer. (The changing pitch of a whistle of a passing train witnessed by a stationary observer is one of the manifestations of the Doppler effect.) Explained and verified, the Doppler effect served later for the calculation of the speed of stars in fast, radial with regard to the earth and therefore otherwise undetectable, motion. Later still, E. Hubble's work on the comparative speeds of 'near' and distant galaxies [249, p. 117] based on the Doppler effect, fuelled the Theory of the Expanding Universe. Today, Doppler's frequency shift serves more earthly purposes: the *unobtrusive* determination of the velocity of a fluid or gaseous flow in process control, the *out-of-contact* monitoring of the motion of a piece of machinery or a hot metal sheet in industry, or the detection of the speeding motorist on the motorway by laser Doppler velocimetry (LDV).

12.6.1   Physical explanation of the Doppler effect

Consider a coherent radiation directed by the source S onto a reflective target R, with R moving towards S with a velocity v. Think now in terms of wavefronts. Not only does the reflector come towards the incident wavefronts (Figure 12.38), it also chases the reflected ones. The number of wavefronts a hypothetical stationery R would be struck by, every second, is c/λ. The number of wavefronts of a supposedly stationary wavetrain that the moving R would encounter every second would be v/λ. In the real

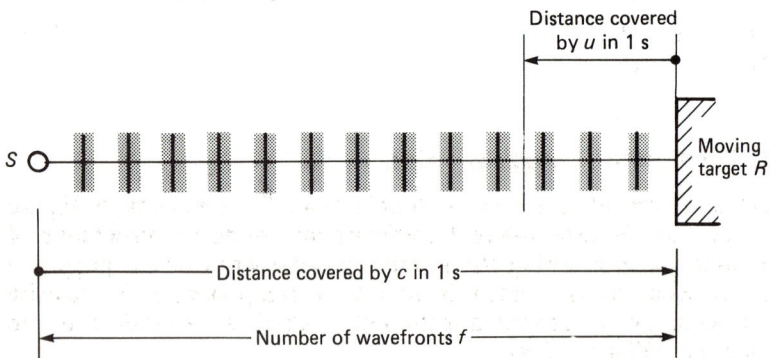

Figure 12.38 Situation after a 1 s observation

situation, with both wavetrain and R moving, the frequency of wavefront/target encounters, f_R, is:

$$f_R = \left(\frac{c}{\lambda}\right) + \left(\frac{v}{\lambda}\right) = \frac{c+v}{\lambda} \qquad (12.4)$$

$$\underset{\substack{\text{immobile} \\ \text{reflector}}}{} \quad \underset{\substack{\text{immobile} \\ \text{wavetrain}}}{}$$

Thus, through its motion, the target gains v/λ wavefronts per second or, frequency-wise, experiences a shift of:

$$\Delta f = \frac{v}{\lambda} \qquad (12.5)$$

If we analysed similarly the effect of this motion on the wavefronts reflected towards a stationary observer (here a photoreceiver) located very closely to S, we would find that owing to 'chasing' they are more crowded in space than were the incident ones. The wavelength has shrunk by a fraction $\Delta\lambda = v/f_R$. The combined effects of the target motion produce an overall frequency shift:

$$\Delta f' = 2\Delta f = \frac{2v}{\lambda} \simeq f\frac{2v}{c}$$

an approximation permitted by the fact that in practice $v \ll c$. (The exact value of the shift works out to be $f\,[2v/(c-v)]$; a relation the enthusiastic reader may wish to check.) Therefore:

$$\Delta f' = 2v\frac{f}{c} \qquad (12.6)$$

Impressing both, the primary and the reflected, frequency-shifted, radiations onto the photoreceiver results in *heterodyne mixing*. Responding to luminous intensity, a very ordinary photodiode will oblige by detecting the beat frequency $2v(f/c)$, or more practically:

$$F_B = \frac{2v}{\lambda} \qquad (12.7)$$

This, in the case of a HeNe laser, gives the target velocity:

$$v = 0.3165 \times F_B \times 10^{-6} \qquad (12.8)$$
$$\text{(m/s)} \qquad\qquad \text{(Hz)}$$

12.6.2 Practical laser Doppler velocimeters

Practical arrangements use oblique illumination. The schematic of Figure 12.39 shows the Goldstein–Kreid configuration, quite representative of systems used for monitoring the motion of solid webs (steel, paper) in industry. In such arrangements it is the $v\cos\alpha$ component of the velocity v, the projection of v onto the direction of c, which is responsible for the frequency shift (Figure 12.40).

Owing to the dual beam arrangement (Figure 12.41), while one of the projections is additive (beam A), the other is subtractive (beam B). As the

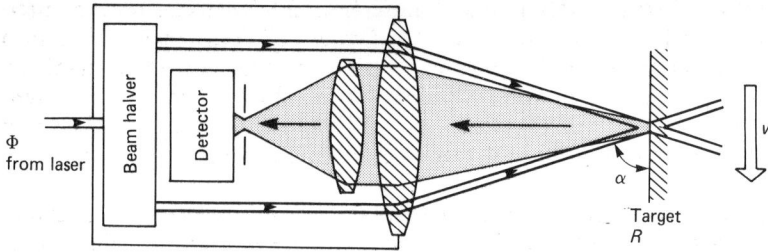

Figure 12.39 Goldstein–Kreid configuration

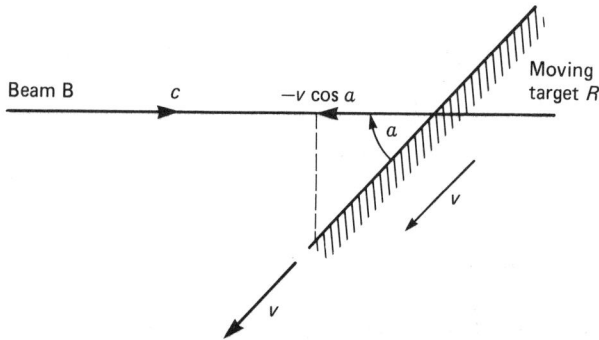

Figure 12.40 In an obliquely moving target $v \cos \alpha$ is responsible for the frequency shift

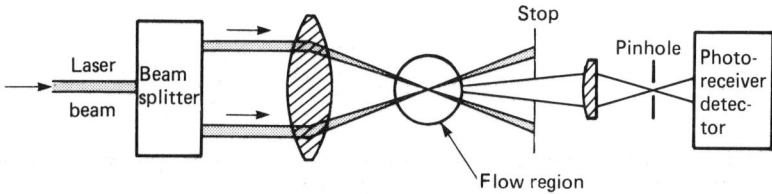

Figure 12.41 A DLV configuration for monitoring liquid or gaseous flow

two Doppler shifted frequencies beat, a heterodyne action takes place. The difference frequency:

$$\Delta f' = \frac{2v \cos \alpha}{\lambda} \qquad (12.9)$$

(within the previously mentioned restriction $v \ll c$, true for all industrial applications) is detected by the intensity-sensitive photoreceiver collecting the backscatter.

One cannot help being impressed by the fact that the arrangement yields output frequencies right down to the audible range, while the carrier has an f value of hundreds of terahertz ($\approx 4.7 \times 10^{14}$ Hz for HeNe). With a HeNe laser, the target velocity in m/s is:

$$v = \frac{0.3165}{\cos \alpha} F_B \times 10^{-6} \qquad (12.10)$$
$$\text{(m/s)} \qquad\quad \text{(Hz)}$$

Velocities as low as 0.001 m/s and as high as 0.5 km/s can be measured (Polytec [255]). The introduction of a Bragg cell frequency shifter into one of the beams A or B makes it possible to detect speed reversal, by introducing a Δf offset, e.g. the TSI Inc (USA) machine. The system differs little for monitoring the flow of liquids or gases, mostly so in that forward scatter and not backscatter is now impressed onto the photoreceiver.

While with solids scatter is produced by surface roughness, liquids and gases are nearly always graced by the presence of small solid particles (e.g. smoke), which oblige in the same way; in any case, some can always be added.

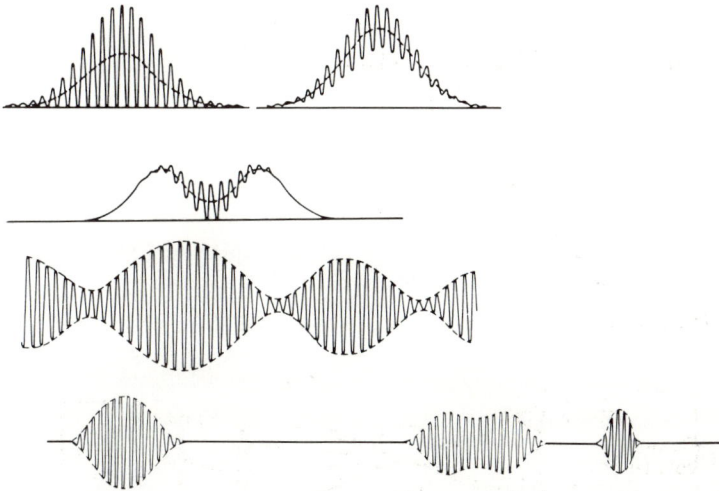

Figure 12.42 Typical patterns of raw (untreated) LDV outputs. Note the presence of d.c. in the two upper traces and of 'drop-outs' in the lower traces

It is important not to underestimate the part electronics play in real LDVs. The untreated Δf signals are weak, sullied by noise, d.c., 'drop-outs', etc. (Figure 12.42). Yet, what the user needs is a clear, unambiguous reading of v. This calls for sophisticated signal processing (see, for example, Ref. 252, chapter 6, for details).

12.6.3 Closing note on the LDV principle

Some readers may find it satisfying to know that the working of the laser Doppler method can be explained very lucidly without invoking heterodyne mixing. Forget about teracycle frequencies and think simply of the interference fringes produced on the target by beams A and B (Figure 12.43). Their pitch works out to be

$$d = \frac{\lambda}{2 \sin \left(\dfrac{\pi}{2} - \alpha \right)}$$

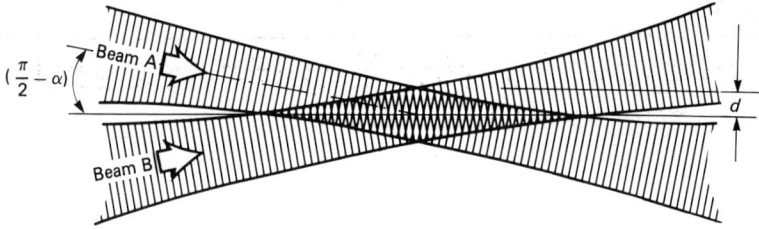

Figure 12.43 Interference fringes produced by the intersecting beams A and B

which is, of course, no other than

$$d = \frac{\lambda}{2 \cos \alpha}$$

Count now the number of fringes a scattering 'high spot' of the target crosses in a second, and you will find it to be

$$f = \frac{2v \cos \alpha}{\lambda}$$

exactly the same answer as that given for $\triangle f$ by Equation 12.9!
  For further reading on LDVs, see Refs 252, 228 and 283.

Chapter 13

Holography, Fourier Transforms and integrated optics

13.1 Three-dimensional imaging by holography

Holography was conceived and initially developed by Dennis Gabor (Imperial College of Technology, London) in 1948, more than a decade before the laser. This achievement (for which he received the Nobel Prize in 1971) would have remained no more than a scientist's interesting contribution to wave optics without the laser. With it, it became an

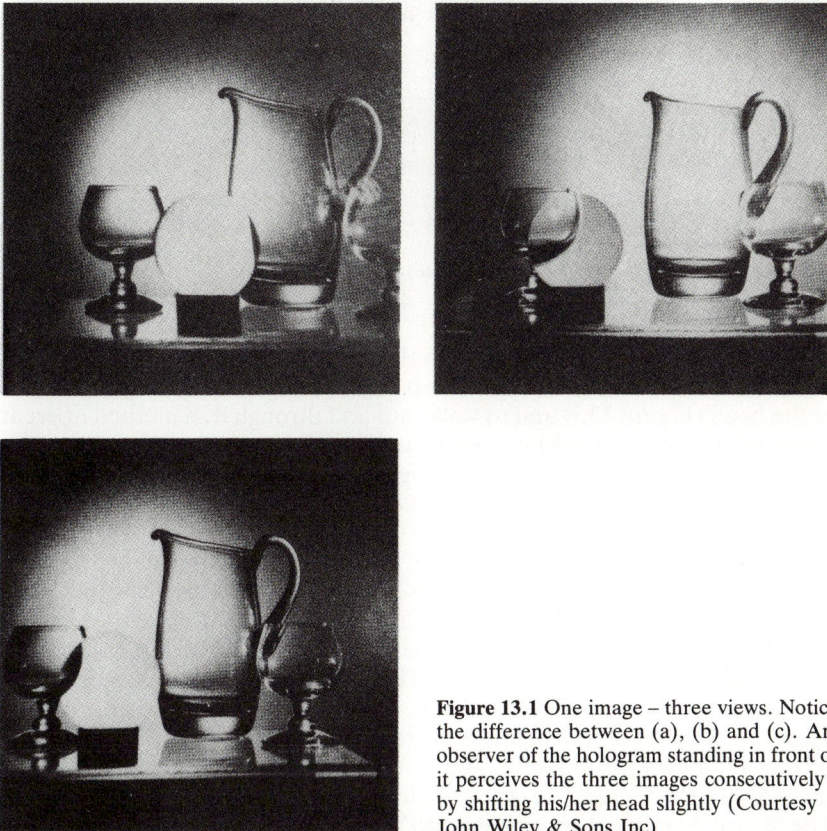

Figure 13.1 One image – three views. Notice the difference between (a), (b) and (c). An observer of the hologram standing in front of it perceives the three images consecutively by shifting his/her head slightly (Courtesy John Wiley & Sons Inc)

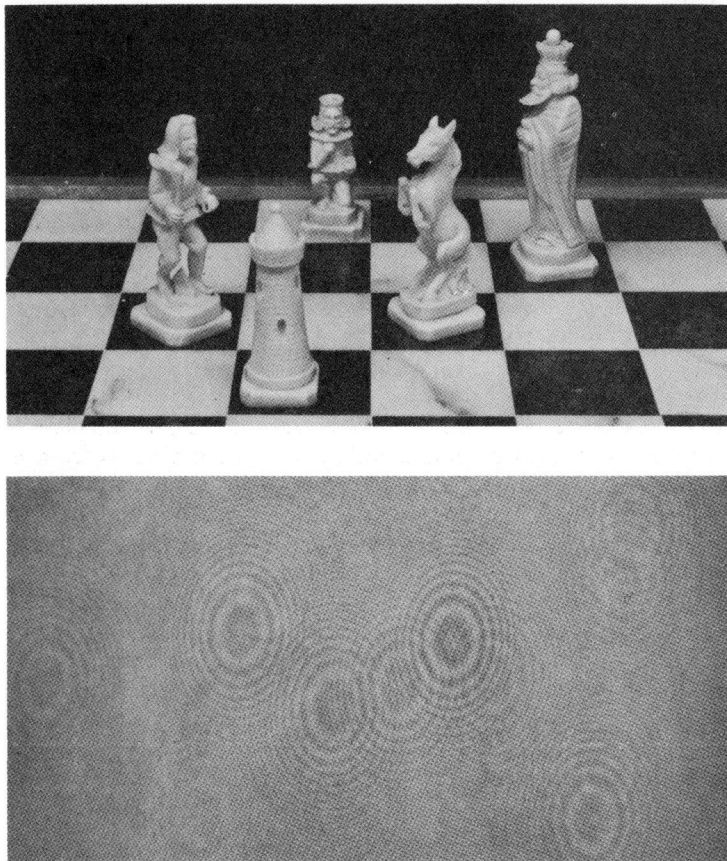

Figure 13.2 Scene (a) and its hologram (b) bear little resemblance

amazing method of three-dimensional imaging with full, rich perspective effects, enabling us to catch sight of an object behind another by a mere tilt of the head (Figure 13.1) and to walk into and through it, a method of great value in pattern recognition and in industrial interferometry, a method with enormous potential for extra high capacity information storage. So astounding are the effects of laser holography that it is often referred to as 'Light fantastic'.

When an object is illuminated by a light beam, much of the light is re-emitted into space, its wavefronts reshaped, redirected and intensity modulated. Photography is only a second-best image recording process, as it performs no more than a two-dimensional mapping of localised directional reflectivities of the object. The phase of the wavefronts, truly representative of the object's shape, is lost. Holography, on the other hand, records (in a coded way) the re-emitted wavefronts, making their subsequent decoding possible. The word 'subsequent' hints at the two-step character of the technique: first encode (to record), then decode (to view). The recording is made on a photographic plate. Without the decoding step, however, a hologram of a complex, irregular object is – or, rather, appears to be – no more than a shambles of light and dark areas (Figure 13.2).

13.1.1 How is it all done?

A hologram (holographic recording) is an ensemble of interference fringes obtained by projecting onto a photographic plate simultaneously two wavefronts originated by the same laser: one directly, the other after its reshaping by the object. This is Step 1 – encoding. The photographic plate will have responded to the local irradiances, I. These are proportional to the square of the sum of the fields E_o (object) and E_r (reference) (see Figure 13.3).

The viewing (Step 2 – decoding) relies on diffraction. The ensemble of interference fringes of the recording – the developed photographic plate – are used as a diffraction grating, illuminated by the (or a similar) reference beam in a darkened room (Figure 13.4). At this stage, all is admirably revealed: viewed through the recording, a faithful replica of the object is seen as through a window brightly, with every tilt of the head changing the scene slightly, as if through a real window. The code is unscrambled. The wavefronts of the object have been resuscitated.

It becomes clear, now, why Gabor had to wait for lasers for good holography: wide monochromatic beams with good coherence (spatial and temporal), and sufficient power were required and no pinhole-and-arc combinations could provide them. Note that the holographic process as such is *lensless* and whatever lenses might be used with it fulfil ancillary functions only, such as beam expansion. As for the name, it stems from the Greek word *holos*, meaning 'whole', in recognition of the fact that the process records both *amplitude* and *phase* of the reflected wavefront, as opposed to amplitude alone, in photography. In a hologram, the disturbance created *wholly* characterises the disturbant.

(a) Two basic examples

To make holography more tangible, two examples of producing a recording and reconstructing the image will now be given. Let us start with just a single point in space, its hologram and its reconstruction. We begin by looking at Figure 10.19 and remembering that the radiation issuing from a point – whether self-luminous, an illuminated aperture or a scattering centre (Figure 13.5) – produces spherical waves. To make a hologram we project onto a photographic plate, simultaneously with these spherical waves, plane waves derived from the same source (say by using a beam splitter and a set of mirrors, not shown for simplicity's sake). An interference pattern forms on the plate – in this case a set of concentric, ever-increasing circles, with a reducing growth pitch – as can be clearly seen from the equiphase intersections of the two waves. The thickness of the growing circles reduces too. (Such plates are of some importance in optics. When made up of full circles they are called Fresnel zone plates and were, of course, known well before holography. For a hologram of a point to yield a genuine Fresnel zone plate, the object must be placed on axis of the reference beam.) This pattern is uniquely representative of the point

262

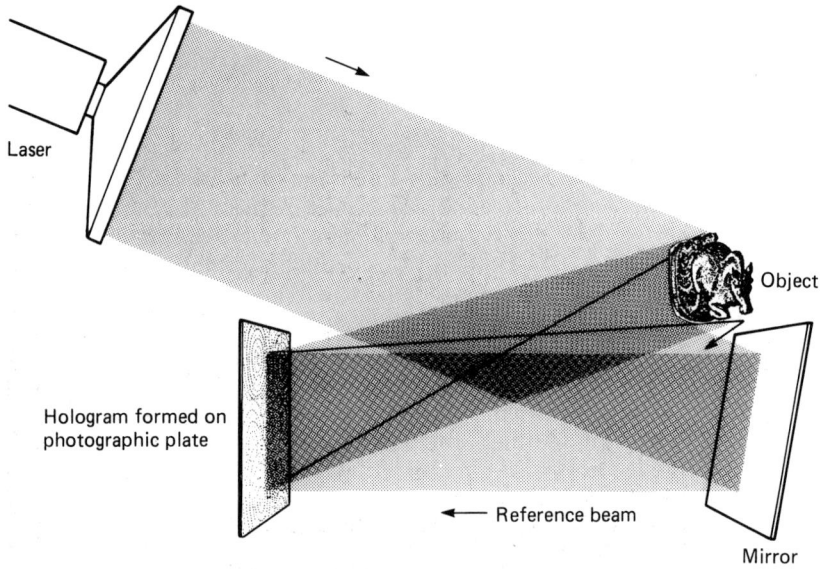

Figure 13.3 *Step 1*. The hologram is an ensemble of interference fringes formed on a photographic plate by projecting onto it simultaneously two wavefronts originated by the same laser – one directly, the other after its reshaping by the object (After Schawlow, 1969)

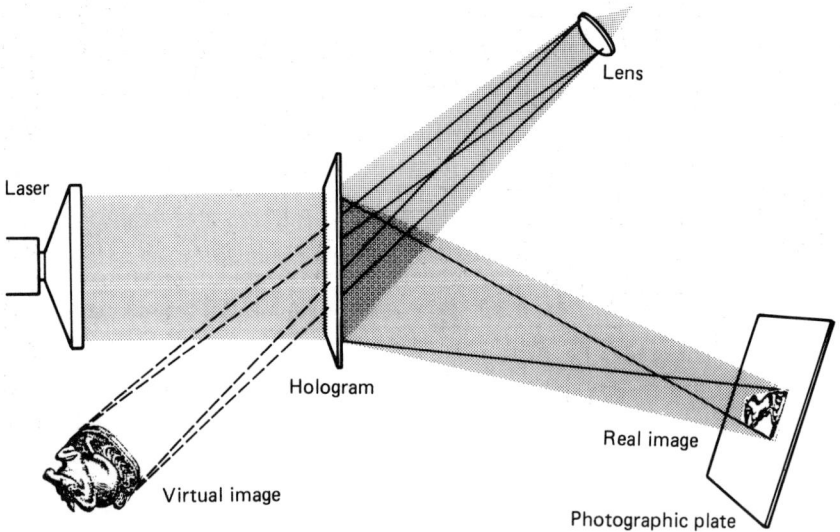

Figure 13.4 *Step 2*. The viewing relies on diffraction. The hologram is illuminated by the same (or similar) reference beam in a darkened room (After Schawlow, 1969)

Photographic plate, irradiated and developed. Shown in black are areas of constructive interference (maximum irradiance) of cross section X – X of plate.

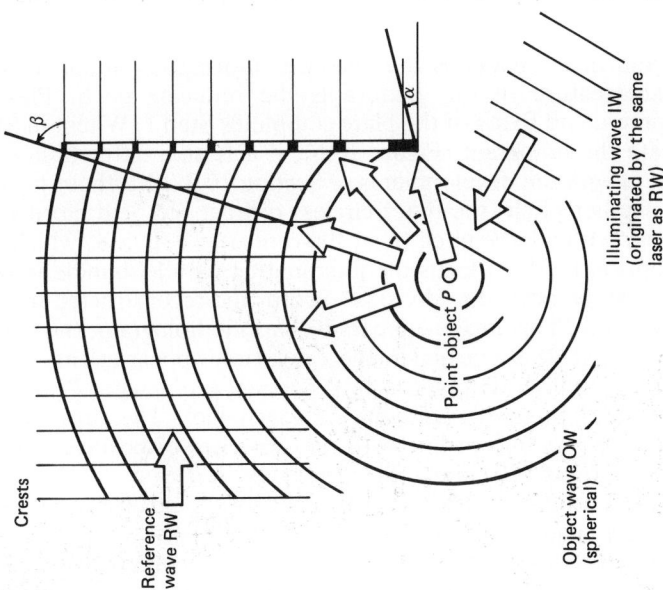

Crests

Reference wave RW

Point object P O

Object wave OW (spherical)

Illuminating wave IW (originated by the same laser as RW)

Photographic plate after development

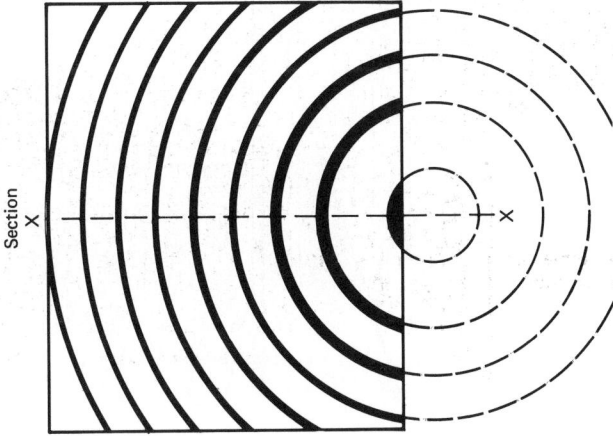

Section X

Resulting hologram

Figure 13.5 Hologram of a point. Step 1 – the making of a hologram. The diffraction of a plane wave by a point object produces spherical waves. Interfering with a reference plane wave the spherical wave produces rings on a photographic plate

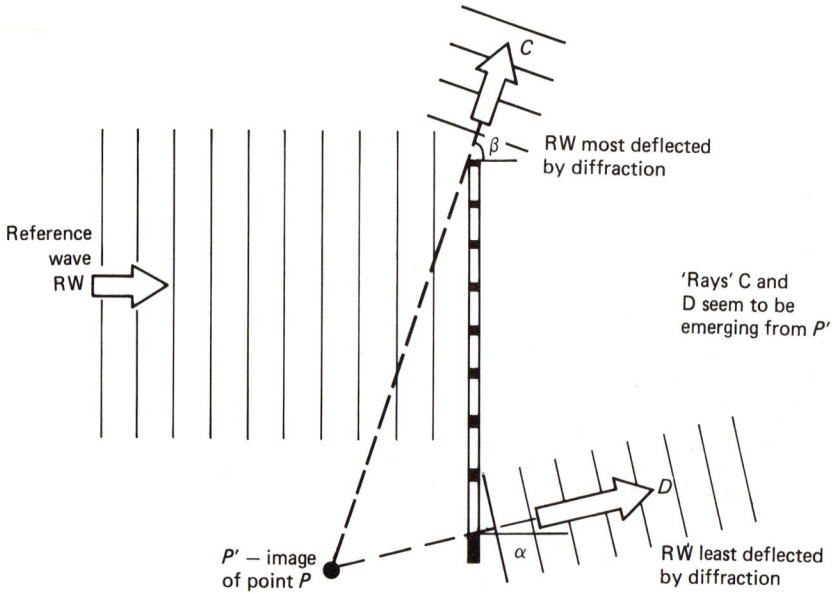

Figure 13.6 Hologram of a point. Step 2 – reconstructing the image. Diffraction of the plane wave RW by the hologram produces in P' the image of point P

object and hence contains the encoded information about it. The most important feature of this pattern is the reducing pitch. Photographic development and fixing of the plate completes Step 1. When, in Step 2, we illuminate the hologram (Figure 13.6), it acts as a diffraction grating. In agreement with the developments of Section 10.3.2(b), those portions that have a smaller pitch (the outer circles) diffract the field more vigorously than those with a large pitch (near the centre), i.e. $\beta > \alpha$, which, in terms of ray optics, tends to focus the illuminating parallel bundle into a single point. It can be proved to do so. P_1 is thus the reconstructed image of the point object P. This image appears *behind* the hologram, the 'window' of our earlier text. It is a virtual image – rays only appear to emerge from P_1; it will not be obliterated by touch. A second, *real* image, P_2, is formed in front of the hologram. This one is 'touchable'. The formation of two conjugate images falls in line with the existence of positive and negative diffraction orders of Section 10.3.2(b).

The above example of holographic imaging should clarify the working of the *hologon* beam deflector of Section 10.3.2(c).

Let us consider now the hologram of a plane by taking the idealised case of a large (to avoid edge effects), flat, first surface mirror. This hologram will, of course, be a row of equispaced opaque bars on a clear background, resulting as it does from the interference of two coherent plane waves (Figure 13.7(a)). More precisely, it will be a sinusoidally modulated transmittance plate as shown here and as per Figure 10.20(b), and thus, effectively, a regular diffraction grating. The reconstructed wave, in this case, is a plane wave (Figure 13.7), in accordance with Section 10.3.2(b) on

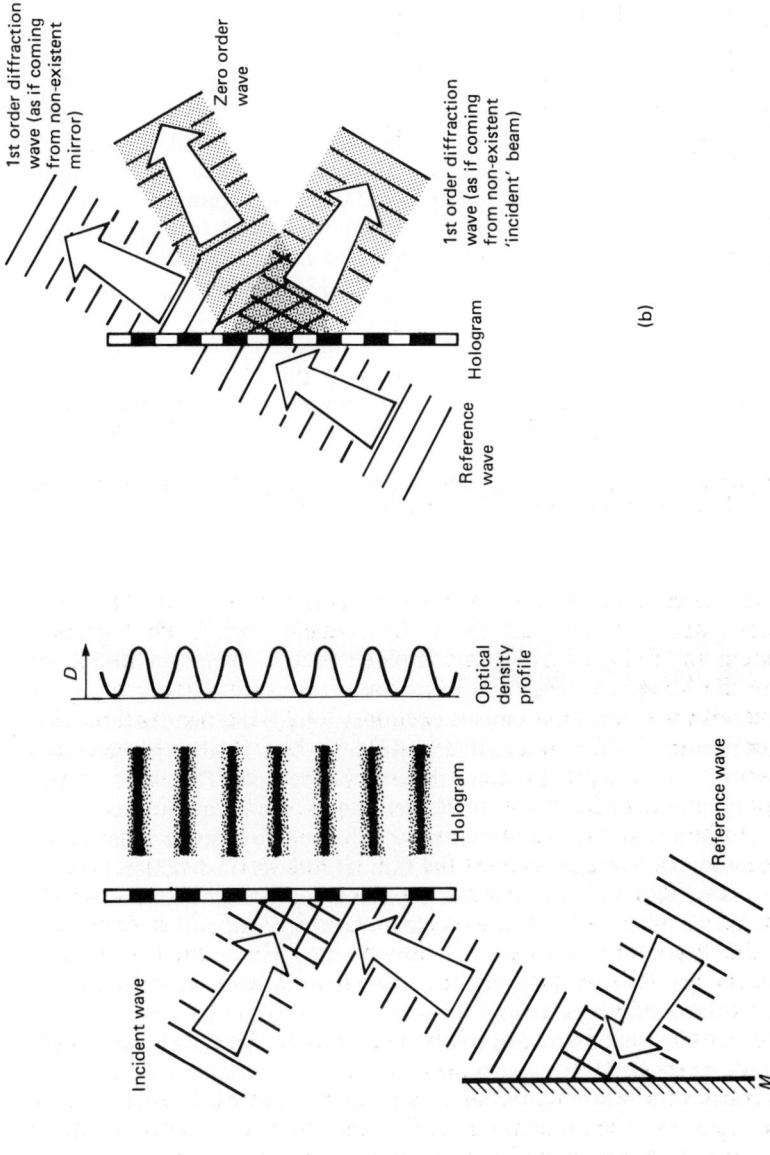

Figure 13.7 Hologram of a plane (a) and its reconstruction (b)

diffraction. As in beam deflectors, the sinusoidal character of transmitt-ance of the hologram helps to eliminate the higher diffraction orders, and the reconstruction yields an image of a plane surface, located identically to the original object, plus its conjugate.

(b) 'The real thing'

To understand how holograms of three-dimensional objects seemingly create concreteness out of thin air, we can extend mentally the example of a point object by taking all the elementary points it is composed of, one by one. From Figures 13.5 and 13.6 it will be seen that the virtual image of a point P_1 farther away from the plate than point P will appear in the reconstruction farther away from it too, and will give a genuine impression of depth by appearing to the right of P, when viewed from position K, disappearing behind P when viewed from point L and appearing to the left of P to a viewer positioned in M (Figure 13.8). Obviously the same reasoning applies to parallel and perspective effects in the 'vertical' direction (orthogonal to the plane of the drawing). At this stage it is proper to mention that the holographic image of a three-dimensional object, called 'real' in the above type of reconstruction, is in fact less real than its conjugate, the so-called 'virtual' one, inasmuch as it is a 'turned-out-glove' or a 'hollow-cast' representation of the truth: the image of an empty cup looks like a seaside sand pie made with it (a *pseudoscopic* image to the experts). The virtual image alone looks like an empty cup. (A little reflection will show why this is so.) A rearranged geometry of the process makes it possible, however, to make the 'real' image look genuinely real.

13.1.2   Aspects of holography

One of the many astounding aspects of holography is the fact that a broken hologram retains most of its original usefulness. This feature makes it into a much safer preservation medium than a photograph. The light diffused by each elementary region of the object covers the *entire* surface of the hologram – quite unlike in photography, where every single point of the object illuminates a single point of the film. In the reconstruction process, an *incomplete* hologram will produce a *full image*, though with the penalty of a certain loss of definition. Should, therefore, a portion of the hologram accidently break off and be lost, the consequence will be far less dramatic than in the case of a photographic plate from which, for example, a member of the family or a whole wing of a house could be gone for ever.

The size of the reconstructed image depends on the wavelength of the readout wave. Indeed, in agreement with the developments of Section 10.3.2(b) and more particularly with Equation 10.11, the diffraction angle of the incident 'rays' of the reconstructed wave will be proportional to λ. It follows that a reconstruction obtained with a red laser ($\lambda_{HeNe} = 633$ nm) of a hologram recorded with a blue laser ($\lambda_{Ar} = 488$ nm) will be approx 1.3 times larger than the object.

Building *secrecy* into a hologram is quite easy: make the reference wave other than plane and no meaningful reconstruction will be obtained unless the readout wave is the same. We have a key-to-fit-the-lock situation.

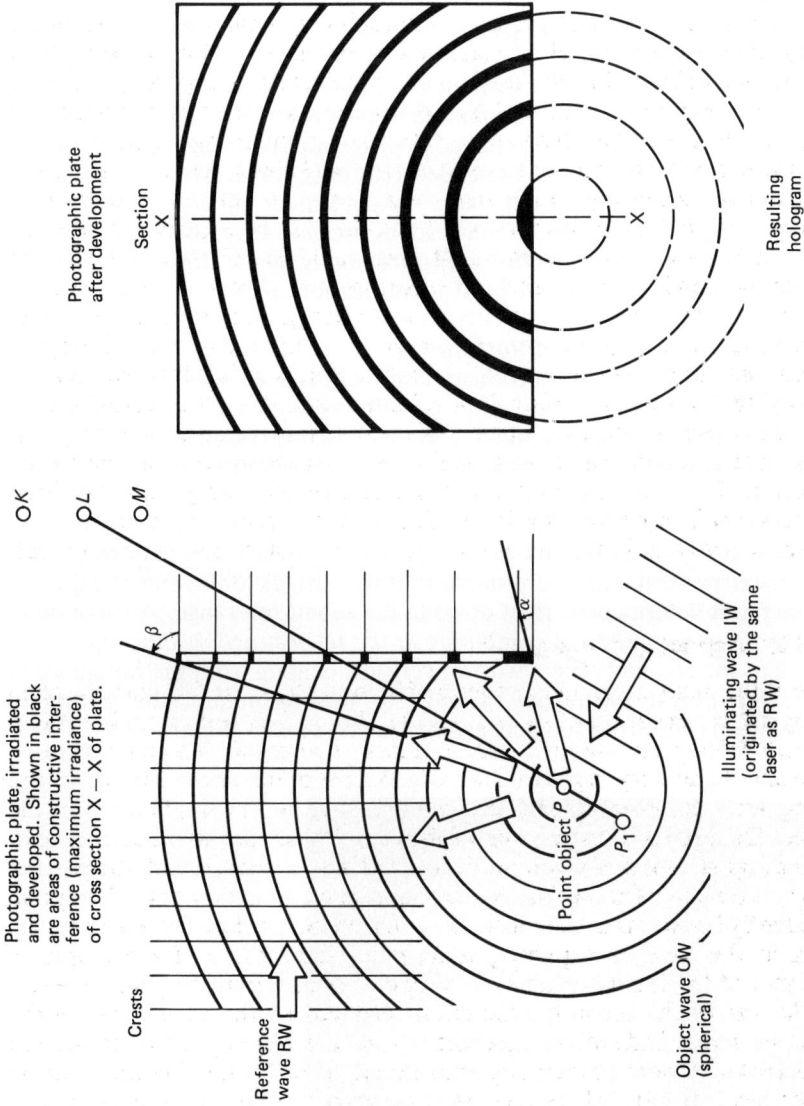

Photographic plate
after development

Section
X

Resulting
hologram

Photographic plate, irradiated
and developed. Shown in black
are areas of constructive interference (maximum irradiance)
of cross section X – X of plate.

Crests

Reference
wave RW

Point object P

Object wave OW
(spherical)

Illuminating wave IW
(originated by the same
laser as RW)

Figure 13.8 Creating the impression of depth

Wavefront shaping code marks can be quite complicated and the password (almost) impossible to imitate.

13.1.3   Applications

(a)   Associative record identification

The above way of thinking paves the way to a holographic technique of grouping objects which share a common characteristic. Here – astonishingly again – we do holography without even an external reference wave. That portion of the object which is common to the group (Ref. 223 p. 104 quotes the example of a specifically shaped tail of a variety of pigeons) acts like a code mask by generating the encoded reference wave. The light scattered by a similar tail of any other pigeon – and a similar tail alone – will reveal all the birds of the species, in the readout process. From here, a short step takes us to *associative record identification* in general, such as optical character recognition, finger print identification, etc.

(b)   Making holograms interfere

Small deformations can produce large effects, thanks to *interference holography*. Even complex objects with nearly matt surfaces can be investigated this way. Take the hologram of an unstressed object, stress the object *in situ* and take a second hologram, superimposed on the first one. The readout of this double-exposure hologram will reveal an interference pattern, the fringes of which are witnesses of the stress-caused deformations. A fringe count and a little experience will tell not only where the stress was greatest and where smallest: it will also yield the numerical value of the finest deformation, right down to the submicron range. A variation of this technique consists of combining *at the reconstruction stage* the wavefronts of the *stressed* object with wavefronts coming from the hologram of the same object *unstressed*. Thanks to this technique, the effects of stressing an object (or subject) can be viewed *in vivo* (Figure 13.9). This is

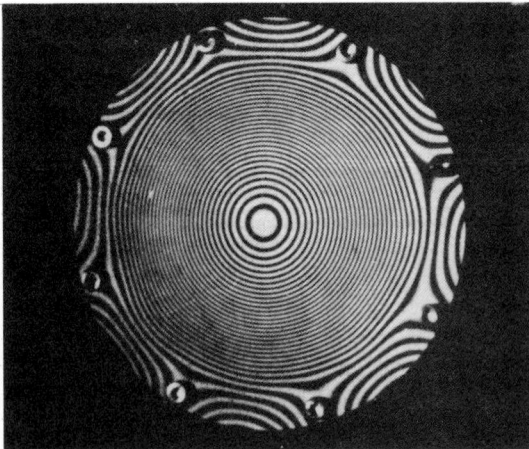

Figure 13.9 Deformation of circular membrane visualized

real-time holographic interferometry. Other refinements of the technique permit the observation *in vivo* of vibrating objects, such as parts of musical instruments or turbine blades [192], to reveal existing or potential crack areas.

Holographic interferometry constitutes today a proved, accepted method of engineering inspection.

13.1.4 Volume holograms

For all the applications so far mentioned, fine grain *thin* photographic emulsions are required in order to record the many intermingled interference fringe fields, pitched down into the micron region, produced by intricate three-dimensional objects. *Thick holograms* (called also *volume holograms*) not only exist, but hold an important place of their own. Here, too, fine-grain emulsions are required. A thick hologram emulsion is several times thicker than the fringe pitches. It records interference layers transversely. The thick hologram is reminiscent of the electro-acoustic deflector cell (Section 10.3.2(b)), although, in the former, the diffracting planes are permanent. Here too the Bragg condition needs to be satisfied – at the readout stage – for good brightness. Meaningful readouts will be obtained for only that wavelength used in the recording process. This makes us suspect that lambda multiplexing might be possible. Indeed it is. By illuminating several objects successively, with different lambdas, multiple holograms can be obtained, to be read out selectively. By illuminating the same object successively, with say red, green and blue lasers, we can record three holograms of the object in the same thick emulsion and reconstruct a *colour* (three-dimensional) image by using a simultaneous 3-lambda readout. Claims are even made to obtaining such readouts with white light sources [223, p. 107]. (For more in-depth reading on volume holography, see, for example, Ref. 193, pp. 27–32.)

13.1.5 Holography and computer technology

Computer technology can both benefit from and be beneficial to holography. This interplay is so interesting and has so much potential that it simply cannot be omitted from a text like this.

(a) Helping computers

Optical *static* (as opposed to *rotary*) mass memories with short access time can be obtained by means of two-dimensional arrays of small (2–5 mm diameter) holograms, storing some 10^5 bits each. The desired hologram is addressed by a beam deflector, for both writing and reading. In the write mode, both the object (data) and reference beams are directed onto the selected area while an on/off beam modulator imprints the ones and zeros onto the recording medium. In the read mode the reference beam is used alone. Access to data is then parallel, with a real image of the reconstructed object being optically magnified and projected onto a two-dimensional array of photodiodes. Apart from the very high storage density with high readout rates, data storing holograms present advantages of low sensitivity to scratches, relatively low positional tolerances and the potential of volume holography multiplexing. The weakness of the scheme lies in the processing of the transparencies. At the time of writing it is this that limits the applicability of holographic data storage to Read Only Memories (ROMs). Much materials research, however, is in progress to overcome this difficulty [192, 223].

(b)   Being helped by computers

The second aspect of the computer–holography give-and-take relationship concerns Holographic Optical Elements (HOEs). This, probably the youngest branch of holographic optics at present, bears upon the fabrication of diffractive optical components.

Remember that the hologram of a point, the equivalent of a Fresnel zone plate, is capable of making a parallel beam of light converge. It is, in fact, an HOE equivalent to an extra-thin positive lens made of extremely high refractive material ($n \simeq 100$, lambda dependent). Other diffractive optical elements can be configured capable of fulfilling many useful functions in optical systems. Various methods of fabrication are possible (e.g. by mechanical, chemical, electron beam or ion beam pattern formation). Computer-generated HOEs (CGHs) are first either projected on a CRT or put out on a digital plotter and then optically reduced. The flexibility of computer programming is such that not only can CGHs be produced with a great variety of single functions, but they can also commit a single layer of film to *multiple* functions. A *thin* CGH with the resolution of some 200 lines/mm can already be produced and 2000 lines/mm are hoped for. Some *thick* HOEs have been reported with up to an incredible 550 addressable (by variation of the angle of incidence of the RW) 'pages', though these particular HOEs were not computer generated.

The list of materials for HOEs is long; thermoplastics look promising. Even reflective single surface write/read HOEs have been obtained with them. The manufacturing problems, however, are still very great. The most promising among the single components is the large collimator. (For more details on HOEs and CGHs see Ref. 222 on which this section is based.)

Hopes are high for 'whole optical systems designed from the keyboard' (to quote Ref. 222), even for HOEs that would work with incoherent light.

13.1.6   Concluding notes

For the mathematics of the basic holographic process, see Refs 17, 192, 193 and, more specifically, 221, which is a tutorial review of holography written in terms familiar to electrical engineers. It contains 109 bibliographic references.

Holography is a fascinating technique of great intellectual elegance – a technique based on strongly original and highly creative thinking. The applications, as explained, are many. Regretfully, our enchantment is tinged with disappointment for its not having led – yet – to practical, broadly used three-dimensional television or cinema projections. The challenge is open.

13.2   Frequencies and filters both temporal and spatial. Fourier Transforms by lens

13.2.1   Spatial frequency

To most people, the term 'frequency' tacitly implies a *temporal* frequency. This is as true for the common notion of the number of buses per hour as it is for the scientific concept of sinusoidal vibrations or pulses per second. While temporal frequency specifies the number of identical, repetitive events per unit of time, *spatial* frequency, its analogue in space, defines the number of identical repetitive variations of optical density per unit of length (Figure 13.10).

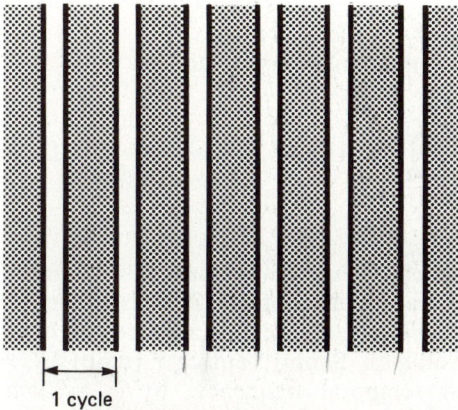

Figure 13.10 Spatial frequency defines the number of identical, repetitive variations of optical density per unit length. (In this example $f = 60$ cycles/m)

The transference of the frequency concept into the spatial domain has equipped the opticist with a powerful analytical tool. The adoption proves itself more useful everyday, in particular in the fields of image evaluation and processing. The resulting inter-disciplinary penetration will not fail to arouse a vivid interest among students of electro-optics. Those with an electronic background will recognise en route familiar extensions of the frequency concept to notions of 'spectrum', 'bandwidth' and 'filtering'. Their experience will be crowned by the realisation that the Fourier Transform is as much at home with diaphragms, lenses and pictures, as it is with pulses, amplifiers and filters.

13.2.2   The Fourier method

We recall that, in electronics, the masterly piece of mathematical wizardry called the Fourier Series enables us to decompose any regularly repetitive waveform of frequency, f, into a number of sinusoidal waveforms with frequencies 2, 3, 4, ..., n times greater than f. (To the mathematical purist the word 'any' is not strictly true but such restrictions will seem far-fetched to the practising engineer.) The all-important numerical values of the individual amplitudes and phases of these harmonic components can be calculated from:

$$A_n = \frac{2}{T} \int_{\Delta T}^{\Delta T + T} f(t) \cos n\omega t \, dt \tag{13.1}$$

$$B_n = \frac{2}{T} \int_{\Delta T}^{\Delta T + T} f(t) \sin n\omega \, dt \tag{13.2}$$

$$\phi_n = \arctan\left(-\frac{B_n}{A_n}\right) \tag{13.3}$$

given that the waveform under investigation is represented by:

$$f(t) = \frac{A_0}{2} + \sum_{n=1}^{\infty} A_n \cos n\omega t + \sum_{n=1}^{\infty} B_n \sin n\omega t \tag{13.4}$$

with $\omega = 2\pi f = 2\pi/T$.

We also recall that the more finely chiselled the waveform or the sharper its transitions, the stronger the high harmonics contents. All this applies as well to spatial variations of optical density, radiance or irradiance as it does to the familiar time-dependent voltages. Simply replace t (time) by x (length) and $\omega = 2\pi/T$ (angular temporal frequency) by $\omega = 2\pi/a$ (angular spatial frequency). Moreover, that extension of the Fourier method which so effectively lifts the restrictions of repetitiveness from the waveform under investigation, called the Fourier Transform, is usable with spatial frequencies too. Here, the frequency spectrum of a once-for-ever occurrence is continuous, as witnessed by the replacement of the series of Equation 13.4 by the integrals:

$$f(t) = \int_0^{\infty} A(\omega) \cos \omega t \, d\omega + \int_0^{\infty} B(\omega) \sin \omega t \, d\omega \tag{13.5}$$

with the spectral amplitudes given by:

$$A(\omega) = \frac{1}{\pi} \int_{-\infty}^{+\infty} f(t) \cos \omega t \, dt \tag{13.6}$$

and

$$B(\omega) = \frac{1}{\pi} \int_{-\infty}^{+\infty} f(t) \sin \omega t \, dt \tag{13.7}$$

or yet by the more elegant equations:

$$f(t) = \frac{1}{2\pi} \int_{-\infty}^{+\infty} F(\omega) e^{-i\omega t} \, d\omega \tag{13.8}$$

with the spectral components given by:

$$F(\omega) = \int_{-\infty}^{+\infty} f(t)\, e^{i\omega t}\, dt \qquad (13.9)$$

in which $F(\omega)$ is the Fourier Transform (FT) of $f(t)$ while $f(t)$ is the *inverse* Fourier Transform (FT^{-1}) of $F(\omega)$. (For a spatial as opposed to a temporal event, the replacement of t by x will obviously be made.) Looking carefully at these equations we see that the transition from Equation 13.8 to Equation 13.9 replaces the temporal (or spatial, with x in place of t) spectrum of an event by its frequency spectrum (Figure 13.11). The FT^{-1} operation performs the domain-to-domain transition in reverse. The

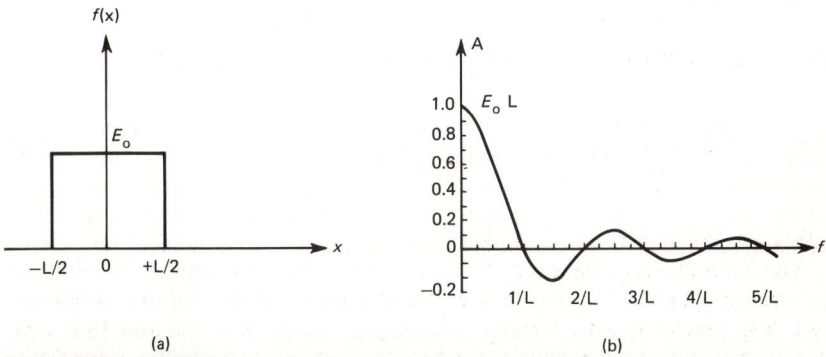

Figure 13.11 Example of a Fourier transformation of a spatial variation of an event: (a) rectangular 'pulse'; (b) its frequency spectrum

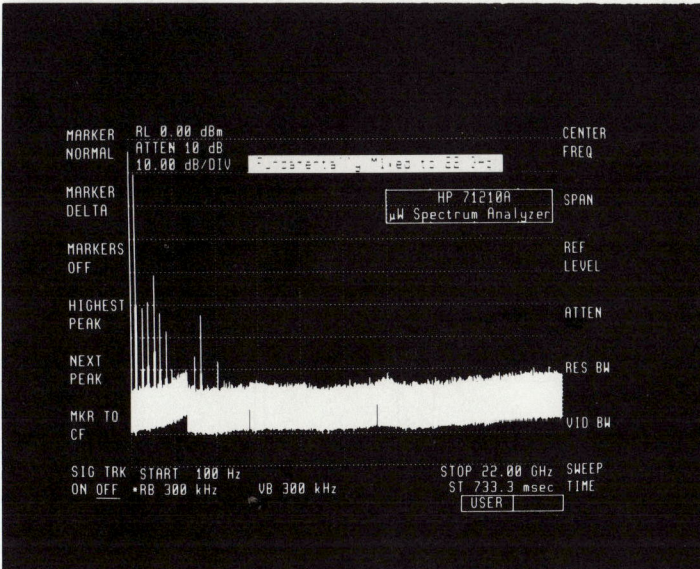

Figure 13.12 Decomposition of an electric waveform into its frequency components (Courtesy of Hewlett Packard)

exponential $e^{\pm i\omega t}$ acts as a hinge in this swinging act. That it should be so is quite natural considering it contains for exponent the product of both ω and t.

Just as *decomposition* into harmonics of an electric waveform (Figure 13.12) can be experimentally verified by means of a spectrum analyser, waveform *composition* can be accomplished by an electronic synthesiser, both thereby proving to the incredulous that Jean Baptist Joseph Baron de Fourier was right.

Let us now turn to optics and discover how images can be decomposed into their constituent spatial frequencies, how some of these frequencies can be either discriminated against or looked for, and what practical uses all this can have.

13.2.3   Using diffraction for frequency sorting

In optics, the Fourier jugglery can be performed for us by our glassware without computation on our part. Figure 13.13 shows how. Section 10.3.2(b) shows how diffraction can extract a single spatial frequency from a transparency. Looking at various figures there and reflecting anew upon the diffraction equation (Equation 10.11) we can see that to each particular 'pitch' – hence to each spatial frequency, its reciprocal – corresponds a particular set of diffraction angles. The spatial modulation used there being of the on/off type, it contains a fundamental frequency, f, plus its harmonics. Had we considered the fundamental alone, we would be left with just one diffraction mode, corresponding to $m = |1|$ and thus only one diffraction angle $|\alpha|$. In other words, a sinusoidally modulated grating would have just this one diffraction mode. Do we not, after all, recognise in this the very principle of the acousto-optic beam deflector?

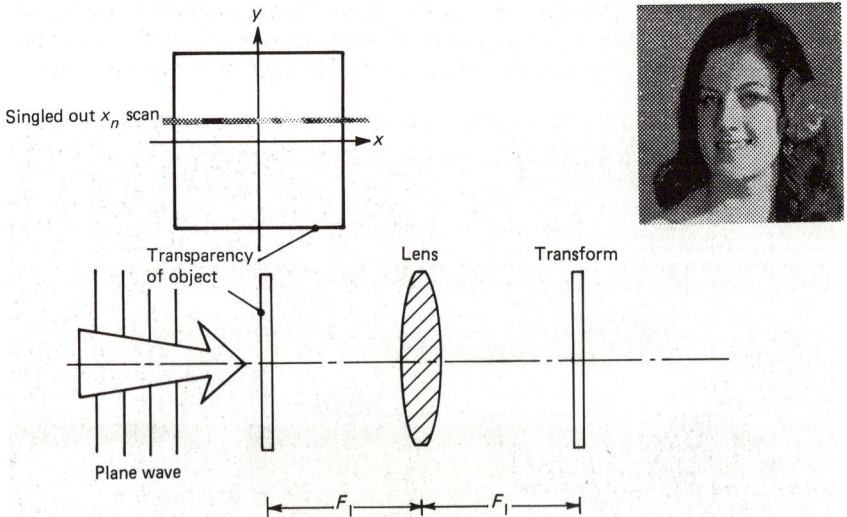

Figure 13.13 Basic set-up for the generation of the Fourier Transform of a single scan of a transparency

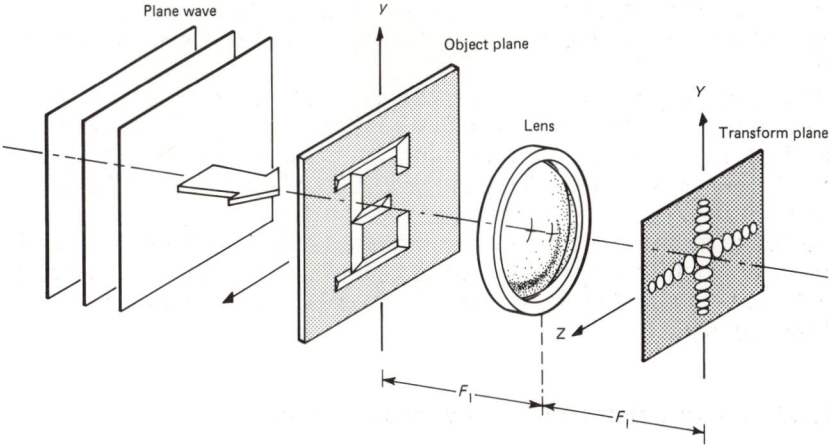

Figure 13.14 Set-up for obtaining a two-dimensional optical Fourier Transform of a transparency

Restricting our thoughts, at this stage, to a single x scan of the transparency, we now see that, whatever the complexity of the opacity variation function of that scan, it can always be decomposed into its constituent spectral components. Observe that the largest α values correspond to the components of the highest spatial frequency. With a set-up that is symmetrical about z, a single y scan will experience a similar spectral analysis. We thus end up with a two-dimensional Fourier Transform of the transparency on the screen (Figure 13.14).

While we now know that the position of each particular bright dot on screen defines the frequency, f_n, of the corresponding component, the question arises: how can we tell the *field* amplitude of the components of the obtained spectrum? The answer is that we cannot. What can we know. however, is the relative *power* levels of the components, as this information is betrayed by the irradiance of each dot through its proportionality to

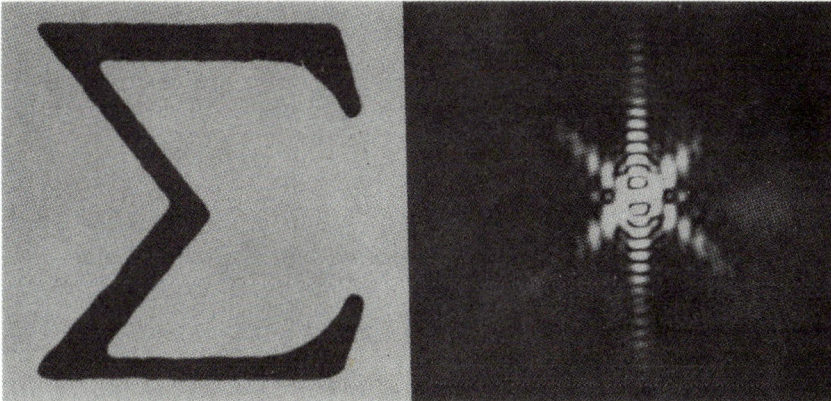

Figure 13.15 This set-up forms a readily accessible power spectrum of spatial frequencies making up the Σ sign (Reproduced from SIRA[194] by permission)

(a)

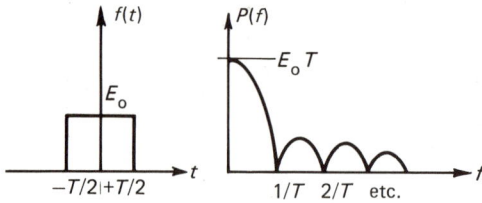

(b)

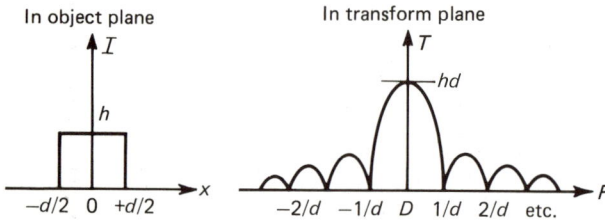

Figure 13.16 Frequency power spectrum (temporal) of a pulse in electronics (a) side by side with the frequency power spectrum (spatial) of a slit optics (b)

the square of the amplitude. Thus, the arrangement has formed for us, in the Fourier Plane (FP), a readily accessible power spectrum of the spatial frequencies of the object (Figure 13.15).

The whole operation was *instant* and of course *parallel*. We conclude by showing side by side the frequency power spectrum (temporal) of a single rectangular pulse in electronics and the frequency power spectrum (spatial) of a long rectangular slit in optics (Figure 13.16).

The amplitude spectra can, of course, be calculated. (For a few examples of FTs of simple geometrical forms, see Ref. 227, p.154.) We shall see later that their virtual presence in the FP can also be utilised.

13.2.4   Sculpting the frequency response

Having to hand the frequency spectrum of an image opens up a range of interesting possibilities. To begin with, we can remove one or more of the spectral components, just as we can, in electronics, block off an unwanted interference, such as a 50 Hz or 100 Hz ripple ('hum'), by inserting a notch filter into the circuit. For this to be of any use we must, of course, reconstitute an image from the Fourier spectrum, i.e. to perform the FT^{-1} operation. Figure 13.17 shows the simplicity with which this can be done. Lens L_2 is our synthesiser. Any mask in the Fourier plane will remove a part of the spectrum. We can see at once that the hardware of an optical notch filter is truly minimal: just two blobs of black paint. Extending the blobs to small oblong sectors can have staggering effects: they can free a captive animal from its cage. In Figure 13.18 a horizontally oriented mask has removed the spatial frequency corresponding to the pitch of the vertically oriented bars and its harmonics. We could, of course, obstruct all

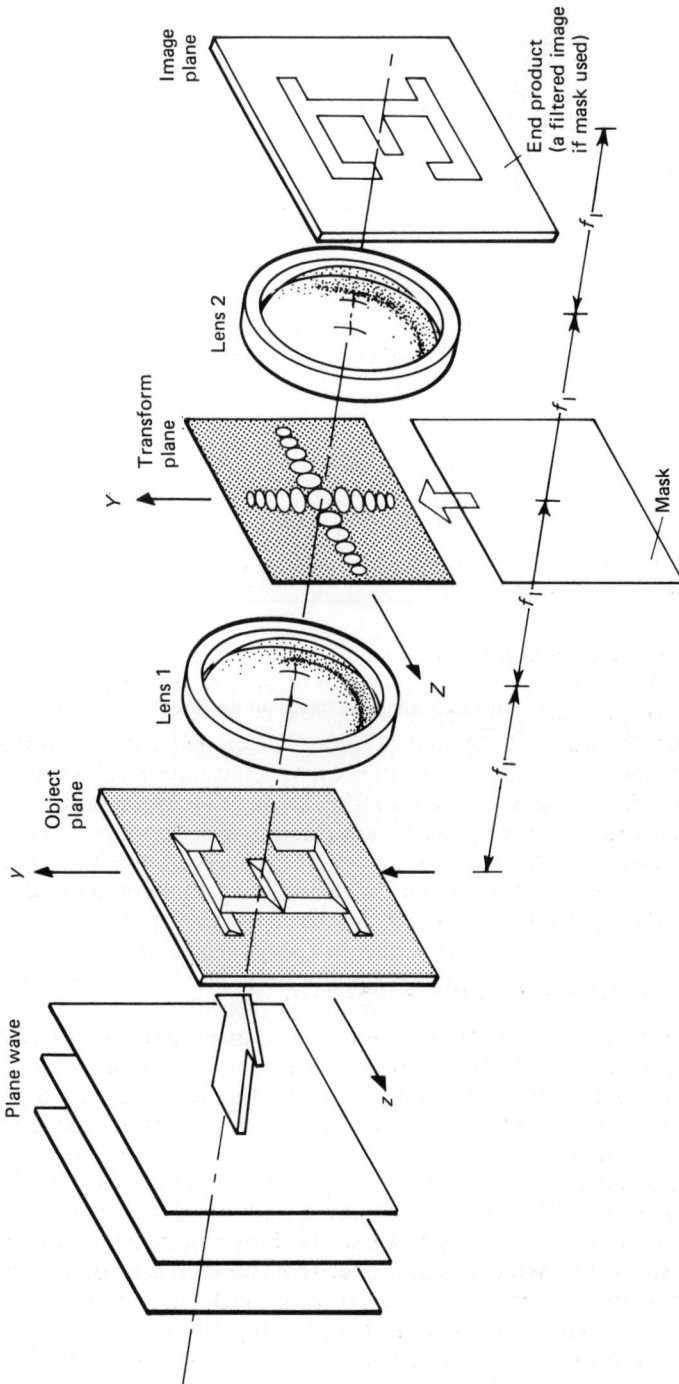

Figure 13.17 From spectrum back to image (FT^{-1} operation). Lens L_2 is our synthesiser. Use of a mask in the FT plane can remove unwanted frequencies

Mask

Luminous power
vs. distance

Luminous power
vs. frequency (spatial)

Before and after
spatial frequency filtering

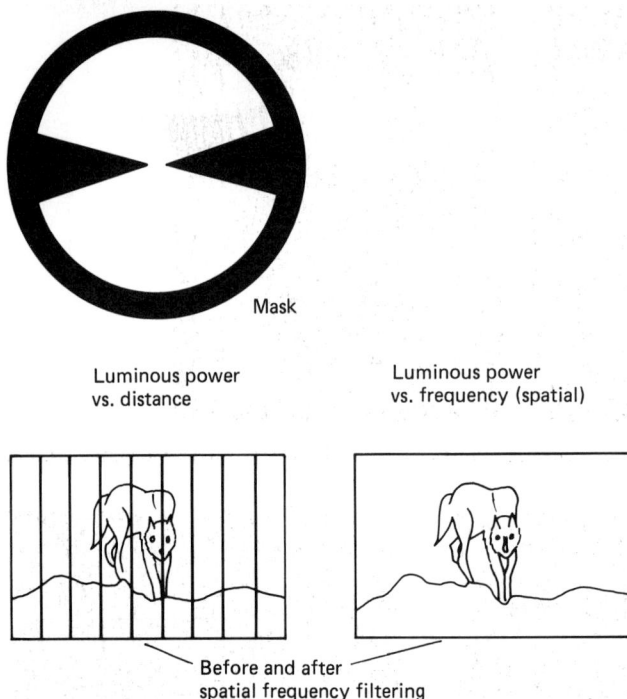

Figure 13.18 'Let the captive out.' Extending the 'blobs' of the filter to oblong sections can free the captive animal from its cage

the remaining frequencies instead, letting through the ones corresponding to the bars (complementary mask) and be left with an empty cage from which the animal has escaped!

This amusing example of much simplified image manipulation demonstrates forcibly the power of *spatial filtering*. The shaping of the frequency response of an optical system can take on a very high degree of refinement, much higher than that of an audio amplifier equipped with a row of slide potentiometers. The applications are many, particularly in the image improvement and image identification fields, in the latter of which the sculpting of the mask can involve holographic techniques for phase sensitivity. Real time applications, from which the waiting for photographic processes has been eliminated, are emerging from laboratories. Some are described below.

(a) Image improvement

Just as an audio amplifier with its tone control adjusted to discriminate against high-pitched sounds produces a sonority more acceptable to most of us (though not necessarily Hi-Fi), a picture can be more pleasant to look at without at least some of its higher frequencies (Figure 13.19), for instance those produced unavoidably in the screening process of much of our printed matter. Spatial filtering can remove them. On the other hand, a high-pass spatial filter will enhance edge contrast (Figure 13.20). This can

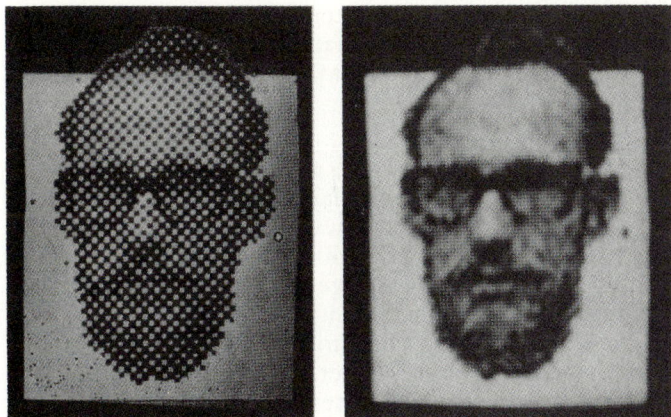

Figure 13.19 Removal of some higher frequencies can make a picture more pleasant to look at

Figure 13.20 Removal of lower frequencies can facilitate the detection of 'fine' (sic!) details (Reproduced from SIRA[194] by permission)

be required to facilitate the detection of fine details in a uniform large background.

NASA makes frequent use of spatial filtering techniques to remove the seams from surveyance photographs obtained by jointing individual strips of film in order to improve their reconnaissance value. In scientific research selective removal of some spatial frequencies can greatly enhance the intelligibility of microscopic views, bubble chamber track records (e.g. Ref. 17, figure 14.14) or films with a marked granularity.

Amplitude filters, attenuating some frequencies more than others in accordance with a selected $T(f)$ law, can improve the faithfulness of a picture. We thus see that the accessibility of the Fourier Plan (FP) enables us to re-sculpt the spectrum of an image so that it forms, through the FT^{-1} transformation, a more useful or a more pleasant representation than the original itself.

(b)   Pattern recognition

The univocal correspondence between a graphic pattern and its FT induces us to think that two identical (or nearly identical) FTs are indicative of two identical (or nearly identical) images. This, in turn, suggests a solution to the problem of finding which one of a large number of transparencies matches (or nearly matches) a specific model. The case can be quite real when a criminal has to be identified by comparing fingerprints found at the scene of the crime with those of a large number of suspects on file. In a less sinister vein, we may wish to detect the presence of a specific type of microorganism in a large number of microscope smear slides in a hospital, compare a signature on a cheque with a number of specimens held in a bank (*temporal* spectral analysis can also assist in spotting *real-time* signature forgeries), or endow a data handling system with a character recognition capability. Comparing spectra is much easier and lends itself far better to mechanisation than comparing pictures. This ease is further enhanced by the fact that the FT is independent of horizontal and vertical shifts of the original (a little reflection will show why this is so). Not only does this relax the demands for the registration accuracy of the samples under test, but it also allows us to look for details or small objects of a specific shape randomly located within a larger workframe, such as chromosomes in sperm or specific microbes in blood.

While this type of pattern identification by comparative FP anlaysis for fingerprints, signatures or blood cells is interesting, the finest technique in object recognition by Fourier optics relies on the application of *matched filters*.

(c)   Matched filters

We remarked earlier that the photographic process used to record the spectrum in the FP entails the loss of phase information. Exploiting the holographic principle, Vander Lugt has evolved in the mid-1960s a technique of producing spatial filtering masks, in which the phase information is preserved. If we wanted to detect, by his method, the presence of 2p pieces among a variety of coins of a transparency, we would first produce a hologram of the 2p piece alone, by means of an apparatus represented schematically in Figure 13.21. We would then use this hologram as a spatial filter – a mask – in a slightly modified version of the set-up of Figure 13.17. Now, the screen in the image plane will reveal the presence of any 2p piece contained in the transparency of the object plane! The operation relies on the fact that a mask produced in the above way represents the complex conjugate of the object wave S (denoted S^*, pronounced 'S star'). (Of course, coherent light is used throughout.) A mathematical treatment of the matter shows that the screen gets a signal $Af(x) \times S^{(*)}(-x)$ (where $f(x)$ is the input, A a constant) which reaches a real value for $f(x) = S$, i.e. in our case, the 2p piece (see, for example, Ref. 225, section 8-6). Figure 13.22 shows that the matched filter detects input selectively.

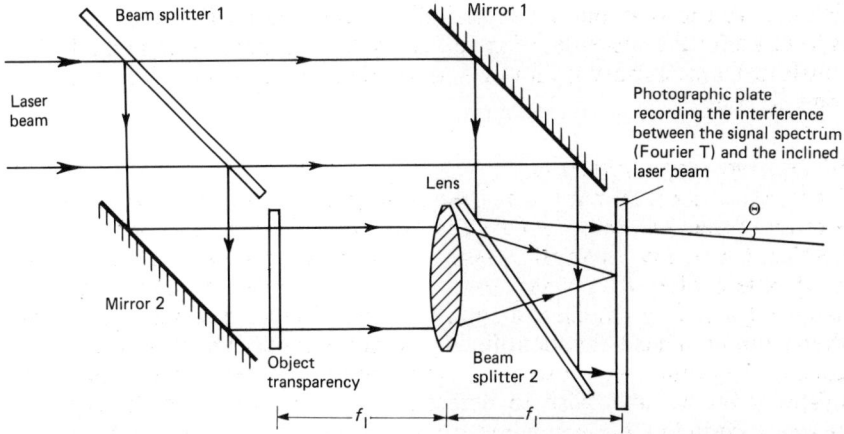

Figure 13.21 Apparatus for producing phase preserving spatial filter masks. The inclination θ of the interfering laser beam is essential in the process of making the filter (see Ref 76, section 7.42 for details)

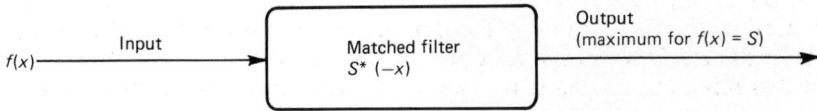

Figure 13.22 Matched filter detects input selectively

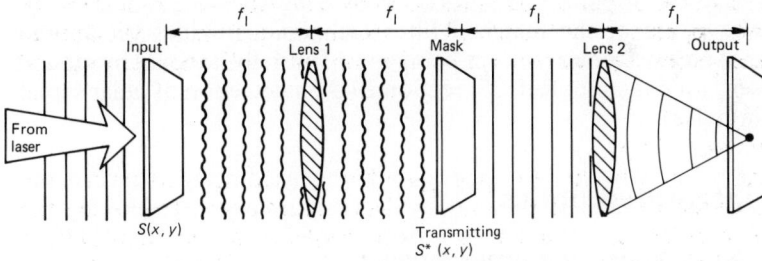

Figure 13.23 Field curvatures are 'straightened out' by the matched filter, enabling lens 2 to form a bright spot on the screen

In physical terms, what happens is that whatever field curvatures have been inflicted onto the plane wave of the analyser by the object transparency are 'straightened out' by the composite conjugate filter (Figure 13.23), but *only if it 'matches' it*. The wave, made again straight plane, will of course be focused in a very bright spot on the screen in the image plane. A return to the mathematical treatment reveals that this is the case of the largest output signal, i.e. the brightest spot. Figure 13.24 shows how this has worked for detecting the letter g (notice the white spot bottom right) from among 24 alphanumerics. Observe that, whilst unburying it from noise, a matched filter does not recover a signal: it merely *detects its presence* and produces an acknowledgement.

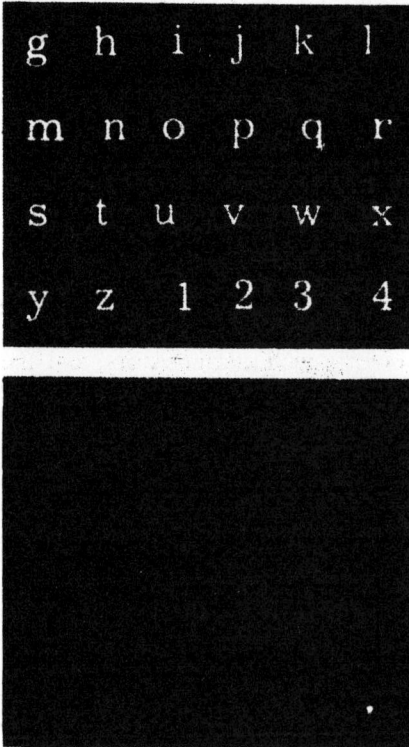

Figure 13.24 Detection of the letter g from 24 alphanumerics (Reproduced by permission of IEEE Inc.)

The above use of the matched filter technique provides yet another example of an electronics concept being most fruitfully adopted by modern optics (see, for example, Ref. 37, section 29-5 on detection of radar signals buried in noise).

13.3 Integrated optics

13.3.1 Aims, hopes and achievements

At its inception, integrated optics (IO) was intended to become to electro-optics what integrated circuits (ICs) are to electronics. Cheap, small, mass produceable, easily interconnectable units performing a whole range of EO functions accomplished hitherto by discretes: lasers, lenses, beam splitters, path switches, modulators, photodiodes, etc., were hoped for. Over a decade later they still are. Despite this disappointing overall picture, IO developments have led to interesting achievements, in addition to strengthening awareness of and familiarity with coherent optic physics.

13.3.2 Materials used

The base materials used in IO fall into two categories: dielectrics and semiconductors. While lithium niobate ($LiNbO_3$) is the most used dielectric, the dominant semiconductors are III/V ternary and quaternary compounds, mostly GaAlAs. Waveguiding is obtained through *raising* locally the refractive index, n, of the base material. To this effect, $LiNbO_3$ will have, typically, titanium (Ti) channels diffused into it, while III/V compounds will either undergo a localised control of the aluminium content or be subjected to mask controlled carrier concentration processes. The choice of $LiNbO_3$ stems from its high electro-optic (Pockel's) effect and piezoelectric coefficients, its very high anisotropy and the ease with which it yields sufficiently large crystals. The unique attractions of III/V compounds are: their combined optical and semiconductor properties which hold promise of a true integration of both optical and electronic functions *in monolithics*, lending themselves as these compounds do to both photon and electric carrier confinements, let alone to the fabrication of light emitting and receiving junctions, as well as transistors; and the existence of an enormous technological know-how of various processing techniques, as these materials are already widely used in the micro-electronics industry.

13.3.3 The physics behind the devices

(a) The evanescent field

IO uses coherent, most often monomode ($m = 1$) radiation. Like in optical fibres, light ducting relies here on total internal reflection. (The reader may find it helpful to revise Section 7.2.) Unlike the FO technology,

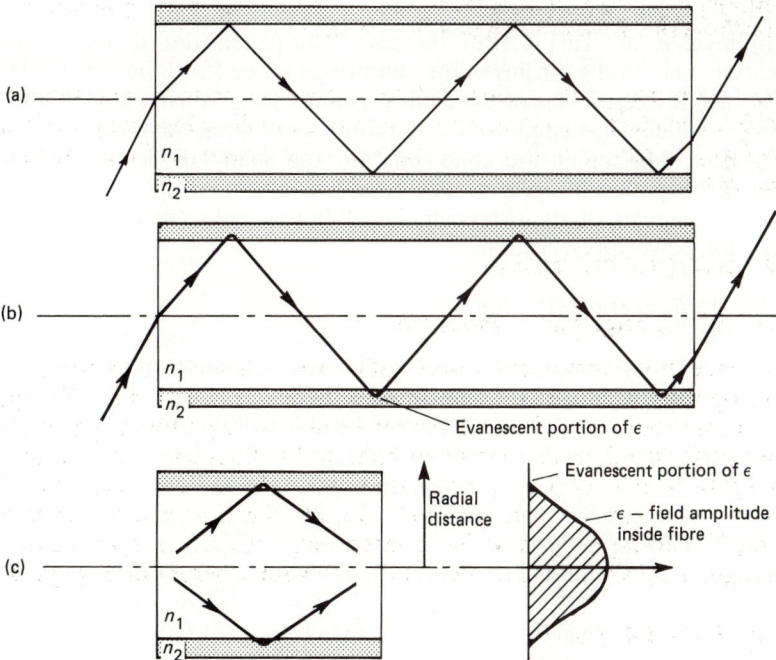

Figure 13.25 The evanescent field in optical fibres. One mode only illustrated. (a) For simplicity, not mentioned in earlier chapters; (b) now shown; (c) field profile in a monomode fibre

however, IO makes frequent use of that portion of optical energy which penetrates slightly into the lower index guiding layers (the cladding, in OFs), an effect which, partly because of its smallness, partly for simplicity's sake, has not been mentioned in the earlier chapters. Figure 13.25 shows how the *evanescent field* (so named because of its very rapid decay) carries this minute overflow of energy. Its presence is physically necessary to satisfy the boundary conditions. IO uses it for coupling purposes. Light guiding is obtained by raising the n of the substrate locally, by diffusion or by topping it with a higher n material. Figure 13.26 shows that the 'upper'

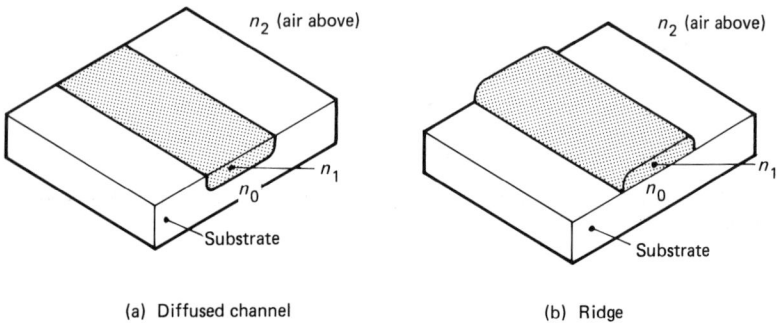

(a) Diffused channel   (b) Ridge

Figure 13.26 Examples of strip waveguides. Note that $n_1 > n_0 > n_2$

guiding layer is air. This is often the case. The penetration depth of the evanescent field in the air lies in the submicron range (for a hundred-fold field decay); the thickness of the central guiding layer is usually of a few, say two, in-guide wavelengths of the radiation used for monomode ($m = 1$) propagation. Making it too thick enables too many modes; too thin disables propagation altogether. The cross-section, for simple ducting, is usually rectangular or quasi-rectangular. Bending radii are in excess of $1000 \lambda_g$ for power preservation.

(b) Controlling n electrically
The electro-optic (Pockel's) effect (see Section 10.3.3(c)) is often exploited in IO. As interelectrode spacing can be made small, voltages not exceeding 30 V are sufficient to produce functionally adequate fields. The Pockel's effect is used for the control of n, either in forked light steering junctions or in interferometric on/off switches and modulators based on phase retardation.

(c) Transverse gratings
Guiding layers with surfaces corrugated transversely to the direction of wave propagation (Figure 13.27) can be used in IO in two ways: as diffractive in and out coupling elements (Figure 13.28) and as reflective elements (Figure 13.29(a)). The latter, also called *Bragg reflectors,* are particularly useful in integrated lasers, as replacements of the usual cleaved end reflectors. They can also be used as reverse coupling elements (Figure 13.29(b)). It can be shown that, for Bragg mirrors to work, the corrugation

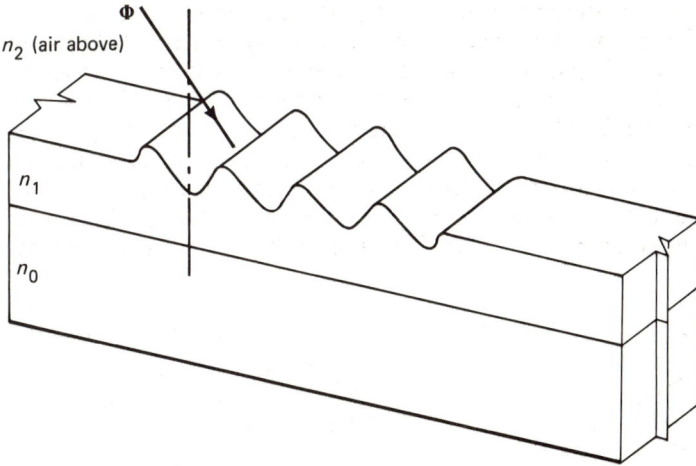

Figure 13.27 Guilding layers with surfaces corrugated transversely to the direction of wave propagation

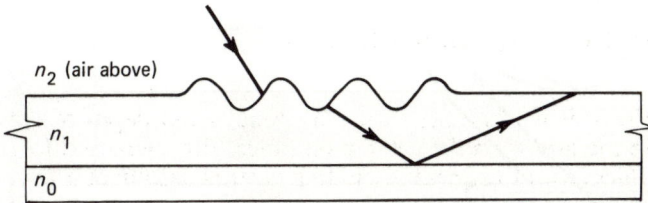

Figure 13.28 Corrugated layers as diffractive in and out coupling elements

pitch must be smaller than the wavefront pitch in the medium ($p < \lambda_1$) while for diffractive coupling the necessary condition is ($p > \lambda_1$)[36].

(d) Surface acoustic waves

Diffraction grating can be produced without making corrugations. Remember the Bragg cell AO laser beam deflector of Section 10.3.2(b). Exploiting the piezoelectric properties of an IO substrate, two interdigitated comb-like electrodes deposited on its surface will generate, when a.c. powered, acoustic waves (Figure 13.30(a)). These penetrate very little into the substrate and are therefore called Surface Acoustic Waves (SAWs). The word 'acoustic' is slightly misleading, as the frequencies used in IO lie in the megahertz–gigahertz region. The best EA conversion efficiency takes place at a frequency for which the comb pitch equals the acoustic wavelength ($p = \Lambda$). This is of some importance when considering a variable pitch electrode ('chirped') SAW generator (Figure 13.30(b)). The *dynamic* gratings produced by SAW generators result from stress-induced localised changes of the refractive index of the substrate.

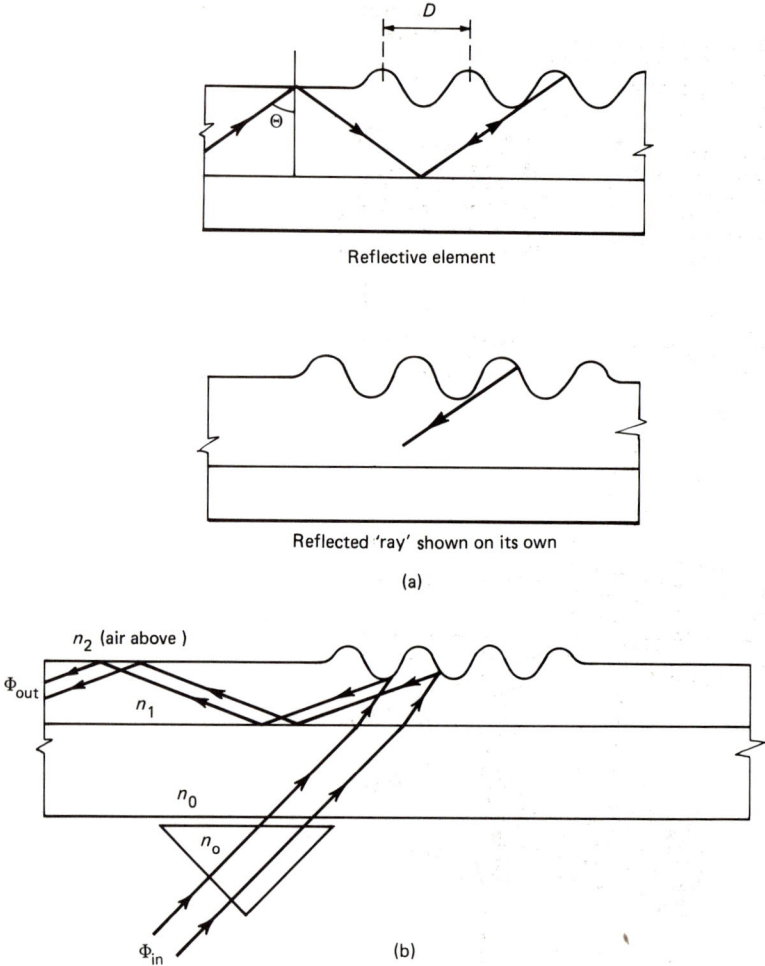

Reflective element

Reflected 'ray' shown on its own

(a)

Figure 13.29 Corrugated layers as (a) reflective elements and (b) reverse coupling elements

13.3.4   The devices

The following lines give a few illustrative examples of IO devices performing some of the fundamental functions of signal switching, routing, modulating, filtering, etc.

(a)   The beam splitter

The 'bulk optics' solution to the problem of splitting an optical beam into two (Figure 13.31) consists of passing it through a partly transmitting, partly reflecting surface, for example, the thinly metallised interface of the two halves of a cube beam splitter. The use of the Y coupler in fibre optics (Section 7.6.1) comes a step closer to the IO solution of the problem (Figure 13.32) especially when fusion-jointed single fibres replace the bundles in (a).

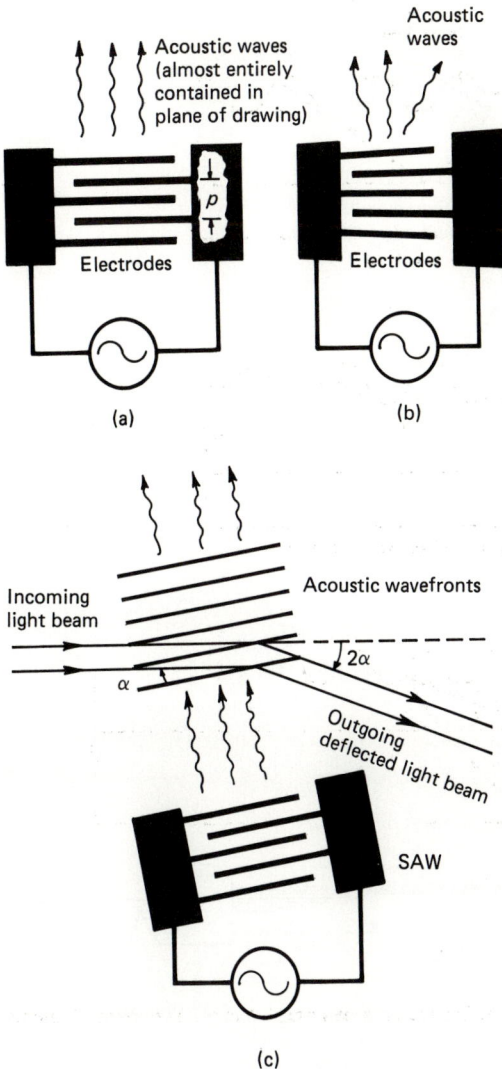

Figure 13.30 Surface acoustic wave generators: (a) straight; (b) 'chirped' – substrate is piezoelectric. (c) A guided IO beam can be deflected by the SAW equivalent of the 'bulk optics' Bragg cell

The design and execution of such a directional Y coupler must be carried out with care, if a small excess loss is to be obtained: the guide width must be of a few wavelengths only, all edges must be smooth and the bifurcation angle must be of no more than a fraction of a degree. The implication is that beam splitters take up lots of room, as sufficient leg length must be allowed to give good free access to the extremities. The Y splitter can be used, of course, in reverse, i.e. as a beam combiner (used here at a cost of a 3 dB loss).

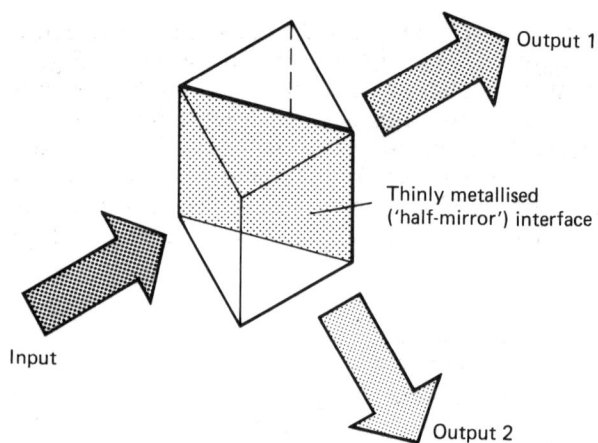

Figure 13.31 'Bulk optics' solution to beam splitting

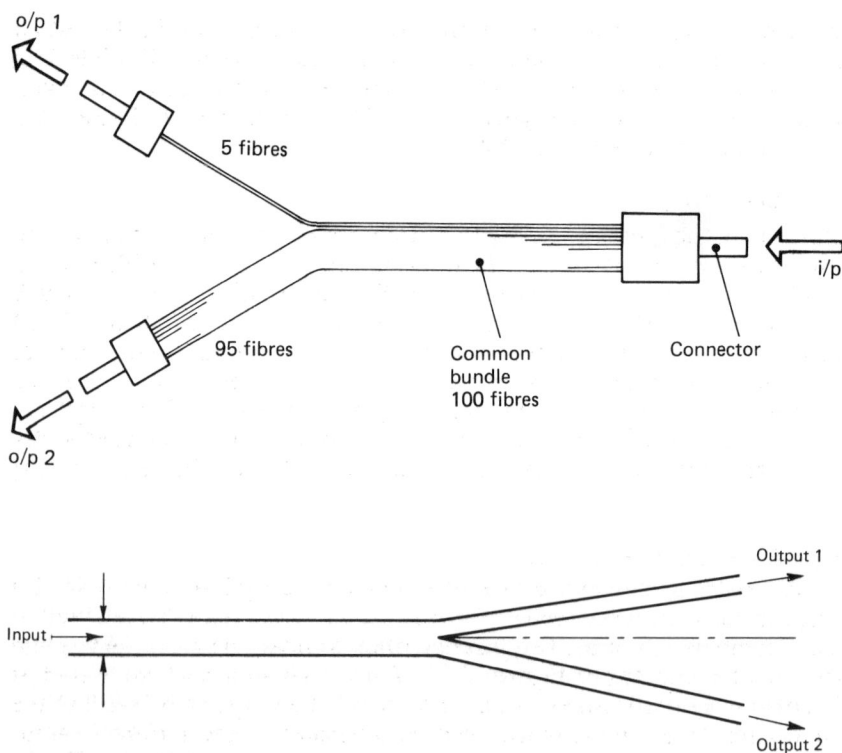

Figure 13.32 Beam splitting in fibre optics (a) and in IO (b)

(b)   On/off switch

Combine a direct and reverse Y splitter on the same substrate, add a couple of electrodes and you get the IO electro-optic switch of Figure 13.33. Remember from Sections 10.3.2(c) and 10.3.3(c) that the presence of an electric field in an EO material changes its refractive index, n. The frequently used $LiNbO_3$ substrate exhibits a very high Pockel's effect (see, for example, Ref. 192, table 3.1) which, with the large L/D ratios obtainable in IO, is conducive to large phase retardations being obtainable with only a few volts applied to the electrodes. For a voltage

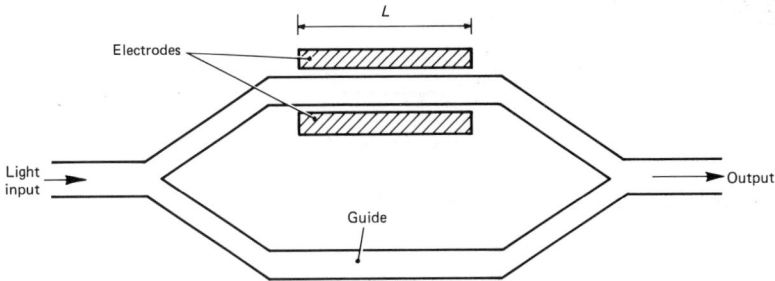

Figure 13.33 Electro-optic switch

producing a 180° phase shift in one arm, the interference in the output guide will be *destructive* and the beam will be switched *off*, while in a zero-voltage situation the constructive interference creates an *on* condition. The reader will recognise in this EO switch the Mach–Zehnder interferometer of Section 12.2.4.

(c)   Modulators

Basic amplitude modulation can be achieved with the architecture of the on/off switch just described. By drawing on the analogies with the EO modulator for open-air laser beams of Section 10.3.3(c), the reader will deduce how the Mach–Zehnder interferometer recognised in Figure 13.33 can convert phase modulation into quasi-linear amplitude modulation. A possible refinement consists of adding a third electrode to produce push–pull changes of the refractive index in the two arms of the interferometer, thus lowering the voltage requirements of the modulator. Modulation bandwidths of 1 GHz have been claimed with this arrangement [233].

(d)   Beam routing device

When confronted with the task of designing the optical equivalent of a single-pole changeover switch (Figure 13.34(a)), one might be tempted to use a combination of the two just described devices, the beam splitter and the on/off switch as per Figure 13.33. While such an arrangement is most likely to work, its lossiness would most surely be in excess of 3 dB, as the energy reaching the off branch will be dissipated in the substrate. A far better solution consists of an *active* version of the beam splitter (Figure 13.35), in which an electric field 'encourages' the beam to 'get in lane' just

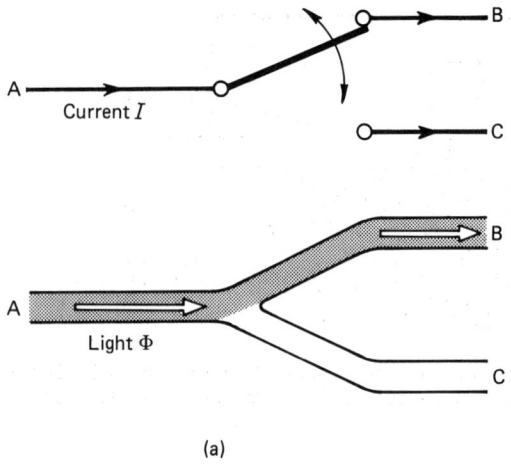

(a)

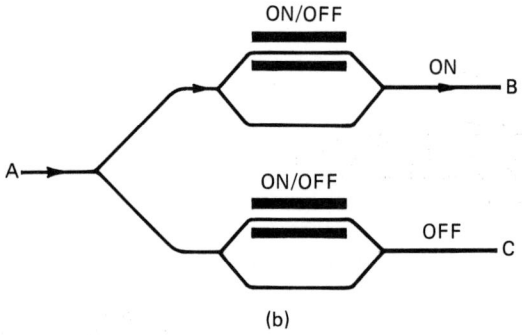

(b)

Figure 13.34 (a) Electrical single-pole changeover switch and its optical analogue; (b) possible, though not very practical, EO realisation

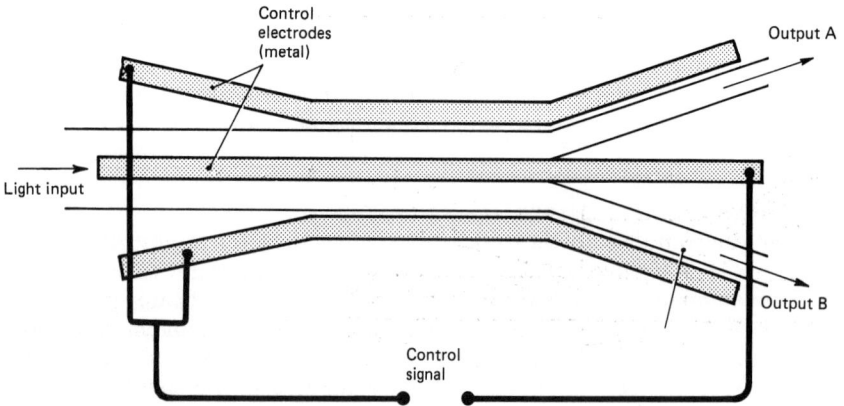

Figure 13.35 Real EO single-pole changeover switch

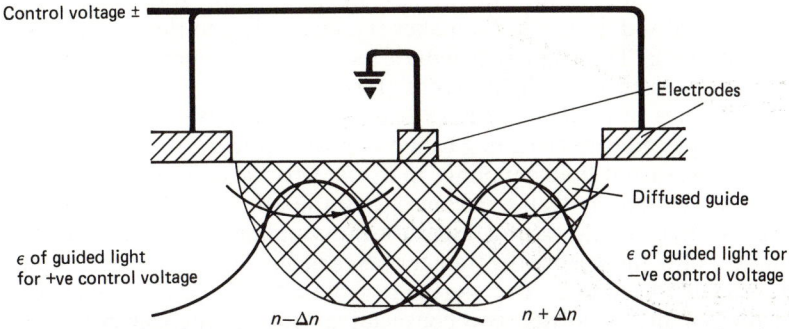

Figure 13.36 Electrically controlled localized increase of n makes the beam cling to the chosen route. While the arrows and localised variations of n are shown for a positive voltage applied, ϵ is shown for both signal polarities

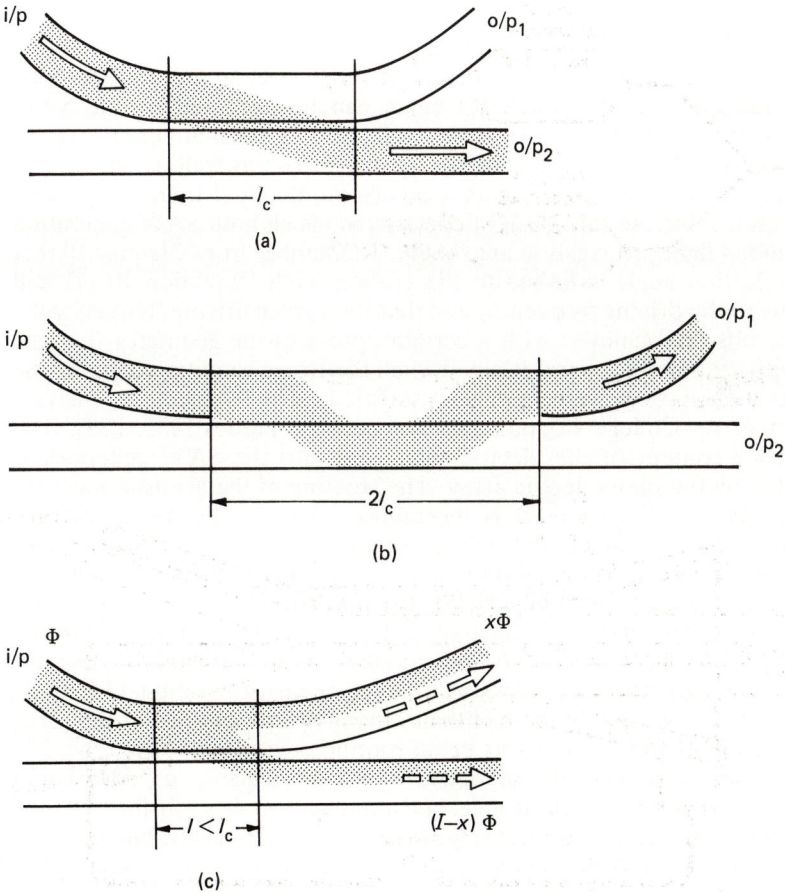

Figure 13.37 The COBRA switch (guiding electrodes not shown)

prior to the forking point. Figure 13.36 shows how the electrically controlled, localised increase of n makes the beam cling to the chosen side of the input guide. The EO coefficient of the substrate is maximised in the direction transverse to the light guides. Insertion losses of such active Y junction switches of under 1 dB have been reported [233].

Another type of beam switching called COBRA (Figure 13.37) relies on controlled evanescent field coupling. In it, the very close proximity of two guides causes an energy transfer from one to the other. In the absence of electrical fields, a critical value l_c exists for the coupling length l for which a total guide 1 into guide 2 energy transfer takes place. For $l > l_c$ some of the energy returns to guide 1. The working of the COBRA device relies on electrically halving l_c, so that two consecutive total energy transfers take place, when a total return of the beam to guide 1 is required [284].

Finally, multiple output beam routing could, conceivably – though expensively – be achieved by means of the IO spectrum analyser described in the next section.

(e)   Acousto-optics based devices

Our last example of function-orientated IO devices is the acousto-optic spectrum analyser. A guided IO beam can be deflected by the SAW equivalent of the bulk optics Bragg cell used in open-air laser systems (Section 10.3.2(b)). LiNbO$_3$, being a piezoelectric as well as an electro-optically active material, is again a highly suitable and therefore a most often used substrate for AO IO deflectors in which both SAW generation and guided light propagation are sought. Remember from Chapter 10 that the deflection angle is linked to the grating pitch (Equation 10.11) and thereby to the driving frequency, and that for a given driving frequency the best results are achieved with a certain opto-acoustic geometry (i.e. the best diffraction efficiency is attained when the Bragg condition defining the optical-to-acoustic wavefront angle is satisfied). The IO spectrum analyser of University College London exploits both properties [232, 235]. The frequency content of the electric signal fed into the SAW generator is revealed by the photoelectric array. The steering of the acoustic wave (to satisfy the second property) is ingeniously achieved by means of 'tilt-chirping' the interdigitated SAW generator electrodes. In Figure 13.38 note the high degree of integration of the entire device: the holographically produced expander (Section 13.1.5), the flat Fresnel lens (strictly speaking, Fresnel zone plate – see Section 13.1.1(a)) and the waveguide fan-out used for activating the photoreceiver array. (The expander hologram, formed on a photoresist, works as a Bragg reflector, Section 13.3.3(c).) Worth observing, too, is the fact that a hypothetical removal of the array would leave us with a multiport beam routing device!

Spectrum analysers of the above type are capable of covering a bandwidth of a few hundred megahertz centred on an f_o in the 300–400 MHz range. Channel resolutions yielding at least 125 resolvable spots (or addresses, for the hypothetical routing device) have been claimed [45]. Their potential lies in the parallel – as opposed to swept frequency – operation.

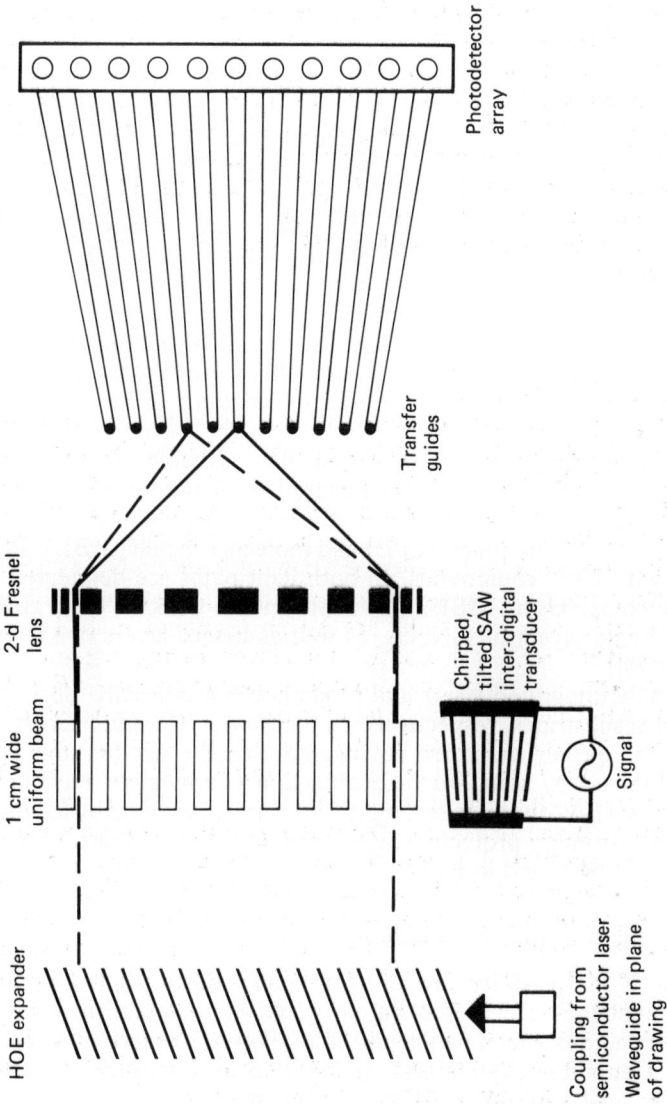

Figure 13.38 IO spectrum analyser of University College London. The frequency contents of the electrical signal fed into the SAW generator are revealed by the photodetector array. A device of this kind could be used for multipath beam routing

13.3.5 Working the substrate

The technique of making IO devices draws heavily on the know-how of micro-electronics: planar layer fabrication, liquid phase epitaxy, photo-lithographic masking prior to selective diffusion or etching, etc.

Lightguiding is obtained either by raising locally the refractive index n of the substrate itself or by topping it with a thin planar layer of a similarly higher n material, the unwanted parts of which are destined to be subsequently etched away.

All n-raising methods (diffusion, ion exchange, proton bombardment) lend themselves to localised action through a preliminary masking off of selected parts of the substrate. Of the two topping methods, epitaxy and sputtering, the former alone can grow waveguiding ridges.

In *n-raising*, typically for IO, $LiNbO_3$ substrates are used with titanium diffusion doped light guiding channels. These are obtained by first evaporating a thin layer of titanium onto the substrate through a photo-resist mask and then heating the unit. Ion exchange techniques are used on glass. For example, the replacement of sodium ions by silver ions in sodium glass (by a wet, hot process [36]) raises the n value of the outer surface. Aluminium masks are used for waveguide delineation. Proton bombardment can be used for producing required n variations in monolithically integrated GaAs components.

Topping by sputtering can be achieved on glass, sapphire or $LiNbO_3$. It deposits a thin 'icing' of a higher n material onto the whole substrate making subsequent pattern fabrication a necessity. Topping by epitaxy falls into two categories: liquid phase (LPE) and molecular beam (MBE). Both can be used with III/V compounds. In both light paths are delineated by subsequent etching. Using MBE, C.W. Pitt and co-workers have grown waveguiding homo-layers of $LiNbO_3$, as well as hetero-layers of $LiNbO_3$ on sapphire [234].

Monolithically integrated lasers and photodiodes are usually fabricated on GaAs and Si substrates, respectively. In the former, the end mirrors are replaced by the earlier described corrugated Bragg reflectors and light guiding is obtained by standard heterojunction confinement techniques (see Section 9.4.1). With the latter, light feed can be obtained by localised doping of the customary protective SiO_2 layer. (Very clear illustrations of both devices will be found in Ref. 192, figures 9.31 and 9.33.)

13.3.6 Making the right connections

Connections to IO modules are of two kinds: electrical and optical. Electrical connections can be fabricated by means of established micro-electronics techniques: metal pad deposition followed by thermocompression bonding. Coupling light into and off an IO device is a different matter. Butt-coupling produces efficient energy transfer only at the cost of troublesome guide-end preparation and painstaking alignment. The so-called prism-coupling [36], relying on evanescent field energy transfer, is even less practical, save for measurements. The 'sandwich ribbon' coupling (see Ref. 230 and the bibliography in Ref. 36) comes nearest to resembling electrical thermobonds, especially as the layered connecting waveguide is flexible (Figure 13.39). Also reminiscent of microelectronics

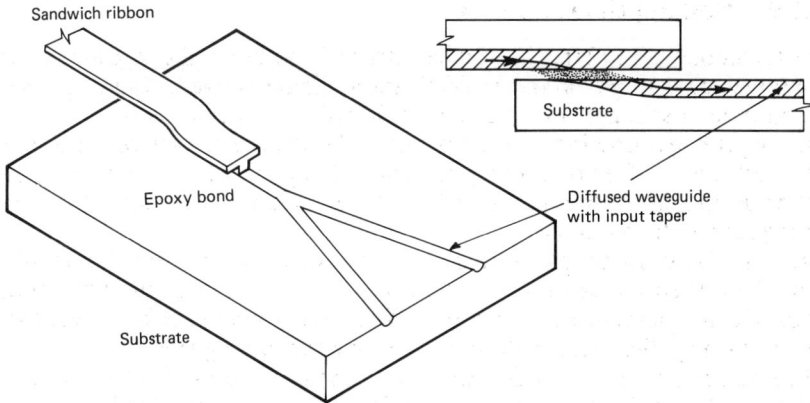

Figure 13.39 The 'sandwich' ribbon coupling comes nearest to resembling electrical thermobonds. The connecting light guide is flexible

are diffused in high *n* overlay bridges, used for interconnecting different components of the same monolithic. Beam expanders and beam focalisers for in and out coupling can use holographic optical elements (HOEs, see Section 13.1.5) as, for example, in the integrated microwave spectrum analyser described in Section 13.3.4(e). They might, possibly, some day use 'geodesic' lenses, formed by a filled-in depression in the substrate [36, 235].

13.3.7 Conclusion

IO has produced, in the laboratory, a number of intellectually pleasing, technically promising EO units, capable of performing a few basic signal processing functions such as switching, routing, modulating, and in-parallel spectrum analysing. Operating voltages are conveniently low (from a few volts to just under 30 V), bandwidths suitably high (approx. 1 GHz). However, all units need coherent light feeds and most are tied to single mode waveguiding. Although monolithic integration of light sources (lasers) and receivers (photodiodes) on GaAs substrates is now achievable too, large or just larger scale monolithic integration has still a long way to come.

Despite a considerable deployment of ingenuity and a great deal of hard work, backed up by as much enthusiasm, research workers have not yet succeeded in developing the long hoped for self-contained family of cheap, small, easily interconnectable IO signal processing modules, even remotely reminiscent of electronic families of ICs. The big breakthrough is still to come.

Chapter 14

The electro-optics curiosity shop

A number of rather unexpected aspects of modern optics are little known outside the esoteric circles of highly specialised opticists. Some of these aspects are so astonishing that one is almost tempted to call the aggregate of such exotic species the Opto Zoo. This final chapter considers, in a rather lighter tone, a few of the more intriguing ones.

14.1 Lenses by the length

The tiny cylindrical elements, millimetres or tens of millimetres long, have properties that are drastically dependent on their length. A typical Selfoc element is some 5–20 mm long and 1.5 mm in diameter (Figure 14.1). (Selfox is a registered trade name® of the Nippon Sheet Glass Co Ltd, Tokyo, Japan.) The full-length unit will produce at one end a real, erect image of an object presented to its other end. Cut it down to three-quarters of its length and it will either collimate a point source, or form a real erect image of an 'at-infinity' object. If you choose to cut it down further to half of its original length, L, it will act as a unity inverter. Cut it again by half

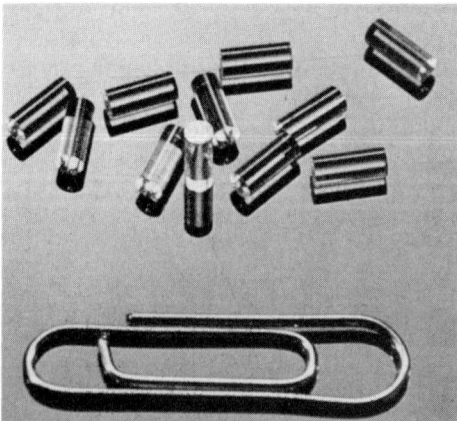

Figure 14.1 Typical length-dependent lenses are 5–20 mm long

(a)

(b)

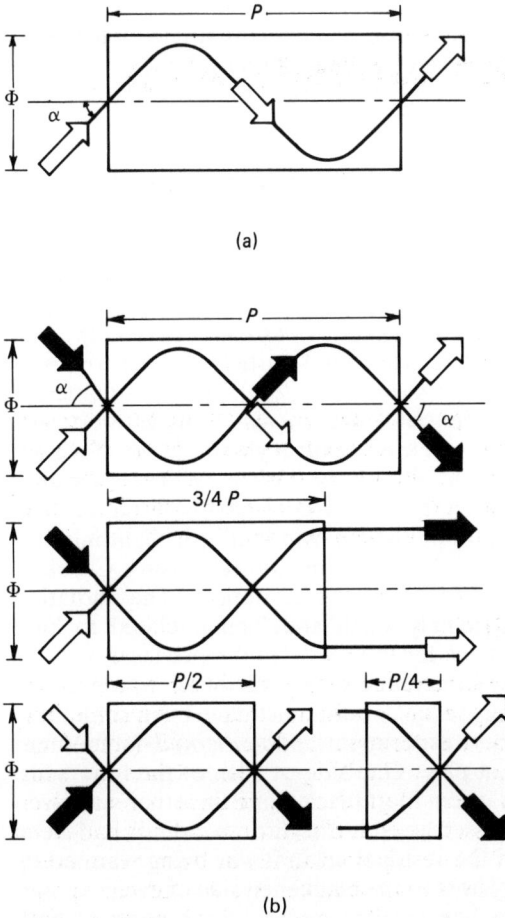

Figure 14.2 (a) Ray guiding properties of the graded index lens and (b) their effect on length-dependent lens action. A progressive reduction of the length of this lens profoundly changes its properties

and you get an inverter of an 'at-infinity' object. You have been controlling the f_1 of this unlenslike lens by cutting it to length! This arcane behaviour of the little glass rod is explained by the presence of a radial variation of its refractive index, strongly reminiscent of graded index transmission fibres (see Section 7.3.3). Figure 14.2 shows the ray guiding properties of the Selfoc rod. P is the 'pitch'. The $N = f_{(r)}$ variation is parabolic

$$n_r = n_o \left(1 - \frac{A}{2} r^2 \right) \qquad (14.1)$$

with n_o the on-axis refractive index and A a constant. (For the meaning of A and for other geometric optics relationships in Selfoc lenses, see the manufacturer's data sheets.) The generic name of these rods is GRIN lenses.

Applications range from fibre splitter/coupler through LED or APD-to-fibre coupler to bar code and optical disc readers (for single element) to 1 : 1 imaging for copying machines (for Selfoc arrays).

For further reading on GRIN optics, see Refs. 3(e), 103 and 238.

14.2   Optically activated levitation, rotation and flexural vibration

Can light produce direct *mechanical* action, as opposed to illuminating, imaging, transmitting and processing information, sensing, etc? In the 26 January 1970 issue of *Physical Review Letters*, A. Ashkin reported 'the first observation of acceleration of freely suspended particles by the forces of radiation pressure from CW visible laser light' [260]. By the summer of the following year he wrote: 'In our experiment, a single, vertically directed focused CW laser beam was used to lift a glass sphere off a glass plate and stably levitate it' [261]. A century earlier, William Crookes succeeded in putting into motion and then continually sustaining, by luminous energy, the rotation of a miniature 'windmill' (or 'lightmill'?).

There is an essential difference between the forces causing Crookes' rotation and those responsible for Ashkin's levitation. The radiation pressure (Ashkin's case) is extremely small and, being related to light intensity ($\mathscr{P} = I/c$) would have been far too small, in the pre-laser days, to compensate the frictional forces encountered by Crookes' windmill, let alone to overcome its launching inertia. There must have been something else there. Indeed, in Crookes' experiment, *photothermal* (sometimes called *radiometric*) action was at play. One side of each of the four 'sails' was covered with radiation-absorbent matt black while the other was given a reflective shine. The glass vessel housing the little windmill had been incompletely evacuated so that the residual air in it was being warmed by the incident radiation – but only next to the blackened side of each exposed vane. The resultant differences between the molecular moments of both sides of the exposed vane created a torque (pushing the 'hot' side), causing Crookes' little weathercock to turn. It was not, after all, the 'photonic wind' of radiation pressure that turned this windmill. We shall see in a moment that photothermal forces play an important role in today's light activated flexural resonant sensors too. In contrast, optical levitation relies on genuine radiation pressure.

14.2.1   Optical levitation and its uses

You may have seen in a fair a ping-pong ball suspended mid-stream in an ornamental water fountain. The similarity with Ashkin and Dziedzic's glass sphere trapped in a vertical laser beam is striking (Figure 14.3). The experiment is also reminiscent of the well known Millikan's balancing act with charged oil droplets in an electric field. The two Bell Lab physicists used a 0.25 W TEM_{00} 0.515 μm laser beam to support a 20 μm sphere in mid-air [261]. If the sphere 'sits' stably, for hours at rest, in an on-axis position, it is because transverse forces, caused by the difference between the refractive indices of glass and air, push it towards the maximum light

Figure 14.3 Similarity between a ping-pong ball suspended amid a stream of water and a glass sphere trapped in a vertical laser beam

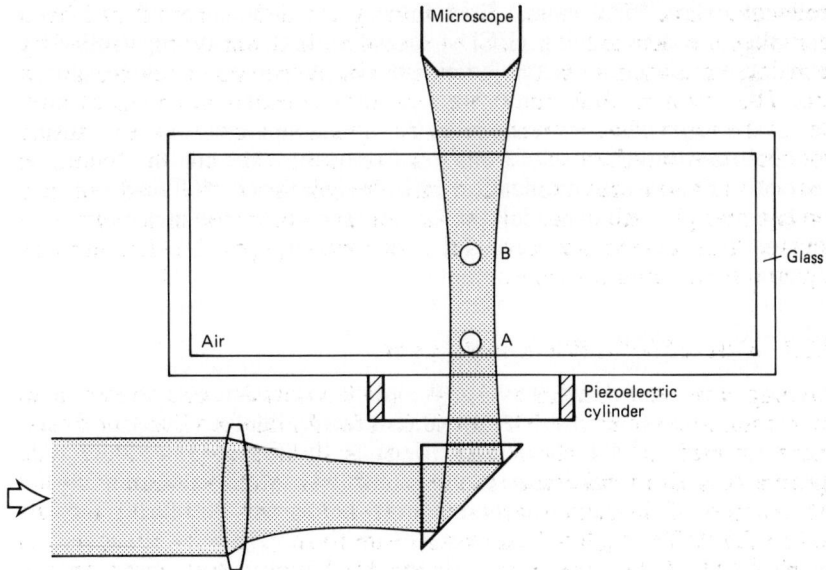

Figure 14.4 Schematic of apparatus for lifting the sphere from position A and upholding it through the 'potential well' effect in position B

intensity. These forces create a genuine *potential well* (Figure 14.4). At first demonstrated theoretically [260] the existence of the potential well has been later confirmed experimentally, in a most convincing way. The manoeuvrability of suspended spheres, in singles, doubles or larger aggregates, by beam position control, provided the proof [262]. The position of the moved around spheres, too transparent and too small for direct viewing, can be observed by the scattered light of the supporting beam itself or of that of an ancillary, very low power laser. Beautiful the levitation experiments certainly are, but can they also be useful? In their 1971 paper the authors of Ref. 261 speculated on using this support for micromanipulation, for the study of light scattering by single, small, yet larger than lambda, particles (called MIE scattering), for accelerometers and for absolute optical power measurement. In 1980 [262] they showed that this refined manipulative technique had matured sufficiently to fulfil at least one of their expectations: using two laser beams, aggregates of two, three and four spheres could be assembled mid-air and thereby further the study of MIE scattering by non-spherical particles. The technique also made high-accuracy (10^{-5}) size determination of small spheres possible. Highest on the list of possible applications of today lie studies of atmospheric physics and fabrication and holding of laser fusion targets.

14.2.2 Optically activated mechanical vibrations

In Section 12.1 optical fibre sensors were praised for their immunity to electromagnetic interference and for their inherent safety. It was pointed out, however, that much of the goodness of OFSs could be spoiled by the non-repeatability of measurements and mentioned that elegant compensational schemes were being revised to minimise or eliminate the adverse effects of its principal cause: light intensity variations. Such variations could be caused by ageing (LEDs, lasers, fibres), tarnishing (reflective surfaces), reconnections (cable breakages or extensions), etc. One scheme proposes to ignore much of intensity variations by relying for the measurements on the *mechanical resonance*, F_r, of a small vibrating structure – fork, slab or bridge – excited *and read* optically. The method could be likened to FM radiocommunications, where signal strength fluctuations are largely ignored by the demodulator. In this scheme it is the measurand that changes the resonant frequency, F_r, of the sensor.

Figure 14.5 shows an experimental lever bar from University College London [266], destined to measure force. The resonant frequency of the unstressed quartz slab, F_o, lies in the 14 kHz region. The stress sensitivity of this non-optimised device (adapted from a piezoelectrically driven vibrator) is $\triangle F \simeq 1$ Hz/g. The 4 mW optical excitation ϕ comes from an infrared laser ($\lambda = 830$ nm). A maximum flexural deflection A of 50 μm and a Q-factor of 2000 are claimed. Because the device is predominantly *photothermal* it exhibits some residual sensitivity to the value of ϕ, as the 'd.c.' component of the thermal energy bends the structure, thereby shifting F_o slightly.

A much smaller, optically driven, optically read structure (Figure 14.6), named a 'bridge' by its Strathclyde University designers [267], was made to oscillate at its resonant frequency F_o of 260 kHz by a mere 0.05

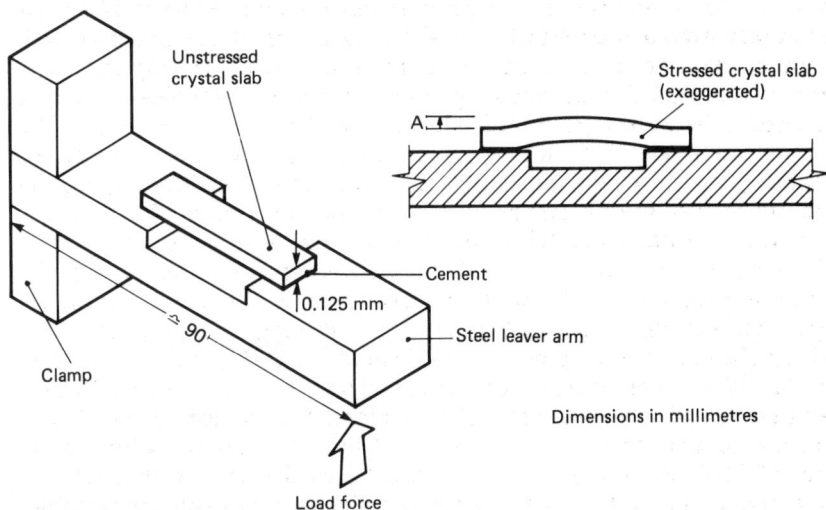

Figure 14.5 Schematic of an experimental force measuring resonant lever bar from University College London

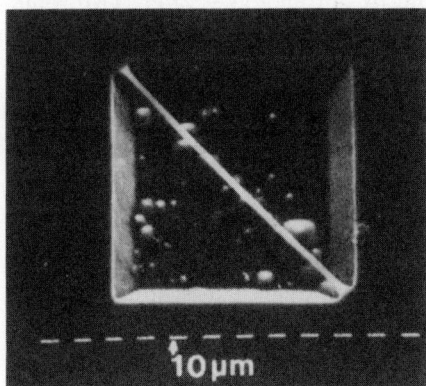

Figure 14.6 Resonant silicon oxide bridge of Strathclyde University (Courtesy of Dr S. Venkatesh)

mW of HeNe laser power ($\lambda = 633$ nm). Resonant deflection amplitudes of a few to a few tens of nanometres are reported. Optical heterodyne detection is used for reading F_r.

The main attraction of this device lies in its smallness (200 μm $\times$ 5 μm $\times$ 1 μm), achieved by silicon microfabrication techniques. With this sort of size, integration into biological OFS of the invasive *in vivo* type might be possible, as silicon is bio-compatible. The use of FO links would make remote powering and interrogation feasible.

Although Ref. 267 does not say it explicitly, it would appear, judging by the power levels involved, that this device too relies predominantly on photothermal effects and not on genuine radiation pressure.

14.3   Changing colours, reversing time and flip-flopping through non-linearities

Usually a nuisance, non-linearities can nevertheless sometimes be put to good use. Electronics is crowded with examples. When, in optics, the internal electrical polarisation of a material stops growing proportionately with an increasing light input, or when n becomes intensity dependent, harmonics generation can be observed, real-time holography can be achieved and bistability can be made to appear.

14.3.1   Changing colour through harmonics generation and frequency mixing

Frequency doubling and trebling has been known to radio engineers for decades. Should an optical frequency v double or treble, a change of colour will, obviously, result as $\lambda = c/v$. This can in fact be observed. Infrared radiation can be converted into a shorter-wave visible one, e.g. $\lambda = 1.06$ μm into $\lambda = 0.53$ (green), or ultraviolet obtained from visible. The question this raises straight away is: where does the photon energy surplus come from, as a 'blue' photon hv_2 carries twice as much energy as an 'infrared' one hv_1? The puzzle is solved by putting $hv_2 = 2hv_1$. Two weak long-wave photons combine (and die in the process) to produce a single strong, short-wave one ($\lambda_2 = 0.5\lambda_1$). This coalescence has been corroborated experimentally. (C.K.N. Patel mentions hole–electron pairs in semiconductors corresponding to a two-photon energy [7, p. 275]. Conversion efficiency, initially very low ($\simeq 10^{-6}$) grew, by the late 1960s to an impressive 20%, under specified crystal orientation [17, p. 504].

Frequency mixing, long known in radiocommunications, also has its counterpart in optics. Both sum ($v_1 + v_2$) and difference ($v_1 - v_2$) terms can be obtained by two photon mixing of coherent v_1 and v_2 radiations. Optical heterodyning is thus with us. Photonic interpretation for both sum and difference generation exists, the latter a little less straightforward than the former (see 'further readings' at the end of this section). Together with harmonics generation, frequency mixing opened up impressive opportunities for the creation of laser-like radiations at frequencies at which direct laser action cannot be obtained.

Other optical effects than those already mentioned use the language of electronics without the slightest embarrassment.

Optical rectification pertains to the presence of a constant, non-oscillating ('d.c.') electric polarisation in a strongly irradiated non-linear crystal, analogous to the d.c. component created by the passing of a sinusoidal electrical signal through a non-linear component (such as a diode), or network (such as an overdriven amplifier). Optically generated d.c. voltages engendered by this rectification can, in fact, be measured.

Parametric amplification, exploited in magnetic and microwave amplifiers for a considerable time, is achievable in optics too. In it, a power transfer takes place from the higher v local oscillation (pump) to the weaker, lower v external signal. Valuable low-noise amplification of very weak signals can thus be achieved.

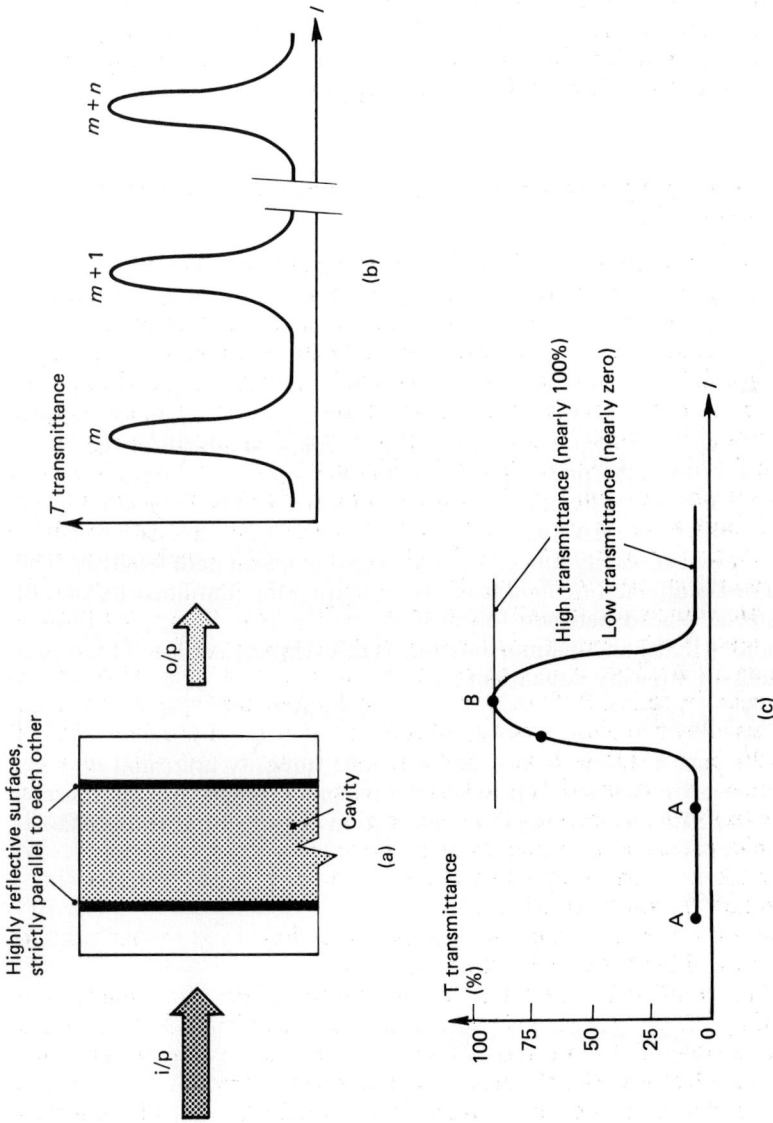

Figure 14.7 The Fabry–Perot etalon: (a) basic structure; (b) generalised resonance curve (Airy function) for $m, m+1, m+2, \ldots, m+n$; (c) transfer curves of interest $T_{\lambda = \text{const.}} = f(l)$

14.3.2   Doing arithmetic or the digital ray

Non-linear signal transfer and hysteresis are the two pillars of electronic digital computing. If non-linear light transmittance and luminous hysteresis were to be found, *optical* digital computers could be built. Laboratory work of the early 1980s has brought about devices with intensity-dependent transparency of the right kind, as well as devices exhibiting hysteretic, two-valued transfer characteristics. Although no optical digital computers are yet in existence, their potential is growing, with research work being fuelled as much by the expectations of speeds well in excess of what best can be expected of electronics, as by the hopes of *parallel* operation. Here is, in a nutshell, how the budding basic optical computing elements, the switch and the bistable, work.

(a)   The all-optical switch for logics

Remember how positive optical feedback helps with building up and maintaining a strong field within a laser cavity, and that, thanks to constructive interference, the field is strongest when the optical path length of the cavity equals an integer multiple of half-wavelengths, $l = m\lambda/2$, i.e. at resonance. Such cavities, called for historical reasons Fabry–Perot etalons (FPEs), and their transmittance curves $T = f(\lambda)$ ($l = $ const) and $T = f_T(l)$ ($\lambda = $ const) were known well before the birth of the laser. Figure 14.7 shows the one of interest. The operation of the optical switch relies on the FPE resonance. Working with a fixed-frequency source, the switch is thrown open (working point A, cavity detuned) or closed (working point B, cavity tuned) by changing the optical path length, l. The control of l is effected *by light itself* by exploiting the non-linearity of n of the material placed between the mirrors, as $l = n \times l_{\text{geometrical}}$. Indium antimonide (InSb) and gallium arsenide (GaAs) are examples of materials exhibiting an intensity dependent n:

$$n = n_o + n_1 \left| E \right|^2 \tag{14.2}$$

Illuminate such an FPE feebly, say with one intensity unit, and $n = n_o$, leaving the cavity detuned. Throw onto it two units, and n goes up, thereby tuning the FPE. Clearly, an AND-gate action (Figure 14.8) is obtainable.

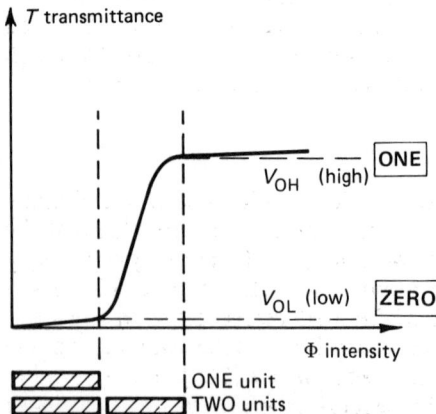

Figure 14.8 Optical AND gate

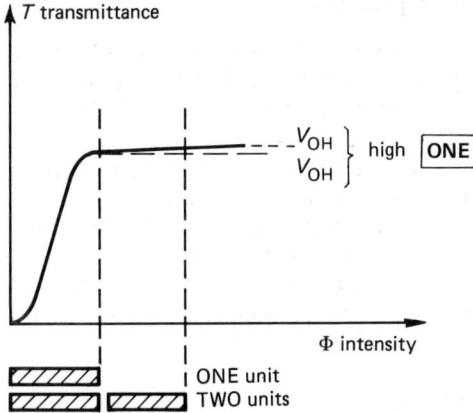

Figure 14.9 Optical OR gate

Furthermore, the non-linear 'filler' of the FPE has transformed its transfer curve so that its upper part now lends itself to the performance of the OR-gate function (Figure 14.9). Logical negation can also be obtained [257].

Remember that the gating of light is achieved by purely optical means, in the total absence of electrical connections.

(b)   The optical bistable
The single valued S-shaped transfer curve of Figure 14.9 is but a special case of the general hysteretic curve of Figure 14.10, the breadth of which is a matter of design. The consequence of hysteresis is great: it endows the device, clearly, with bistability. Figure 14.10 shows the two stable working points for the same value of incident intensity, ϕ: L with a low and H with a high transparency. Like any other bistable, e.g. ferrite core or transistor flip-flop, the device is a memory cell, capable of storing a 0 after activation by a 0 input, a 1 after activation by a 1. The hysteresis, responsible for the bistability, stems from the combined effects of non-linearity and optical positive feedback. Thanks to the latter, the field within the FPE can exceed significantly that of the incident beam. Such a condition corresponds to the presence of in-cavity standing waves at resonance. The electronics engineer will perceive in it the presence of a Q-factor, reminiscent of that of a microwave cavity or an LC circuit. Once thrown into the H state by a high A signal, the FPE will stay there after A is removed, quite content with the minimal, loss-compensating bias, B.

(c)   The optical signal amplifier
Prior to closing this section, I ought to mention that it is not only arithmetic that the non-linear FPE can do. Figure 14.12 shows that, provided that the working point is properly chosen, a small change of the input light intensity can cause a large change of the output. The Heriot-Watt University research team, responsible for much of the work connected with lightwave logics, see in a device exhibiting such amplifying behaviour the optical analogue of the transistor and call it a *transphasor* [256]. The operating speed of a transphasor is, potentially, several orders of magnitude higher

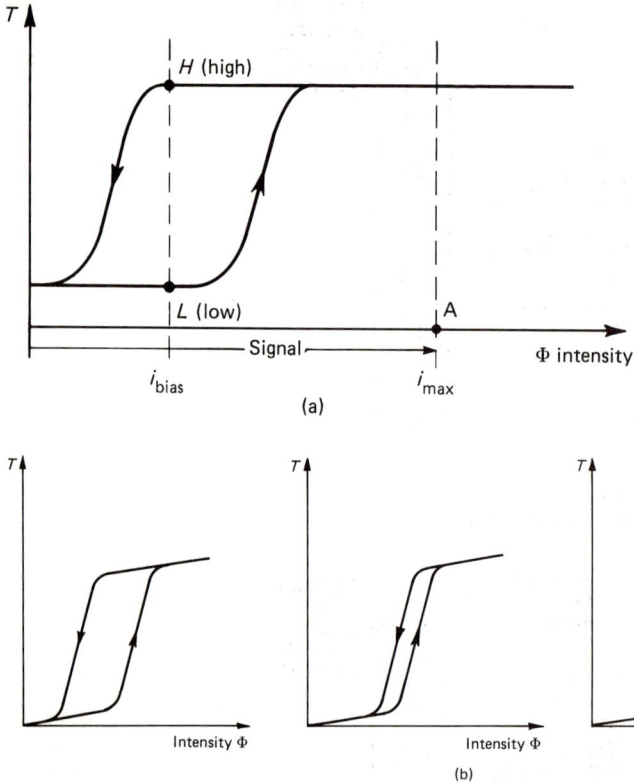

Figure 14.10 (a) Optical bistability. (b) The single valued S-shaped transfer curve of Figures 14.8 and 14.9 is but a special case of the general hysteresis curve of (a). Shown above are three typical loop widths

than that of the fastest possible transistor. Room temperature operation for non-linear FPEs, worked with hitherto cryogenically, appears today within reach.

(d)   Other optical flip-flops
Optical flip-flops and kindred pulse technique devices based on the use of liquid crystal light valves have been proposed, designed or even built. As such, they are intrinsically slow, with operating speeds of milliseconds rather than picoseconds. Since the subject of liquid crystals has not received a coverage in this work, we refer readers with a specific interest in such flip-flops to the bibliographic references 243 and 259.

14.3.3   Reversing time through optical phase conjugation

The term Phase Conjugate Mirror (PCM) applies to a reflector that makes a ray striking it retrace its original path in reverse. Such a mirror will send back every single ray of a bundle coming from the object, so that object and image become one. (Check for yourself that the best retroflectors of

(a) Original rays
    Ancillary beams not known

(b) PCM 'reflected' rays
    One ancillary beam shown

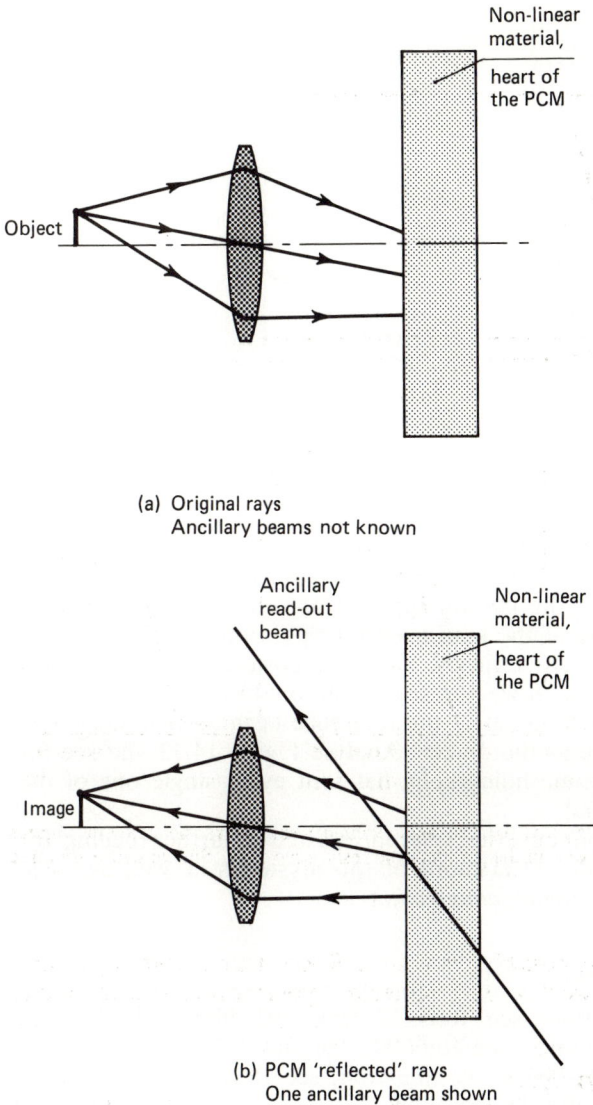

Figure 14.11 A phase conjugate mirror (PCM) will make a striking ray retrace its original path in reverse

Section 12.3.1(b) will not do this.) Figure 14.11 shows that the strange behaviour bestowed on rays by a PCM changes a diverging beam into a converging one and vice versa! Logically this implies that *any aberration introduced by the object–mirror space*, e.g. the spherical aberration of an interposed lens, *will be automatically corrected* (see Ref. 76, pp. 496–498, for details). There are further implications of fields propagating in a time-reversed sense, some of which are potentially useful but, rather than listing them, the mechanism of these almost bewitching mirrors will be described.

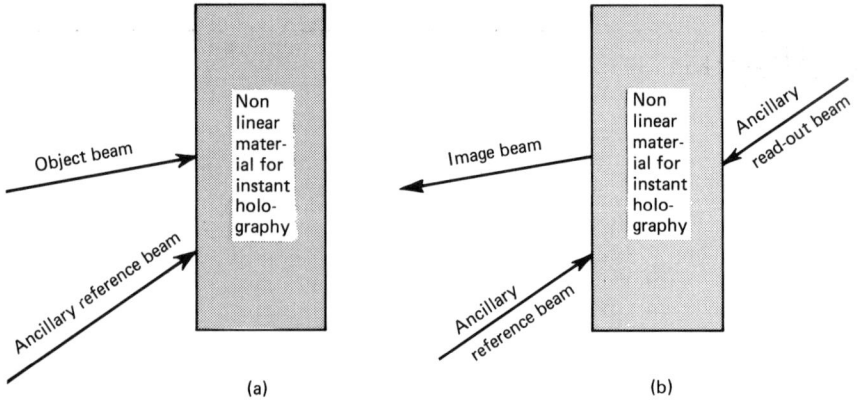

Figure 14.12 It is instant holography (a) combined with a 'read-out beam' at 180° to the reference beam (b) that makes PCMs work

Let us assume monochromatic illumination of the kind introduced in the section on holography, remembering that the latter is a two-step process. Imagine that instead of the photographic plate you now use a plate of non-linear material, say of the kind used for opto-optic beam deflectors (Section 10.3.2(c)). Given sufficient luminous power, such a plate will form a hologram *instantly*. If we now use a read-out beam pointing at 180° to what we called in Section 13.1.1 the reference beam, a *real* image will form in the same place as the object. Analyse Figure 14.12 and see for yourself that this real-time holography has sent every single one of the incident rays back home!

If this has whetted your appetite, references listed in 'further reading' for Section 14.3 below will reveal more of the mysteries as well as some potential uses of *optical phase conjugation*.

14.3.4 Further reading

For harmonic generation, see Refs 17 (pp. 502–505) and 192 (pp. 117–122). For digital optics, see Refs 90, 256 and 257, and 268 and its bibliography. For optical phase conjugation, see Refs 269, 236 (and its bibliography), 76 (pp. 494–496) and 286.

Epilogue

Much of the optical and electro-optic research work goes on on both sides of the Atlantic (and of the Pacific). By the time this book is published, some of this work will have led to the discovery of new phenomena, of new properties of both old and new materials and of dreamt of, and undreamt of, applications. Inquisitive readers will wish to delve into theoretical optics and electro-optics, and to keep abreast of ongoing research by browsing from time to time through specialised magazines. They may find the periodicals listed in Appendix 13 useful.

Appendix 1

Decibels and optical density

Table A1.1 Electrical power ratios expressed in decibels (dB)

Power ratio	dB	Power ratio	Power ratio	dB	Power ratio
1.0000	0	1.000	0.3162	5.0	3.162
0.9772	0.1	1.023	0.3090	5.1	3.236
0.9550	0.2	1.047	0.3020	5.2	3.311
0.9333	0.3	1.072	0.2951	5.3	3.388
0.9120	0.4	1.096	0.2884	5.4	3.467
0.8913	0.5	1.122	0.2818	5.5	3.548
0.8710	0.6	1.148	0.2754	5.6	3.631
0.8511	0.7	1.175	0.2692	5.7	3.715
0.8318	0.8	1.202	0.2630	5.8	3.802
0.8128	0.9	1.230	0.2570	5.9	3.890
0.7943	1.0	1.259	0.2512	6.0	3.981
0.7762	1.1	1.288	0.2455	6.1	4.074
0.7586	1.2	1.318	0.2399	6.2	4.169
0.7413	1.3	1.349	0.2344	6.3	4.266
0.7244	1.4	1.380	0.2291	6.4	4.365
0.7079	1.5	1.413	0.2239	6.5	4.467
0.6918	1.6	1.445	0.2188	6.6	4.571
0.6761	1.7	1.479	0.2138	6.7	4.677
0.6607	1.8	1.514	0.2089	6.8	4.786
0.6457	1.9	1.549	0.2042	6.9	4.898
0.6310	2.0	1.585	0.1995	7.0	5.012
0.6166	2.1	1.622	0.1950	7.1	5.129
0.6026	2.2	1.660	0.1905	7.2	5.248
0.5888	2.3	1.698	0.1862	7.3	5.370
0.5754	2.4	1.738	0.1820	7.4	5.495
0.5623	2.5	1.778	0.1778	7.5	5.623
0.5495	2.6	1.820	0.1738	7.6	5.754
0.5370	2.7	1.862	0.1698	7.7	5.888
0.5248	2.8	1.905	0.1660	7.8	6.026
0.5129	2.9	1.950	0.1622	7.9	6.166
0.5012	3.0	1.995	0.1585	8.0	6.310
0.4898	3.1	2.042	0.1549	8.1	6.457
0.4786	3.2	2.089	0.1514	8.2	6.607
0.4677	3.3	2.138	0.1479	8.3	6.761
0.4571	3.4	2.188	0.1445	8.4	6.918

	−dB+			−dB+	
	← →			← →	
Power ratio	dB	Power ratio	Power ratio	dB	Power ratio
0.4467	3.5	2.239	0.1413	8.5	7.079
0.4365	3.6	2.291	0.1380	8.6	7.244
0.4266	3.7	2.344	0.1349	8.7	7.413
0.4169	3.8	2.399	0.1318	8.8	7.586
0.4074	3.9	2.455	0.1288	8.9	7.762
0.3981	4.0	2.512	0.1259	9.0	7.943
0.3890	4.1	2.570	0.1230	9.1	8.128
0.3802	4.2	2.630	0.1202	9.2	8.318
0.3715	4.3	2.692	0.1175	9.3	8.511
0.3631	4.4	2.754	0.1148	9.4	8.710
0.3548	4.5	2.818	0.1122	9.5	8.913
0.3467	4.6	2.884	0.1096	9.6	9.120
0.3388	4.7	2.951	0.1072	9.7	9.333
0.3311	4.8	3.020	0.1047	9.8	9.550
0.3236	4.9	3.090	0.1023	9.9	9.772

Table A1.2   Optical density (D) versus percent transmittance (T), $D = \log_{10}(1/T)$

D	T	D	T	D	T	D	T	D	T
0.00	100.0	0.48	33.11	0.96	10.96	1.44	3.631	1.92	1.202
0.01	97.72	0.49	32.36	0.97	10.72	1.45	3.548	1.93	1.175
0.02	95.50	0.50	31.62	0.98	10.47	1.46	3.467	1.94	1.148
0.03	93.33	0.51	30.90	0.99	10.23	1.47	3.388	1.95	1.122
0.04	91.20	0.52	30.20	1.00	10.00	1.48	3.311	1.96	1.096
0.05	89.13	0.53	29.51	1.01	9.772	1.49	3.236	1.97	1.072
0.06	87.10	0.54	28.84	1.02	9.550	1.50	3.162	1.98	1.047
0.07	85.11	0.55	28.18	1.03	9.333	1.51	3.090	1.99	1.023
0.08	83.18	0.56	27.54	1.04	9.120	1.52	3.020	2.00	1.000
0.09	81.28	0.57	26.92	1.05	8.913	1.53	2.951	2.05	0.891
0.10	79.43	0.58	26.30	1.06	8.710	1.54	2.884	2.10	0.794
0.11	77.62	0.59	25.70	1.07	8.511	1.55	2.818	2.15	0.708
0.12	75.86	0.60	25.12	1.08	8.318	1.56	2.754	2.20	0.631
0.13	74.13	0.61	24.55	1.09	8.128	1.57	2.692	2.25	0.562
0.14	72.44	0.62	23.99	1.10	7.943	1.58	2.630	2.30	0.501
0.15	70.79	0.63	23.44	1.11	7.762	1.59	2.570	2.35	0.447
0.16	69.18	0.64	22.91	1.12	7.586	1.60	2.512	2.40	0.398
0.17	67.61	0.65	22.39	1.13	7.413	1.61	2.455	2.45	0.355
0.18	66.07	0.66	21.88	1.14	7.244	1.62	2.399	2.50	0.316
0.19	64.57	0.67	21.38	1.15	7.079	1.63	2.344	2.55	0.282
0.20	63.10	0.68	20.89	1.16	6.918	1.64	2.291	2.60	0.251
0.21	61.66	0.69	20.42	1.17	6.761	1.65	2.239	2.65	0.224
0.22	60.26	0.70	19.95	1.18	6.607	1.66	2.188	2.70	0.200
0.23	58.88	0.71	19.50	1.19	6.457	1.67	2.138	2.75	0.178
0.24	57.54	0.72	19.05	1.20	6.310	1.68	2.089	2.80	0.158
0.25	56.23	0.73	18.62	1.21	6.166	1.69	2.042	2.85	0.141
0.26	54.95	0.74	18.20	1.22	6.026	1.70	2.000	2.90	0.126
0.27	53.70	0.75	17.78	1.23	5.888	1.71	1.950	2.95	0.112
0.28	52.48	0.76	17.38	1.24	5.754	1.72	1.905	3.00	0.100
0.29	51.29	0.77	16.98	1.25	5.623	1.73	1.862	3.05	0.089

D	T	D	T	D	T	D	T	D	T
0.30	50.12	0.78	16.60	1.26	5.495	1.74	1.820	3.10	0.079
0.31	48.98	0.79	16.22	1.27	5.370	1.75	1.778	3.15	0.071
0.32	47.86	0.80	15.85	1.28	5.248	1.76	1.738	3.20	0.063
0.33	46.77	0.81	15.49	1.29	5.129	1.77	1.700	3.40	0.040
0.34	45.71	0.82	15.14	1.30	5.012	1.78	1.660	3.60	0.025
0.35	44.67	0.83	14.79	1.31	4.898	1.79	1.622	3.80	0.016
0.36	43.65	0.84	14.45	1.32	4.786	1.80	1.585	4.00	0.010
0.37	42.66	0.85	14.13	1.33	4.677	1.81	1.549	4.25	0.006
0.38	41.69	0.86	13.80	1.34	4.571	1.82	1.514	4.50	0.003
0.39	40.74	0.87	13.49	1.35	4.467	1.83	1.479	4.75	0.0018
0.40	39.81	0.88	13.18	1.36	4.365	1.84	1.445	5.00	0.0010
0.41	38.90	0.89	12.88	1.37	4.266	1.85	1.413	5.25	0.0006
0.42	38.02	0.90	12.59	1.38	4.169	1.86	1.380	5.50	0.0003
0.43	37.15	0.91	12.30	1.39	4.074	1.87	1.350	5.75	0.00018
0.44	36.31	0.92	12.02	1.40	3.981	1.88	1.318	6.00	0.00010
0.45	35.48	0.93	11.75	1.41	3.890	1.89	1.288	From Ref.3(a),	
0.46	34.67	0.94	11.48	1.42	3.802	1.90	1.259	p. 227	
0.47	33.88	0.95	11.22	1.43	3.715	1.91	1.230		

Optical density, D, can be regarded as the analogue of electrical attentuation, A, in decibels, with opacity, O (the reciprocal of transmittance, $O = 1/T$) becoming the analogue of the numerical value of the electrical output-to-input power ratio.

Essential lens formulae*

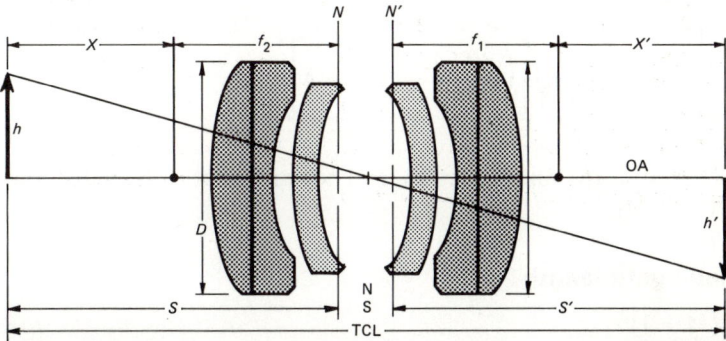

Figure A2.1

Definitions

OA is the optical axis, a line joining the object, image and centre of curvatures of each element

S is the object distance

S' is the image distance

TCL is the total conjugate length (object to image distance)

D is the apparent diameter of the aperture stop, as seen through the long conjugate of the lens.

M is the ratio of image height to object height.

h is the height of the object above the OA

h' is the height of the image below the OA

f is the effective focal length of the lens

NOTE: For a lens used in air, $f_1 = f_2 = f$

X is the distance from the second focal point (f_2) to the object plane

X' is the distance from the first focal point (f_1) to the image plane

N,N' are the nodal points of the lens

NS is the distance between the two nodal points, N and N'

NOTE: NS may be negative (crossed nodes)

EA is the effective aperture

*By permission of JML Optical Industries Inc.

Formulae are based upon thin lens calculations. The same formulae are accurate for thick lens calculations if nodal point separation is considered.

Formulae

1 Magnification

$$M = \frac{h'}{h} = \frac{S'}{S}$$

$$M = \frac{X'}{f} = \frac{S - f}{f}$$

$$M = \frac{f}{X} = \frac{f}{S - f}$$

$$M = \frac{\left(\dfrac{\text{TCL}}{f} - 2\right) \pm \sqrt{\left[\left(\dfrac{-\text{TCL}}{f} - 2\right)^2 - 4\right]}}{2}$$

Note: Positive root is the magnification, and negative root is the reduction factor R. $(R = 1/M)$

2 Total conjugate length

$$\text{TCL} = \frac{f(M + 1)^2}{M}$$

$$\text{TCL} \quad X + X' + 2f + \text{NS} = S + S' + \text{NS}$$

3 Focal length

$$\frac{1}{f} = \frac{1}{S} + \frac{1}{S'}$$

$$f = \frac{M(\text{TCL})}{(M + 1)^2}$$

$$f = \sqrt{(XX')}$$

$$f = \frac{S}{1 + \dfrac{1}{M}} = \frac{S'}{(M + 1)}$$

4 Lens to image distance (S′)

$$S' = f(M + 1)$$

$$S' = M(S)$$

$$S' = \frac{f(S)}{S - f}$$

5 Lens to object distance (S)

$$S = f \left(1 + \frac{1}{M} \right)$$

$$S = \frac{S'}{M}$$

$$S = \frac{f(S')}{S' - f}$$

6 f-number (with infinite object distance)

$$f^{\#}_{\infty} = \frac{f}{D}$$

7 Effective aperture (at a particular magnification)

$$EA = f^{\#}_{\infty} (M + 1)$$

Note: If EA is held constant between lenses, the resulting image plane illuminance is also constant.

Note

This appendix does not use exactly the same notation and sign convention as the main text. Figure A2.1 and a simple sketch on graph paper supporting the calculations should remove all doubt regarding image reversal.

Radiation flux within a solid angle of an emitter, calculated from its polar diagram

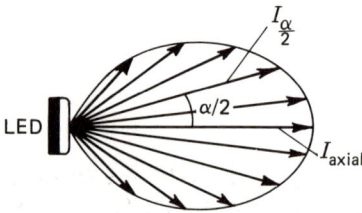

Figure A3.1 Polar diagram of LED

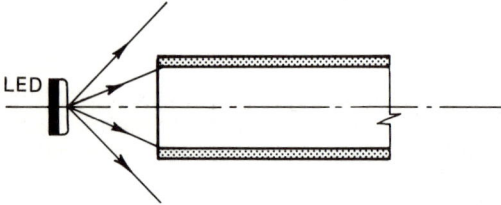

Figure A3.2 LED-to-lens coupling

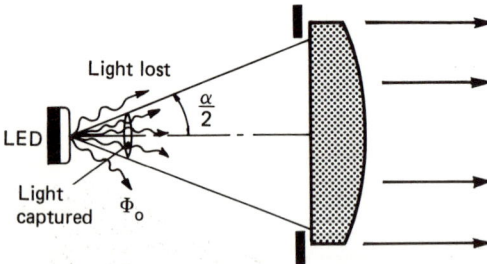

Figure A3.3 LED-to-fibre coupling

The optimisation process of the TXA optics requires the knowledge of the ϕ_θ versus $\alpha/2$ variation. The $\phi_\theta = f(\alpha/2)$ function will also help to determine the efficiency of an LED-to-fibre coupling, for a given NA of the fibre (Figures A3.2 and A3.3).

Prior to the calculations, the following assumptions are made:

1. The spatial distribution of radiant intensity is symmetrical about the main axis of the emitter, i.e. it can be represented by a volume of revolution formed by rotating the polar diagram around its principal axis.

2. We have at hand, in addition to the polar diagram, either the on-axis radiant intensity (in W/sr) or the total radiant power ϕ_T (in W).

The logics of the procedure

To calculate the flux contained within the solid angle θ, called here ϕ_θ, we divide this angle into elementary 'zones', $\triangle\theta$, sufficiently small for the intensity within each one of them to be practically constant. (Here, each zone is formed by the revolution of a flat half angle of 5°.) We then calculate the elementary flux $\triangle\phi_\theta$ radiated by each zone, which is, of course, $I_\theta \times \triangle\theta$.

Summing all the elementary contributions $\triangle\phi_\theta$ contained within the solid angle θ yields the answer to the problem, namely:

$$\phi_\theta = \Sigma \, \triangle\phi_\theta$$

The calculation

The formula for finding the space angle θ resulting from the revolution of a flat half-angle $\alpha/2$ is derived geometrically (Figure A3.4). The elementary area of revolution $\triangle A$ is:

$$\triangle A \simeq \left(2\pi R \sin \frac{\alpha}{2}\right) dl$$

However, as $dl = Rd(\alpha/2)$, dropping the $\simeq$ sign, we get:

$$\triangle A = 2\pi R^2 \sin \left(\frac{\alpha}{2}\right) d \left(\frac{\alpha}{2}\right)$$

The solid angle of revolution:

$$\theta \equiv \frac{\text{Area of spherical cap } A}{R^2}$$

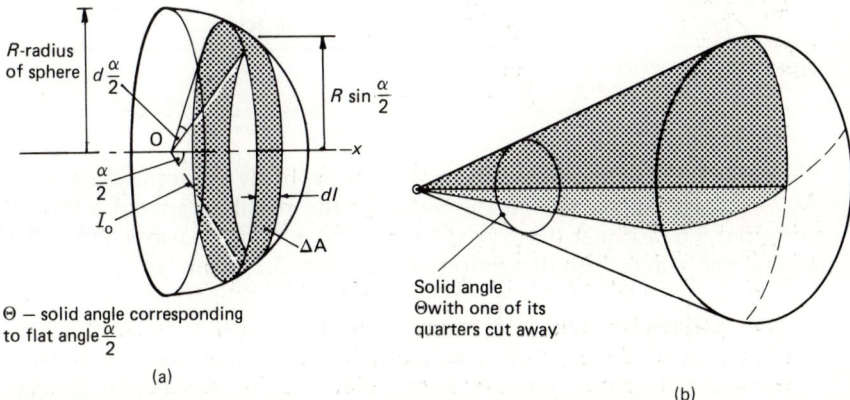

R-radius of sphere

Θ – solid angle corresponding to flat angle $\frac{\alpha}{2}$

(a)

Solid angle Θ with one of its quarters cut away

(b)

Figure A3.4 Elementary 'zones'

Hence

$$\theta \equiv \frac{2\pi}{R^2} \times R^2 \int_0^{\alpha/2} \sin\left(\frac{\alpha}{2}\right) d\left(\frac{\alpha}{2}\right)$$

$$= 2\pi \int_0^{\alpha/2} \sin\left(\frac{\alpha}{2}\right) d\left(\frac{\alpha}{2}\right) = 2\pi\left(1 - \cos\frac{\alpha}{2}\right)$$

Thus, the full solid angle θ, expressed in steradians is:

$$\theta = 2\pi\left(1 - \cos\frac{\alpha}{2}\right) \qquad\qquad\qquad (A3.1)$$

This relation is tabulated in Table A3.1 and plotted in Figure A3.5.

Table A3.1

$\dfrac{\alpha}{2}$		$\Delta\,\dfrac{\alpha}{2}$	$\cos\dfrac{\alpha}{2}$	$\theta = 2\pi\left(1 - \cos\dfrac{\alpha}{2}\right)$	$\Delta\theta$
(degrees)	(mrad)	(mrad)		(sr)	(msr)
Col. 1	Col. 2	Col. 3	Col. 4	Col. 5	Col. 6
4	69.81	} 17.5	0.998	0.01531	} 8.6
5	87.27		0.996	0.02391	
10			0.985	0.0946*	
15			0.966	0.2141	
20			0.939	0.379†	
25			0.906	0.589	
30	523.6	} 17.5	0.866	0.84178	} 55.7
31	541.1		0.857	0.89744	
35			0.819	1.136	
40			0.766	1.470	
45			0.707	1.840‡	
50			0.643	2.244	
55			0.574	2.679	
60			0.500	3.142	
65			0.423	3.628	
70			0.342	4.134	
75			0.259	4.657	
80			0.174	5.192	
85			0.087	5.735	
89	1553.3	} 17.5	0.017	6.173	} 110.0
90	1570.8		0.000	6.283‡	

*Approximation: $\theta = \pi\sin^2(\alpha/2)$ yields $\theta = 0.09473$
†Approximated as above yields $\theta = 0.367965$
‡$\theta = \pi$ sr

Columns 1, 3 and 6 of the table point out explicitly the influence of the *position* of a 1° flat angle on the value of the resulting spatial angle of revolution: 8.6 msr near the x-axis, 55.7 msr at 30° off the x-axis and 110.0 msr near the y-axis. The derivation of Equation 3.1, namely:

$$\frac{d\theta}{d\left(\frac{\alpha}{2}\right)} = \frac{d\left[2\pi\left(1 - \cos\frac{\alpha}{2}\right)\right]}{d\left(\frac{\alpha}{2}\right)} = 2\pi \sin\left(\frac{\alpha}{2}\right)$$

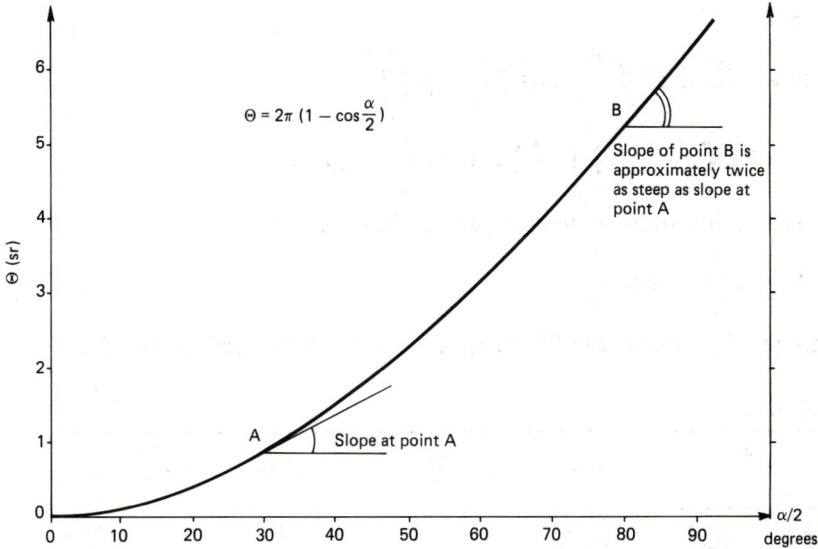

Figure A3.5 Spatial angle θ versus flat angle $\alpha/2$. Note the value of the flat half-angle corresponding to 1 sr, namely 32.77°

proves the point, also witnessed by the variation of the slope of the graph in Figure A3.5. We bring this point to attention in order to emphasise the deceptive visual power of the *flat* polar diagram.

Readers wishing to calculate θ values for small half-angles $\alpha/2$ (<20°) may find useful the approximation:

$$\theta = \pi \sin^2\left(\frac{\alpha}{2}\right) \tag{A3.2}$$

The error is −3% for 20°, reducing to −0.18% for 10°. Should numerical solutions be required for values of $\alpha/2$ other than those shown, the 12-step program for the Commodore P.50 pocket calculator (Table A3.2) may be of help.

Table A3.2   Calculation of space angle θ resulting from the resolution of a flat half angle $\alpha/2$ on Commodore P.50 calculator

Programme

LRN	Step	Execute
COS	1	GOTO
− (minus)	2	
1	3	00
×	4	ENTER $\frac{\alpha}{2}$
2	5	
×	6	RUN (R/S)
π	7	ENTER $\frac{\alpha}{2}$
+/−	8	
=	9	RUN (R/S)
STOP(R/S)	10	ENTER $\frac{\alpha}{2}$
GOTO	11	
00	12	etc.
LRN		

Finally, Table A3.3 gives a numerical example of the flux versus solid angle evaluation for a specific LED, namely the 1 A 119F device made by HAFO of Sweden.

Table A3.3   Calculation of the accrued flux of the 1 A 119F versus flat angle

1	2	3	4		5	6
Flat angle	Flat angle segment	Corresp. zonal space angle	Mean intensity I_i at angle $\alpha/2$ shown		Radiated flux	Accrued flux
$\alpha/2$ (degrees)	$\triangle(\alpha/2)$ (degrees)	$\triangle\theta$ (sr)	$\alpha/2$ (degrees)	I_i (mW/sr)	$I_i\times\triangle\theta$ (mW)	Φ_θ (mW)
10	0–10	0.094	5	8.8	0.83	0.83
20	10–20	0.283	15	9.0	2.55	3.37
30	20–30	0.465	25	7.7	3.58	6.95
40	30–40	0.628	35	6.0	3.77	10.72
50	40–50	0.774	45	6.0	4.64	15.36
60	50–60	0.896	55	5.5	4.93	20.26
70	60–70	0.985	65	4.2	4.14	24.4
80	70–80	1.052	75	2.7	2.84	27.24
90	80–90	1.092	85	0.7	0.76	28.00*

*Total power (flux) P_{TOT}

Appendix 4

Lambertian radiators and re-radiators

In the visible partof the spectrum an extended radiator is said to be *Lambertian* when its luminance (brightness) is the same in all directions, i.e. independent of the viewing angle. A globe lampshade made of 'milky' glass is a good everyday example: it has the brightness in the centre and near the edges of what in fact looks like a flat luminous disc, despite its sphericity, which causes the viewing angle to be 90° in the first and 10°–20° in the second case.

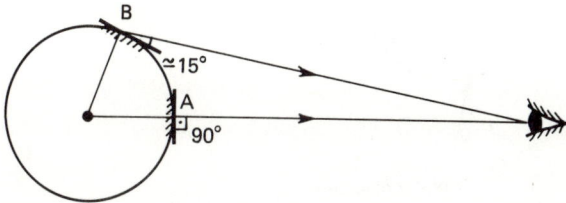

Figure A4.1 Cross-section of a spherical radiator. Portion A viewed at 90°, portion B at 15°

The sun, too, is a Lambertian radiator. The concept is extended to other than visible radiators, e.g. infrared LEDs, which are said to be Lambertian when their *radiance* is direction independent. It is also used in connection with *diffuse* reflectors (i.e. re-radiators) such as good Bristol board.

It will be easily seen that, for a radiator to be Lambertian, its polar diagram must satisfy the relation:

$$I_\alpha = I_o \cos \alpha \qquad (A4.1)$$

and hence be a circle (Figure A4.2) and, three dimensionally, a sphere. Indeed, then and only then will the intensity loss factor $\cos \alpha$ be compensated for, numerically, by the projected area factor $1/\cos \alpha$. (The reader may find it useful to revise Section 4.4 on the definitions of luminance, brightness and radiance.) Indeed, as per Figure A4.3, we have, for oblique viewing, $A' = A \cos \alpha$, so that the luminance, which for normal viewing was $L_o = I_o/A$ becomes, for oblique viewing, $L_\alpha = I_\alpha/A'$ which is $(I_o \cos \alpha)/(A \cos \alpha)$ and therefore L_α equals L_0. QED.

It is worth mentioning that a Lambertian source of area A with a radiance R radiates into a hemisphere ($\theta = 2\pi$) a total flux (power) of

321

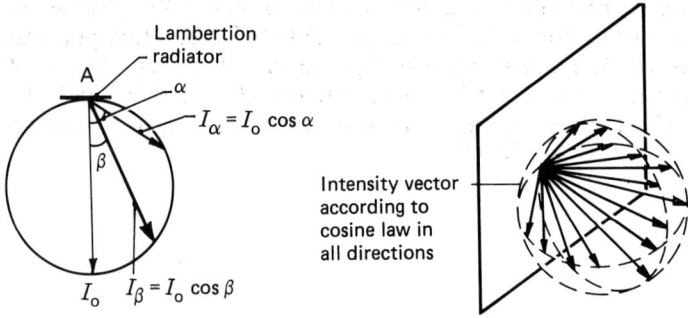

Figure A4.2 The polar diagram of a Lambertian radiator is a circle and three-dimensionally a sphere

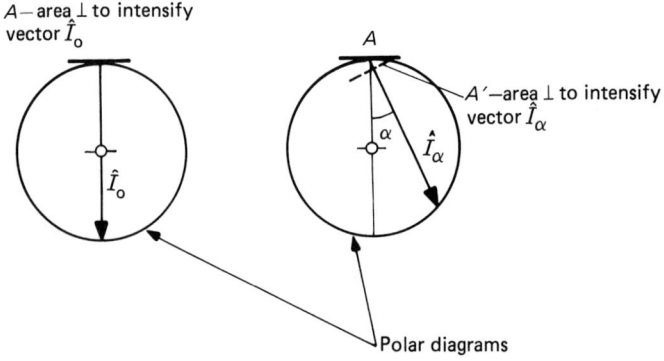

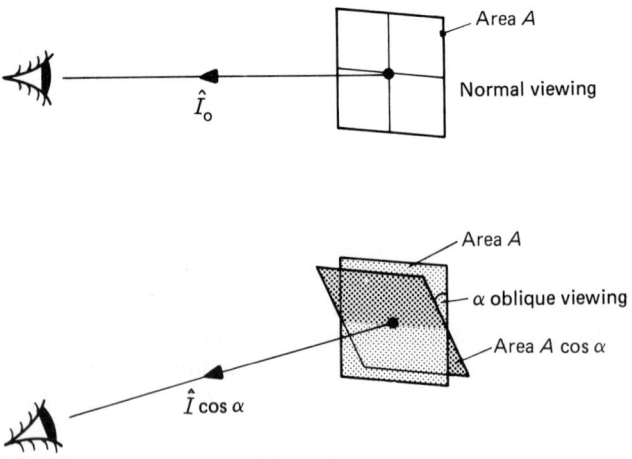

Figure A4.3 Normal and oblique viewing

πRA W and *not*, as one might intuitively expect, of $2\pi RA$ W. (Mathematical derivation can be found in Ref. 2, pp. 184–185.) The flux per unit area, ϕ, is thus πR. It is this result that is probably at the origin of two of the slightly peculiar photometric units, namely the lambert and the foot-lambert. The lambert is the luminance of $1/\pi$ candela/cm^2, and the foot-lambert that of $1/\pi$ candela/ft^2.

On noise in semiconductor diode detectors

No matter how meticulously screened from unwanted electric, magnetic and optical interferences, the detector diode and its load resistor remain the abode of some undesirable currents and voltages. In this context, the term *noise* refers to these, internally generated, residual currents and voltages. Although belonging to two clearly distinguishable kinds, these currents and voltages both owe their existence to the *discreteness* of electrical changes.

In the first kind, the fundamental granularity of electric current causes minute random fluctuations $\triangle i$. This is the *shot noise*. In an obscured diode, i is constituted by the dark current i_d, then alone responsible for the presence of $\triangle i$ fluctuations. When light activated, the light current of the diode i_l introduces its own fluctuations $\triangle i_l$.

In the second kind, the load resistance, R_L, like any resistive element, generates its own tiny noise voltages $\triangle V$ caused by the *thermally induced* random motion of free changes within it. Hence its name: *thermal noise*. The diode detector, being a semiconductor, experiences an ever-present process of carrier generation and recombination. It too causes tiny $\triangle V$ fluctuations, even in darkness. These charge motion produced $\triangle V$ values are independent of the current, unless Joule heating produces a significant temperature rise, in which case its action gets reflected in T, already present in the equation expressing $\overline{\triangle V^2}$. Cooling, of course, makes $\overline{\triangle V^2}$ subside, and finally vanish at 0 K. Quantitatively the noise of these two origins looks as follows.

Shot (also called Schottky) noise

The RMS value $\overline{\triangle i_d}$ of the minute fluctuations of the dark current i_d is given by the relation (derived in Ref. 278, pp. 555–556):

$$(\overline{\triangle i_d^2})^{1/2} = (2ei_dB)^{1/2} \tag{A5.1}$$

The electron charge e equals 1.6×10^{-19} coulomb; B, the bandwidth, is in hertz. We notice that doubling the bandwidth increases the noise current by 41% only. The shot noise is a *white noise*, i.e. independent of the position of the bandwidth B within the total spectrum, which is to say it will have the same value whether looked at through a 1–1001 kHz or a 300–301

MHz filter. Numerically, for a dark current of 100 nA and a bandwidth of 10 kHz, $(\overline{\Delta i_d^2})^{1/2} = 20$ pA and still a mere 0.2 nA for $B = 1$ MHz. In a typical biased PIN photodiode $i_d < 10$ nA at room temperature. Dark current and hence dark current shot noise can be reduced by cooling, but dividends cooling pays are modest. A thermoelectric cooler (the Peltier cell) will lower T_j by some 30 °C, and reduce i_d to $\sim 1/10$ and thereby $(\overline{\Delta i^2})^{1/2}$ to about one-third of their original values. Dark current decreases as the bias voltage is reduced, tending to zero for the SQM (unbiased) case. Ambient illumination produces its own contribution to shot noise $(\overline{\Delta i_a^2})^{1/2}$ giving us yet another reason for keeping it out of the RXA. Adding the dark and light current contributions must be done on the RMS basis $(\overline{\Delta i_s^2})^{1/2} = \sqrt{[(\overline{\Delta i_d})^2 + (\overline{\Delta i^2})]}$ as the two contributions are statistically non-correlated.

Thermal (also called Johnson) noise

In any resistive element the random motion of free changes produces an RMS voltage V_j (j for Johnson) of:

$$\overline{(\Delta V_j^2)^{1/2}} = (4kTRB)^{1/2} \tag{A5.2}$$

with T in kelvins, R in ohms and B in hertz. (An elegant derivation by Johnson and Nyquist can be found in Ref. 278, pp. 551–552.) Obviously, cooling quiets down the thermal agitation and thus reduces V_j, bringing it finally down to zero for absolute zero temperature. k is a physical constant equal to 1.38×10^{-23} J/K, the Boltzman constant, inherited from the kinetic theory of gases.

Viewed through a modern amplifier the Johnson noise is seen a current:

$$\overline{(\Delta i_j^2)^{1/2}} = \left(\frac{4kTB}{R}\right)^{1/2} \tag{A5.3}$$

Numerically, a 1 MΩ resistor at room temperature produces a Johnson RMS noise current $(\overline{\Delta i_j^2})^{1/2}$ of

$$2\sqrt{\frac{1.38 \times 10^{-23} \times 293 \times 106}{1 \times 106}} = 130 \text{ pA}$$

when looked at through a 1 MHz bandwidth filter. The Johnson noise is a white noise.

It is interesting to reflect, in passing, upon the fact that the maximum collectable noise power, P_j, produced by a resistor R is that which would be gathered by a (presumably noiseless) load resistance $R_L = R_1$ with P_j itself becoming:

$$P_j = kTB \tag{A5.4}$$

its value being manifestly independent of the value of R itself! (P_j is obtained by taking a quarter of the open circuit voltage × short circuit current product of the thermal noise generator as per Figure A5.1).

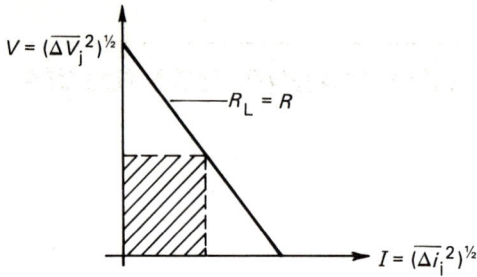

Figure A5.1

Total noise

The shot noise $(\overline{\Delta i_s^2})^{1/2}$ and the Johnson noise $(\overline{\Delta i_j^2})^{1/2}$ contributions add in an RMS manner, being statistically non-correlated. The sum so obtained forms the total RMS noise current i_{NT}:

$$i_{NT_{RMS}} = [i_{s_{RMS}}^2 + i_{j_{RMS}}^2]^{1/2}$$

or

$$i_{NT} = [(\overline{\Delta i_s^2}) + (\overline{\Delta i_j^2})]^{1/2} \tag{A5.5}$$

Choosing R_L

For zero ambient illumination, the two contributions become equal when:

$$2ei_d = \frac{4kT}{R_L}$$

i.e. when $R_L i_d = 2kT/e$ which condition corresponds, for room temperature, to $R_L i_d \simeq 50$ mV.

For $R_L > 50$ mV/i_d shot noise will predominate. In a diode with $i_d = 10$ nA, for example, this will take place for $R_L > 5$ MΩ. How far R_L can be increased in order to reduce i_{NT} will depend on the required speed of response of the circuit. The nomographs of Ref. 275 can be of help in its determination, for a given circuit capacitance. They also give, for a given dark current, the following:

- The value of R_L for which $i_{S_{RMS}} = i_{J_{RMS}}$
- The value of i_{TN}
- The value of NEP for a known R_L

With regard to the influence of the dynamic resistance of the diode R_D, effectively in parallel with R_L, let us say that, with modern diode detectors, it can usually be ignored, as its value runs into hundreds if not thousands of megohms. The exception concerns the odd case of an unbiased SQM(V) diode. Here, the Johnson noise dominates owing to the extreme smallness of the dark current (only picoamps in strength), this smallness being accompanied by a drastic fall of R_D.

Determination of the maximum angle of acceptance

We cover the case of a cladded fibre surrounded by air. The incident ray butt-couples into the fibre core on axis (a 'meridional ray' situation).

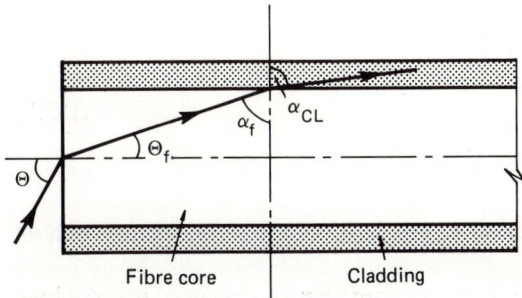

Figure A6.1

Notation

n_0 refractive index of air
n_f refractive index of fibre (core)
n_{CL} refractive index of cladding
θ angle of incidence at input interface
θ_{max} maximum value of θ
θ_f angle of exitance at input interface
α_f angle of incidence at fibre/cladding interface
α_{CL} angle of exitance at fibre/cladding interface

At the fibre/cladding interface we have:

$$\frac{n_{CL}}{n_f} = \frac{\sin \alpha_f}{\sin \alpha_{CL}}$$

Total internal reflection (TIR) will take place for $\alpha_{CL} = 90°$, i.e. for $\sin \alpha_{CL} = 1$. We then have:

$$\frac{n_{CL}}{n_f} = \sin \alpha_f \qquad\qquad (A6.1)$$

At the input interface:

$$\frac{n_f}{n_0} = \frac{\sin \theta}{\sin \theta_f}$$

Noticing that $\sin \theta_f = \cos \alpha_f$:

$$\frac{n_f}{n_0} = \frac{\sin \theta}{\cos \alpha_f}$$

and, in the TIR condition:

$$n_0 \sin \theta_{max} = n_f \cos \alpha_f \qquad (A6.2)$$

Using now the trigonometric identity $\cos x \equiv \sqrt{(1 - \sin^2 x)}$:

$$\sin \theta_{max} = \frac{n_f}{n_0} \sqrt{(1 - \sin^2 \alpha_f)} \qquad (A6.3)$$

Combining Equations A6.1 and A6.3 we finally express θ_{max} as a function of the refractive indices of the air, the fibre core and the cladding:

$$\sin \theta_{max} = \frac{n_f}{n_0} \sqrt{\left[1 - \left(\frac{n_{CL}}{n_f} \right)^2 \right]}$$

or, more practically:

$$\sin \theta_{max} = \frac{1}{n_0} \sqrt{(n_f^2 - n_{CL}^2)} \qquad (A6.4)$$

$\sin \theta_{max}$ is called the numerical aperture (NA) of the fibre, which, with $n_0 \simeq 1.0$, gives:

$$NA = \sqrt{(n_f^2 - n_{CL}^2)} \qquad (A6.5)$$

The particular case of an uncladded fibre, in which n_c is replaced by n_0, gives $NA = \sqrt{(n_f^2 - 1.0)}$, which with the notation of Section 7.2.1 becomes:

$$NA = \sqrt{(n_s^2 - 1.0)} \qquad (7.2)$$

where n_s is the refractive index of the slow medium (glass). The influence exerted by the difference Δn between the indices $n_f - n_{CL}$ upon NA is made explicit by rearranging A6.4 into:

$$NA = \sqrt{\left[\frac{1}{n_0} (n_f + n_{CL}) (n_f - n_{CL}) \right]}$$

giving

$$NA \simeq \left[\frac{1}{n_0} \sqrt{(2n)} \right] \sqrt{\Delta n}$$

where

$$n = \frac{n_f + n_{CL}}{2}$$

leading to

$$NA = const. \sqrt{\Delta n} \qquad (A6.6)$$

The influence of fibre (core) diameter on the number of probable propagating modes

According to C. Sandbank [39, p.3] the number of modes N is given, approximately, by:

$$N \simeq 0.5 \left(\frac{\pi d\, \mathrm{NA}}{\lambda} \right)^2 \tag{A7.1}$$

It is instructive to calculate N for two sets of conditions as follows:

1. 50 μm fibre with a large numerical aperture (NA = 0.4) working at:

   $\lambda = 0.9$ μm ($d = 50 \times 10^{-6}$ m, NA = 0.4, $\lambda = 0.9 \times 10^{-6}$ m)

   The calculation yields $N = 2437$.

2. 4 μm fibre with a small numerical aperture (NA = 0.2) working at

   $\lambda = 1.55$ μm ($d = 4 \times 10^{-6}$ m, NA = 0.2, $\lambda = 1.55 \times 10^{-6}$ m)

   The calculation now yields $N = 1.314$, i.e. only *one* propagating mode.

This also shows that wavelengths in excess of a certain value, λ_c (here 1.78 μm), will simply not propagate along such a thin fibre. λ_c and its corresponding frequency $f_c = c/\lambda_c$ are, respectively, the cut-off wavelength and frequency of the fibre.

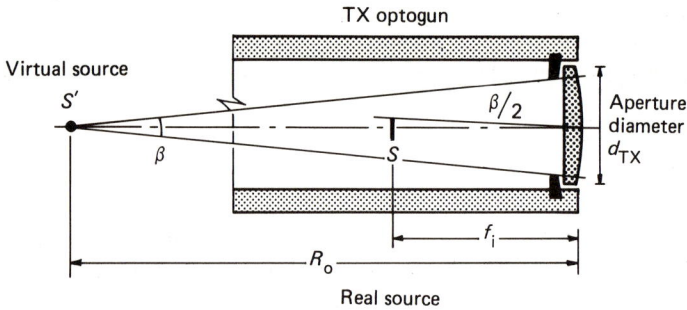

Figure A9.1 The 'back distance' concept

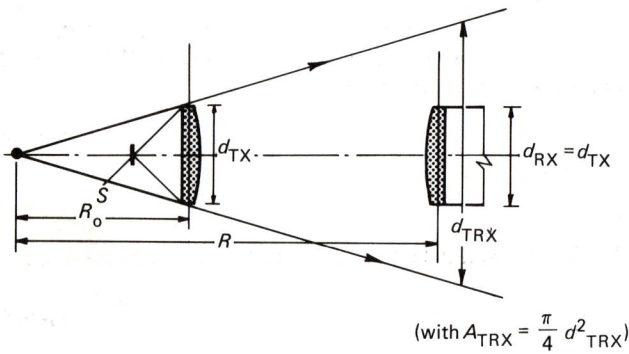

$$(\text{with } A_{\text{TRX}} = \frac{\pi}{4} d^2_{\text{TRX}})$$

Figure A9.2 The 'back distance' concept

$$R_{\text{FS}} = R_0 \, 10 \, \frac{I_{\text{dB}}}{20} \tag{A9.4}$$

This equation stems from the relation:

$$\frac{P_{\text{RXA}}}{P_{\text{TRX}}} = \frac{A_{\text{RXA}}}{A_{\text{TRX}}} = \frac{R_0^2}{R^2}$$

made clear by Figure A9.2.

$P_{\text{RXA}}/P_{\text{TRX}}$ is the proportion of the total flux intercepted by RXA, as the far-field radiation seems to be originating from the virtual source S'. This leads to:

$$I = \frac{R^2}{R_0^2}$$

taking $\log_{10}$ of both sides and multiplying by 10 gives:

$$10 \log I = 20 \log \frac{R_{\text{FS}}}{R_0}$$

$$I_{(\text{dB})} = 20 \log \frac{R_{\text{FS}}}{R_0}$$

from which, finally:

$$R_{FS} = R_0 \, 10^{\frac{I_{(dB)}}{20}} \qquad\qquad (A9.4)$$

Step 4

Calculate range R of link operating in an atmosphere causing an attenuation A dB/km. R is determined by a trial and error method from:

$$\log R + \frac{AR}{20} = \log R_{FS} \qquad\qquad (A9.5)$$

Here is the derivation of this equation: Equation A9.4 can be made to take into account the atmospheric attenuation A dB/km. Simply reduce in it the insertion loss available, I_{dB} (from divergence loss alone) by RA, the path loss due to atmospheric attenuation. We have:

$$R = R_0 \, 10^{(I_{(dB)} - RA)/20} \qquad\qquad (A9.6)$$

This can be usefully reworked into:

$$R \times 10^{RA/20} = R_0 \, 10^{I/20} \qquad\qquad (A9.7)$$

On the right-hand side we recognise R_{FS}; thus, taking $\log_{10}$ of both sides:

$$\log R + \frac{AR}{20} = \log R_{FS} \qquad\qquad (A9.5)$$

Appendix 10

Range calculations for Section 8.4.2(a)

(To be read in conjunction with Appendix 9.)

Assumptions

- Lens diameter $d_{TXA} = d_{RXA} = 0.14$ m
- Divergence $\beta = 0.3$ mrad
- Power radiated $P_{TXA} = 0.5$ mW
- Atmospheric attenuation $A = 0.2$ (in dB/km)
- Required bandwidth BWD = 3 kHz
- Required signal-to-noise ratio, SNR = 12 (in dB)
- NEP = 1.2×10^{-14} WHz$^{1/2}$

Calculations

Step 1

Minimum receiver power:

$$NEP_{RXA} = 1.2 \times 10^{-14} \sqrt{3000} = 0.66 \times 10^{-12} \text{ W}$$

Permissible insertion loss:

$$I_{(dB)} = 10 \log \frac{0.66 \times 10^{-12}}{0.5 \times 10^{-3}} - 12$$

$$= 76.8 \text{ dB}$$

Step 2

Back distance R_0

$$R_0 = \frac{d_{TXA}}{\beta} \qquad \text{(with } \beta < 1°, \frac{\beta}{2} \simeq \sin \frac{\beta}{2}\text{)}$$

$$= \frac{0.14 \text{ m}}{0.0003}$$

$$= 467 \text{ m}$$

Step 3

Range R_{FS} of idealised link operating under $A = 0$ dB/km conditions:

$$R_{FS} = R_0 \ 10^{I_{(dB)}/20}$$
$$= 467 \times 10^{76.8/20} = 467 \times 10^{3.84}$$
$$= 3.230 \text{ km!}$$

Step 4

Range R of link operating under $A = 0.2$ dB/km conditions:

$$\log R + \frac{AR}{20} = \log R_{FS}$$
$$\log R + 0.01R = \log 3230$$

Trial and error calculations (Table A10.1) yield the solution:

$$R \simeq 137 \text{ km}$$

Atmospheric loss, $AR = 0.2 \times 137 = 27.4$ dB

Divergence loss, $10 \log \dfrac{1372}{0.4672} = 49.35$ dB

Total insertion loss $= 76.75$ dB

which is near enough to the total insertion loss available, $I_{(dB)} = 76.8$ dB.

Table A10.1

R (1)	$\log R$ (2)	$0.01R$	Sum (1) + (2)	$\log 3230$	when solved
100	2	1	3.00		
→ 137	2.14	1.37	3.51	3.51	←
200	2.3	2	4.30		

The derivation of coherence length from spectrum width

The derivation below is intuitive rather than rigorous. The term 'spectrum width' of the title refers to the bandwidth broadening caused by the imperfections of coherence. Instead of being given in *frequency* units (Hz), the spectrum width is expressed here in *wavelength* broadening terms (m). It is often called 'linewidth' of the emitted spectrum.

If a waveform remains coherent for a duration t_c, the coherence time, then the corresponding coherence length l_c of the wavetrain in vacuum (or, nearly enough, in air) is:

$$l_c = t_c \times c \tag{A11.1}$$

where c is the speed of light.

Electronics engineers know that the frequency bandwidth $\triangle f$ of the spectrum associated with a time-limited wavetrain is inversely proportional to its duration, t. Thus, we can write:

$$\triangle f_c \propto \frac{1}{t_c} \tag{A11.2}$$

as already stated in Equation 10.3 of the main text.

The width of a spectrum of frequencies – the bandwidth – can be defined in a variety of ways (e.g. between its -3 dB points, between the zero crossings shown in Figure 13.17(b), between $1/e$ amplitude values, etc.), whichever is most convenient for a given range of application. Here, it is highly convenient to define the bandwidth, so that the proportionality factor in Equation A11.2 becomes unity.

This judicious choice turns that equation into:

$$\triangle f_c = \frac{1}{t_c} \tag{A11.3}$$

Combining the latter relation with Equation A11.1 enables us to express the coherence length in terms of bandwidth, namely:

$$l_c = \frac{c}{\triangle f_c} \tag{A11.4}$$

The bandwidth of a laser radiation, however, is usually given in terms of wavelength – the previously mentioned 'spectrum width' or 'linewidth',

$\triangle\lambda$, expressed in units of length (m, nm). We shall accomplish the changeover using the basic relation:

$$f = \frac{c}{\lambda} \tag{A11.5}$$

Differentiating Equation A11.5 gives:

$$df = -\frac{c}{\lambda^2}\, d\lambda$$

$\triangle f_c$ being an extremely small – but finite – fraction of f, with laser light, we can write:

$$\triangle f_c = -\frac{c}{\lambda^2}\, \triangle\lambda \tag{A11.6}$$

Substituting Equation A11.6 into Equation A11.4 yields

$$l_c = -\frac{\lambda^2}{\triangle\lambda}$$

or, more practically, its absolute value:

$$l_c = \left| \frac{\lambda^2}{\triangle\lambda} \right| \tag{A11.7}$$

identical to Equation 10.1.

Equation A11.6 was used for the calculation of the numerical values of the frequency bandwidths of the three types of lasers of Section 10.2.2, from known values of their wavelength, λ, and linewidths, $\triangle\lambda$. For example, for the HeNe laser, we have, in absolute values:

$$\triangle f_c = c\frac{\triangle\lambda}{\lambda^2} = 3 \times 10^8 \frac{1.6 \times 10^{-15}}{(0.632)^2 \times 10^{-12}}$$

$$= \frac{3 \times 1.6}{(0.632)^2} \times 10^5 = 1.2\ \text{MHz}$$

Circular and elliptical polarisation: retardation plates

Some light beams are not plane but *circularly* or *elliptically polarised*. What this means is that the vector **E** of the light produced by them turns all the time (clockwise in some and anticlockwise in other cases) instead of steadily remaining in one plane. The frontal projection of **E** is shown in Figure A12.1

How this can come about will be easily understood from an electrical analogy. Take two sinusoidal voltages V_x and V_y of the same frequency but

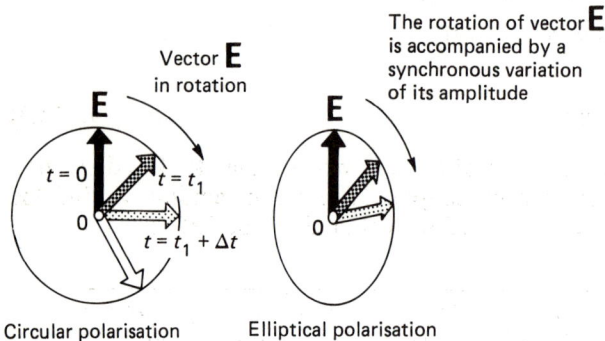

Vector **E** in rotation

The rotation of vector **E** is accompanied by a synchronous variation of its amplitude

Circular polarisation   Elliptical polarisation

Figure A12.1

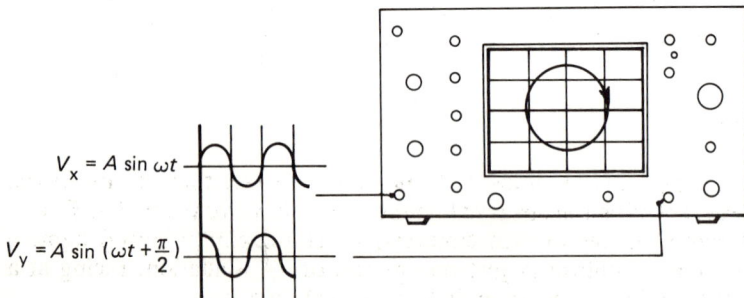

$V_x = A \sin \omega t$

$V_y = A \sin \left(\omega t + \frac{\pi}{2} \right)$

The case of the clockwise circle

Figure A12.2

with an adjustable phase difference ϕ and apply them to the X and Y inputs of an oscilloscope (Figure A12.2). You know what trace to expect:

- For $V_x = V_y$ and $\phi = 0$, a straight line inclined $+45°$.
- For $V_x = -V_y$ and $\phi = 180°$, a straight line inclined $-45°$.
- For $V_x \neq V_y$ and $\phi = 0$ another straight line, with an inclination depending on the value of V_y/V_x.
- For $V_x = V_y$ and $\phi = |90°|$ a circle, traced clockwise or anticlockwise, depending on the sign of ϕ.
- For $V_x \neq V_y$ and $\phi = |90°|$ an upright (if $|V_y|>|V_x|$) or a 'horizontal' (if $|V_y|<|V_x|$) ellipse.
- For $V_x \neq V_y$ and $0° \leqslant \phi \leqslant |90°|$ an inclined ellipse.

The ellipse is the most general case of which the circle is a special case. So is the straight line. The mathematics for a circle is surprisingly simple; with $V_x = V_y = V$ and $\phi = \pi/2$ we can write the values of the X and Y deflections (assuming equal scope x and y deflection sensitivities):

$$X = V_x \sin \omega t = V \sin \omega t \tag{A12.1}$$

$$Y = V_y \sin(\omega t + \frac{\pi}{2}) = V \cos \omega t \tag{A12.2}$$

Squaring and adding both sides we get

$$X^2 + Y^2 = V^2 (\sin^2\omega + \cos^2 \omega t)$$

or

$$X^2 + Y^2 = V^2 = \text{const.} \tag{A12.3}$$

which is the analytical equation of a circle with a radius V. The derivation of the straight-line case is even simpler. The reader anxious to find the derivation of the ellipse is referred to section 8.1.3 of Ref. 17.

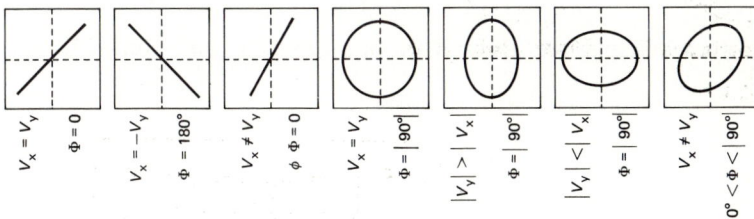

Figure A12.3

The electrical analogy helps us to perceive the fact that the tip of the vector **E** representing the electrical field of the light wave (sum of its $\mathbf{E}_x$ and $\mathbf{E}_y$ constituents) traces out, in circularly or elliptically polarised light, a complete circle or ellipse in just one period of its oscillation, racing at a rate of $60c/\lambda$ rev/min. For green light this worked out at:

$$\frac{3.10^8 \text{ m}}{0.55 \times 10^{-6} \text{ s}} \times 60 \simeq 5.45 \times 10^{14} \text{ Hz} \times 60 \simeq 32\,700\,000\,000\,000\,000 \text{ rev/min}$$

It also helps us to see that horizontally plane polarised light corresponds to the particular case $\mathbf{E}_y = 0$ and the vertically plane polarised to $\mathbf{E}_x = 0$. It further helps us to feel the interesting engineering properties of *quarter-wave* and *half-wave* retardation plates.

Let us strike vertically plane polarised (VPP) light onto a birefringent plate so cut and oriented that the O and E rays travel collinearly and emerge together at point *P* (Figure A12.4) – we already know that within the crystal, and even more so upon leaving it, the phase of one of them, E, lags behind the other – the electrical analogue of the situation can be represented as on Figure A12.5 (making the concession of ignoring the dissipative nature of R).

The *circularly polarised* light results from a phase shift of E with regard to O of 90°. A plate producing such an effect is called, for obvious reasons, a *quarter-wave plate* (Figure A12.6). Looking at both the electrical and the optical sketches of Figure A12.6 makes it clear that a properly oriented quarter-wave plate (VPP vector bisecting the angle formed by the E and O planes) will convert plane polarised light into circularly polarised (Figure A12.7(a)) *and vice versa* (Figure A12.7(b)). It is therefore a most useful component.

A little thought will reveal that, as reflection reverses the sense of rotation of **E**, the addition of a polariser upstream can make a quarter-wave plate into an *isolator,* which could be used for the prevention of back reflections from lenses being returned into the laser in a way which might interfere with its inner working. A half-wave plate will be obtained for φ = 180°.

One of its interesting uses is a controlled (by orientation) rotation of the polarisation plane of a linearly polarised laser beam obtained without turning the laser. The output polarisation will turn by twice the amount of the half-wave plate rotation (Figure A12.8). Another use is for reversing a clockwise circular (or elliptical) polarisation of a beam into an anti-clockwise polarisation or vice versa.

Note that quarter-wave and half-wave plates are *wavelength dedicated* devices.

For further reading see Ref. 3(b), pp. 273–275, and Ref. 17, pp. 246–250.

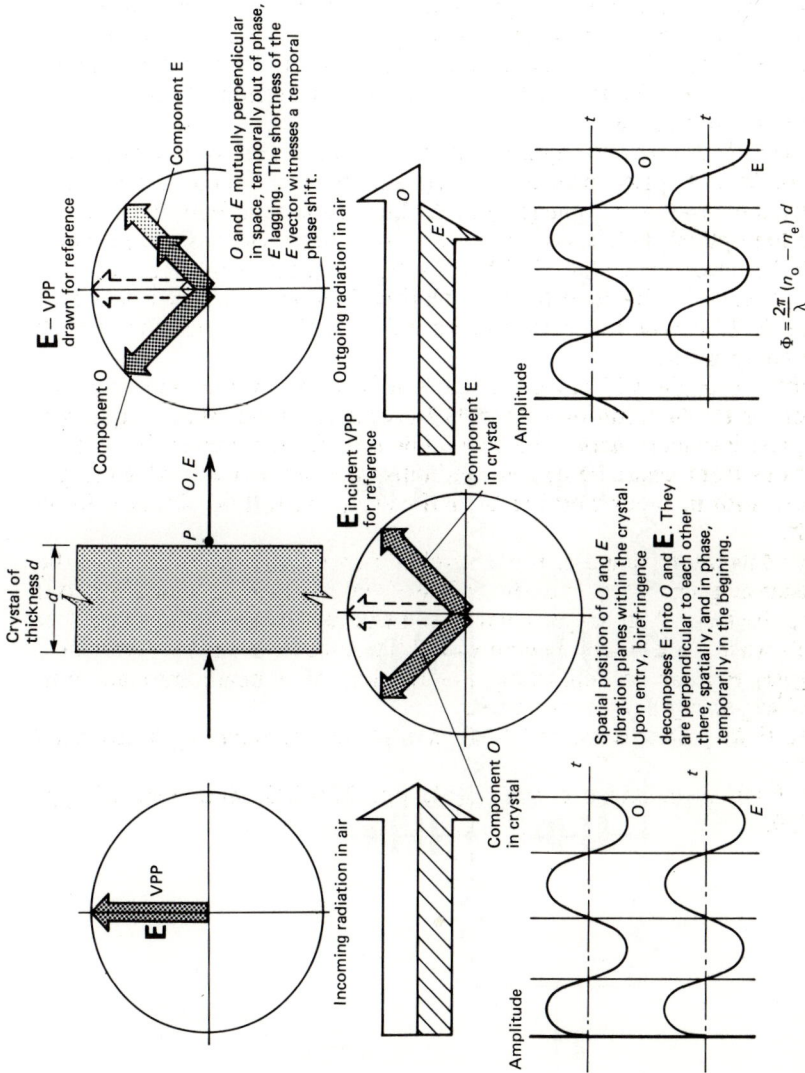

341

Crystal of thickness d

VPP

E ‖ VPP

Incoming radiation in air

Component O in crystal

E incident VPP for reference

Component E in crystal

Spatial position of O and E vibration planes within the crystal. Upon entry, birefringence decomposes E into O and **E**. They are perpendicular to each other there, spatially, and in phase, temporarily in the begining.

Component O

Component E

E – VPP drawn for reference

O and E mutually perpendicular in space, temporally out of phase, E lagging. The shortness of the E vector witnesses a temporal phase shift.

Outgoing radiation in air

Amplitude

O

E

Amplitude

O

E

$$\Phi = \frac{2\pi}{\lambda}(n_o - n_e)\, d$$

Figure A12.4

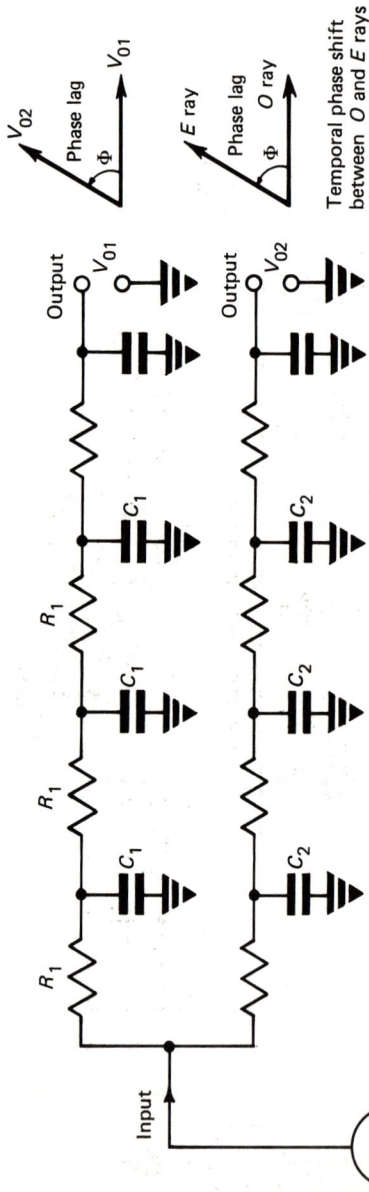

$R_2C_2 > R_1C_1$ causes V_{02} to lag behind V_{01}

Figure A12.5

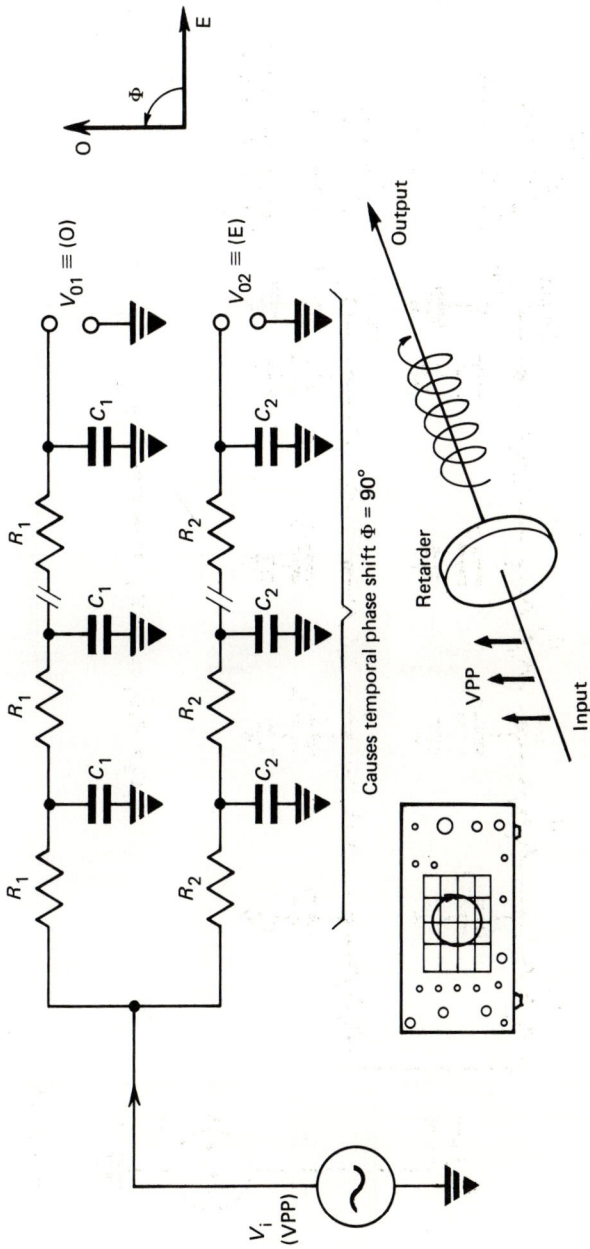

Figure A12.6

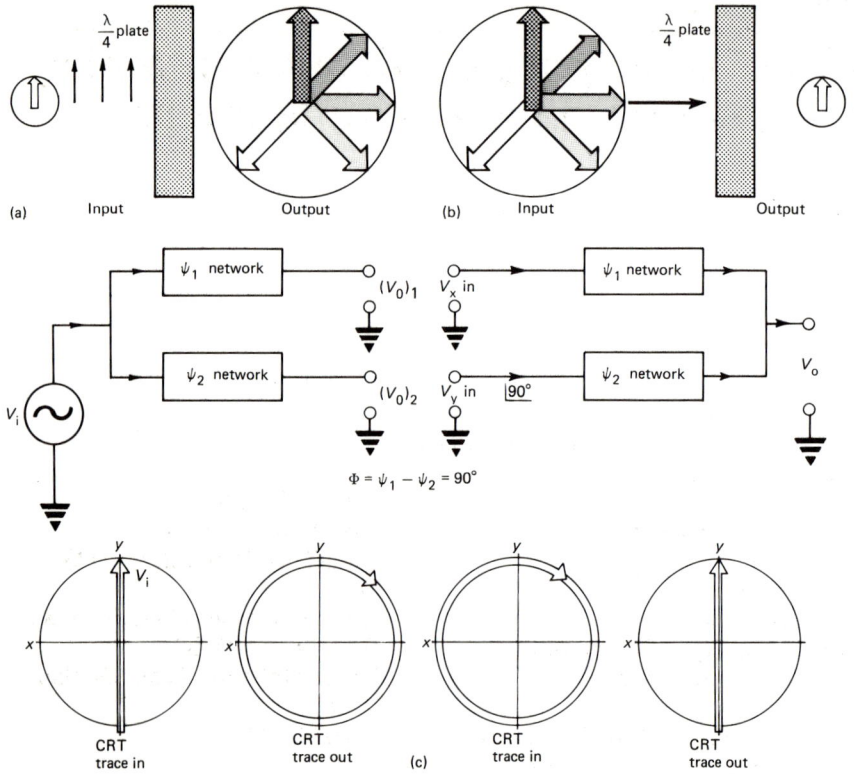

Figure A12.7

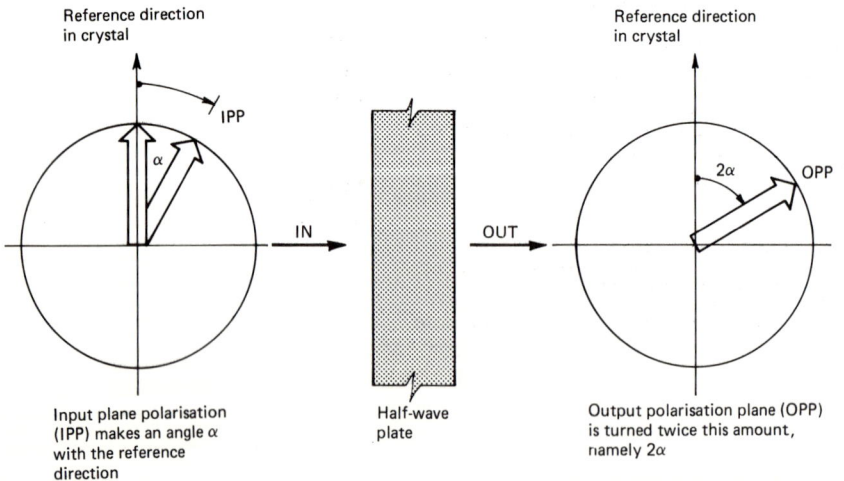

Figure A12.8

List of principal periodicals containing information on EO developments

	Publisher
Applied Optics	OSA
Journal of the Optical Society of America (parts A and B)	OSA
Laser Focus	PennWell (USA)
Lasers and Applications	Opt. Publ. Co (USA)
Photonics	High-Tech Publications (USA)
Optics News	OSA
Telecommunications	Horizon House (USA and UK)
Optics and Laser Technology	Butterworth (UK)
Optics Communications	North-Holland (The Netherlands)
Journal of Light Wave Technology	IEEE (USA)
Proceedings of the IEE, Part J, Optoelectronics	IEE (UK)
Optical Engineering	SPIE (USA)
Optics Letters	OSA
Electronics Letters	IEE

And, to a lesser extent:

IEE Journal of Quantum Electronics	IEEE
Applied Physics Letters	AIP
Journal of Applied Physics	AIP
Soviet Journal of Quantum Electronics (English translation)	AIP
Physics Today	AIP

OSA Optical Society of America.
IEEE Institute of Electrical and Electronic Engineers, USA.
IEE Institution of Electrical Engineers, UK.
AIP American Institute of Physics.
SPIE Society of Photo-optical Instrumentation Engineers, USA.

Bibliography

*denotes commercial promotional literature
†denotes principal books on fibre optics communications

1 Jenkins, F.A. and White, H.E. (1965) *Fundamentals of Optics*, 3rd edn, McGraw-Hill, New York
2 Smith, W.J. (1966) *Modern Optical Engineering*, McGraw-Hill, Maidenhead
* 3 Melles Griot, *Optics Guides*: (a) Issue 1, (b) Guide 2, (c) *Lasers*, (d) *Optics*, 1975 edn, (e) *Laser Diode Optics*, Melles Griot, Irvine, Calif., USA
† 4 Adams, M.J. (1981) *Introduction to Optical Waveguides*, Wiley, New York
5 Chaimowicz, J.C.A. (1976) The choice of emitters, receivers and align geometries in OE com. linkages, *Proceedings of the 2nd European Electrooptics Conference*, Montreux, 1976
6 Bryant, D. (1971) *Physics*, English Universities Press, Sevenoaks
7 Schawlow, A.L. (1969) *Lasers and Light*, WH Freeman, San Francisco
* 8 Chappell, A. (1976) *Optoelectronics, Theory and Practice,* Texas Instruments
* 9 RCA (1978) *Electro-Optic Handbook*, RCA Publication EOH-11
* 10 Gage, S. *et al.* (1977) *Optoelectronics Application Manual*, and Supplement, Hewlett Packard/McGraw-Hill, Maidenhead
11 Messham, S.E. (1977) *Exploring Mind and Body*, HMSO, London
12 Sill, W.B. and Hoss, N. (1968) *Encyclopedia of the Sciences*, Allen & Unwin, London
13 Larch, S. (ed.) (1965) *Optoelectronic Materials and Devices*, Van Nostrand Reinhold, New York
14 Chang, S.S.L. (1963) *Energy Conversion*, Prentice-Hall, Englewood Cliffs, New Jersey
15 Oburns, S. *et al.* (1963) *Discovering the World of Science*, Van Nostrand Reinhold, New York
16 Baldock, R. (1980) Facets of optics, series of articles in *Video*
17 Hecht, E. and Zajac, A. (1974) *Optics*, Addison-Wesley, Reading, Mass.
18 Richtmyer, F. (1934) *Introduction to Modern Physics*, McGraw-Hill, New York
19 Chaimowicz, J.C.A. (1979) CRT's for phototypesetting, *Displays*, July 1979, 111–116
20 SIRA (1967) *Proceedings of the Conference New Developments in Optics*, Eastbourne, 1967
21 Gaskill, J.D. (1978) *Linear Systems, Fourier Transforms and Optics*, Wiley, Chichester
22 Carlson, F.P. (1977) *Introduction to Applied Optics for Engineers*, Academic Press, London
23 Pike, C.A. *Lasers and Masers*, Foulsham and Sams, Slough
* 24 GE Co (USA) Catalogue on Fibre Optics
25 Larousse (1961) *Petit Larousse Encyclopedic Dictionary*, Larousse, Paris (in French)
26 Dance, J.B. (1969) *Photoelectronic Devices,* Iliffe
* 27 Centronic Silicon Photodetectors, Centronic (UK) Publ. PD/065-069/78
28 Brophy, J.J. (1964) *Semiconductor Devices*, McGraw-Hill, Maidenhead

29   Greiner, R.A. (1961) *Semiconductor Devices and Applications*, McGraw-Hill, Maidenhead

30   Rheinfelder, W.A. (1964) *Low Noise Transistor Circuits*, Iliffe

31   Yariv, A. (1976) *Introduction to Optical Electronics*, 2nd edn, Holt, Reinhart and Winston, New York

32   van Heel, A.C. and Velzel, C.H. *What is Light?* World Universities Library, Weidenfeld and Nicolson, London

33   Coulson, C.A. (1961) *Waves*, Oliver and Boyd, Edinburgh

*   34   GE Co (USA) Brochure on FO systems

35   Wallis, D. (1980) The language of lenses. *Video*, Oct. 1980, 32–48

36   Laybourne, P.Y.R. and Lamb, J. (1981) Integrated optics. *The Radio and Electronic Engineer*, **51**, 397–413

37   Westman, H.P. (ed) (1972) *Reference Data for Radio Engineers*, 5th edn, Howard W Sams and ITT

38   Lamont, H.R.L. (1959) *Wave Guides*, Methuen, London

†   39   Sandbank, C. (1980) *Optical Fibre Communications Systems*, Wiley, Chichester

40   Denis-Papin, M. Mécanique et Physique Générales, Dunod, Paris (in French)

41   Kao, C.K. (1982) *Optical Fibre Systems: Technology, Designs and Applications*, McGraw-Hill, Maidenhead

†   42   Barnoski, M.K. (1981) *Fundamentals of Fibre Optic Communications*, Academic Press, London

43   Mossman, P. (1981) Connectors for optical fibre systems. *The Radio and Electronic Engineer*, **51**, 333–340

44   Millarc, C.A. and Mallison, S.R. (1981) *Electronics and Power*, Sept. 81, 337–339

45   Bronowski, J. (1974) *The Ascent of Man*, BBC Publications, London

46   La Rocca, A.V. (1982) Laser applications in manufacturing. *Scientific American*, March 1982, 80–87

47   Sagan, C. (1980) *Broca's Brain*, Coronet Books, Sevenoaks

48   Chaimowicz, J.C.A. (1969) Novel two-way GaAs device for optoelectronic data linkages, *Proceedings of the XVI International Congress for Electronics*, Rome, 1969

†   49   Miller, E. and Chynoweth, A.G. (eds) (1979) *Optical Fibre Telecommunications*, Academic Press, London

50   Seippel, R.G. (1981) *Opto-Elecronics*, Reston/Prentice-Hall, Englewood Cliffs, New Jersey

*   51   Hamamatsu catalogues: Silicon photocells and GaAsP photocells, SC 3–5 (1980) and T–15000 (1985)

52   Klein, H.A. (1967) *Masers and Lasers*, Allen and Unwin, London

53   Millman, J. and Halkias, C.C. (1972) *Integrated Electronics*, McGraw-Hill, Maidenhead

54   Millman, J. and Halkias, C.C. (1975) *Electronic Devices and Circuits*, McGraw-Hill, Maidenhead

*   55   Crystalonics Inc. (USA) catalogue: Fotofet

*   56   RCA Data Sheets

*   57   IPL (Integrated Photo-Matrix Ltd) Data Sheets

*   58   UDT (United Detector Technology Inc., USA) Catalogues and Data Sheets

59   *Security Management*, April 1979

60   Miller, E. (1980) National Semiconductor Inc., Applic. Note 244

61   Grossman, M. (1981) Focus on FO connectors. *Electronic Design*, Nov. 1981

62   Castro, C.J. (1980) Passive FO access coupler systems. *Electronic Engineering*, April 1980

63   Pitcher, J. (1980) Consideration of OF measurements. *Electronic Engineering*, April 1980

64   Lightstone, A. (1979) Coupler for FO communications *Communications International*, October 1979

65   Hudson, C. and Thiel, F.L. (1974) The star coupler. *Applied Optics*, **13**, 2540–2545

66   Mims, F.M. (1980) Alexander Bell and the photophone. *Optics News*, **6**, 8

67   Powell, E.C. (1979) Industry gears up for FS communications, *Optical Spectra*, June 79 and July 79 (in two parts)

68   Boersching, B.A. (1965) A light modulated data link. *SCP and Solid State Technician*, Jan. 1965, 35

69   Price, T.E. (1984) An atmospheric infra red communications link. *Electronic Engineering*, Oct. 1974

70   Richards, D.L. (1981) Optical communications in perspective, *IEE News*, Aug. 1981

71   Ekberg, J. (1969) The focal length of optics in optical communication systems. *Proc. IEE*, Jan. 1969, 88

72   Baker, W.J. (1964) *Lasers, British Communications and Electronics*, p. 412

73   Kock, W.E. (1969) Fundamentals of lasers. *Optical Spectra*, March/April 1969, 65

74   Darmois, E. (1947) *L'electron*, Presses Universitaires de France

75   Holmes, L. (1980) Twenty candles for the laser. *Electronics and Power*, June 1980, 454

76   Charschan, S.S. (1972) *Lasers in Industry*, Van Nostrand Reinhold, New York

77   Printing Industries Research Association, *PIRA Lecture on Lasers, Symposium on Printing*, etc.

*   78   *Acousto-Optic Modulators*, Catalogue of S. Andersen Lans, Inc.

*   79   *Bar Code Reading Made Easy*, Skan-A-Matic Corp. promotional literature

80   Cheng, C.C.K. (1980) Laser scanning systems, etc. *Electro Optical System Design*, June 1980, 41

81   Anon. (1981) Bar codes, *Elektor*, May 1981, 5–22

82   Anon. (1981) Reading bar codes, *Elektor*, May 1981, 43

83   Anon. (1981) Lasers for supermarkets. *Electronic Times*, 24 September 1981

84   Anon. (1979) Bar code scanner, etc. *Design Engineering* Dec. 1979, 13

85   Burrus, C.A., Casey, H.C. and Li, T. (1979) in *Optical Fibre Telecommunications* (eds Miller, E. and Chynoweth, A.G.), pp. 512–514

*   86   Aerotech catalogue

87   Bademian, L. (1980) *Acoustic-Optic Modulation and Deflection*. The Optical Industry Purchasing Directory 1980, Book 2, p. B-823

88   Casasent, D. (1981) Optical signal processing. *Electro Optical System Design*, June 1981, 39

89   Hinds International technical catalog, *The Photoelastic Modulator* (and addendum)

90   Miller, D.A.B. (1982) Bistatic optical devices. *Laser Focus*, April 1982, 79

*   91   Ealing Optics (1981) catalogue

92   Chaimowicz, J.C.A. (1970) Contactless signal coupling. *MCP Electronics World*, **2** (1)

93   Chaimowicz, J.C.A. (1970) A wish come true. *MCP Electronics World*, **2** (3), 2

94   Anafi, D. *et al.* (1975) An overview of harmonic generators. *Optical Spectra*, Dec. 1975, 25

95   Third, B. *et al.* (1978) Paging the new medium is limitless. *Electronics Times*, 19 Nov., 16

96   Editorial (1980) Using differential optical computers. *Design Engineering*, Nov. 1980, 19

*   97   Burr-Brown (1980) Data Sheet 3650–3652

98   Holford, K. (1981) If it moves, microwaves will detect it. *Electronics and Power*, May 1981, 393

99   Malvern Instruments Ltd Catalogue 792 DST (particle sizers)

*   100   Anon. (1972) Using opto-isolators. *Electron*, 27 July

101   Russell, H.T. (1970) Optically coupled isolators in circuits. *Electro Optical System Design*, **2**, 29–33

102   Carlson, F.P. (1977) *Introduction to Applied Optics for Engineers*, Academic Press, London

103   Tomlison, W.J. (1980) Aberrations in GRIN rod lenses. *Applied Optics*, **19**, 1118

*   104   Honeywell Catalogue 110-0155-000, 4/81

*   105   Monsanto Applic. Note AN504

*   106   Monsanto Opto-Isolator Applications Notes, esp. AN505

* 107 Hewlett Packard (1981) Opto Electronics Designers Catalogue
* 108 Litronix (1981) catalogues of optoelectronics devices
* 109 Motorola Catalogue, *Optoelectronics at Work*
* 110 Sahm *et al.* (1976) *Optoelectronics Manual*, The General Electric Co., USA
* 111 Motorola (1981–1982) *Optoelectronic Data Book*
   112 Williams, P. *et al.* (1975) *Circuit Designs*, Vol. 1, W. World Publ
* 113 Parsons R. and Bonham, D. (1972) *Using Optoelectronics*, Texas Instruments
* 114 Smith, G. (1971) App. Note 2, *Application of Opto Isolators*, Litronix
* 115 General Instrument Optoelectronics (1980) *Mid-400 Data Sheet*
* 116 Centronics (1974) *Typical Circuit Applications of Optoelectronic Devices*
* 117 Dionics Inc. (1981) *Iso-Gate DIG-1*
* 118 Mullard (1981) *Mullard Bulletin, Photocouplers CNX 21 (8/1981)*
* 119 Hewlett-Packard Data Sheet 4N55
* 120 Telefunken Data Sheet CNY66
* 121 TRW Optron Data Sheet OP1 125
* 122 Motorola Data Sheet MOC 3020 (in their *Radio Resistor Co. Ltd* catalogue)
* 123 Hewlett-Packard (1978) Data Sheet, *Line Receiver HCPL 2602*
   124 Tomozawa, M. and Doremus, R.H. (1977) *Glass I: Interaction with Electromagnetic Radiation*, Academic Press, New York
   125 Larach, S. (1965) *Photoelectronic Materials and Devices*, D Van Nostrand, New York
   126 Simon, I. (1966) *Infra-red Radiation*, D. Van Nostrand, New York
   127 Chaimowicz, J.C.A. (1970) Immunity to daylight. *MCP Electronics World*, **2**, 6–7
   128 Dionics Data Sheet D1-16V8
   129 Giallorenzi, T.H. (1978) *Fiber Optics Technology II*, IEEE Selected Reprints Series
† 130 Suematsu, K. and Iga, K. (1982) *Introduction to Optical Fiber Communications,* Wiley, New York
   131 Cleobury, D.J. (1982) *Telecommunications*, July 1982, 46-1/52-1
* 132 Mitsubishi Catalogue 5402-20H
   133 Radley, P.E. (1981) Systems applications of optical fibre transmission. *The Radio and Electronic Engineer*, **51**, 377–384
   134 Holmes, L. (1982) Fibre optics in power systems – the Cigré perspective. *Electronics and Power,* Oct. 1982, 673–677
* 135 NKF (1982) Commercial Information
   136 Bradley, N.J. (1980) Fibre optic systems design. *Electronic Engineer*, April 1980, 98
   137 Mossman, P. (1981) Connectors for optical fibre systems. *The Radio and Electronic Engineer*, **51**, 333
* 138 NEC (1981) *Devices for Optical Communications*, Publication E34–031.8106
† 139 Howes, M.J. and Morgan, D.V. (1980) *Fibre Optic Communications*, John Wiley, Chichester
   140 Suzuki, N. *et al.* (1979) Ceramic capillary connector. *Electronics Letters*, **15**, 809
   141 Hensel, P. (1977) Triple ball connector for optical fibres. *Electronic Letters*, **13**, 24, 734–738
* 142 Hellerman-Deutch (1981) *Fibricon FO Systems and Components*, IMO 781
   143 Tebo, A.R. (1981) Fiber-optic coupler directional and otherwise. *Electro Optical System Design*, Nov. 1981, 25–45
* 144 NKF (Philips Group) (1981) *Optical Fibre Cables and Systems* P.150, E5000 010581 NP
   145 Lilly, C.J. (1981) Making light work. *Electronics and Power,* Sept. 1981, 629–634
   146 Eibner, J.A. and Goldberg, N. (1980) *Communication International*, July 1980, 22–23
   147 Maddock, B.J. (1981) *Electronics and Power*, Sept. 1981, 635–636
   148 Turnbull, N. (1980) Fibre optics and cables for fire, *Electrical Equipment*, Nov. 1980, 48–49
   149 Pastelis, A. and Stubbs, G. (1981) Fiber optic techniques work well. *Telecommunications*, Dec. 1981, 16–28
   150 Anon. (1982) New optical fibre record for Britain. *Electronics and Power*, April 1982, 294

151 Moralee, D. (1981) New telecomms venture digs into the past? *IEE News*
152 Editorial (1981) *Electronic Times*, Sept. 1981
153 Anon. (1980) Laser light goes under the sea. *The Electronics Engineer* March 1980, 2
154 Anon. (1981) Planners of next transatlantic cable envisage OF transmission. *Communications International*, Nov. 1981, 7
155 Anon. (1981) Bell plans US–Hawaii fiber cables similar to Atlantic link. *Fiber Optic Technology*, Dec. 1981, 111
156 Anon. (1982) British Telecom's 100 km monomode fibre transmission experiment. *Radio and Electronic Engineeer,* **52** (4), 157
157 Anon. (1980) Boston to Washington by fibre. *Optical Spectra*, Feb. 1980, 41–42
158 Niquil, M. (1981) The wired city of Biarritz. *Optical Spectra*, Aug. 1981, 38–40
159 Bates, C.H. and Tassell, C.H. (1980) Low cost FO data links. *Communications International*, July 1980
160 Anon. (1981) Canada and fiber optics. *Optical Spectra*, July 1981, 37–40
161 Anon. (1981) Japanese technology. supplement to *Electronic Times*, Aug. 1981
162 Verdeyen, J.T. (1981) *Laser Electronics*, Prentice Hall, Englewood Cliffs, New Jersey
* 163 Harvey, W.J. (1983) *STC Free Space Optical Link*, descriptive leaflet of equipment shown at the Fibre Optics Exhibition, The Barbican, London, 1983
* 164 British Hovercraft Corporation (1983) *EEL Free Space Optical Link*, descriptive leaflet of equipment shown at the Fibre Optics Exhibition, The Barbican, London, 1983
165 Keyes, R.J., Quist, T.M. *et al.* (1964) Modulated L-R diode spans 30 miles. In *Opto Electronic Devices* (ed. S. Weber) McGraw-Hill, Maidenhead, p. 86
* 166 American Laser Systems, Inc. (1980) catalogue. Santa Barbara, Calif., USA
† 167 Mims, F.M. (1982) *A Practical Introduction to Lightwave Communications*, Howard W. Sams
168 Swartz, M. (1981) *Information Transmission Modulation and Noise*, McGraw-Hill, Maidenhead
169 Dworkin, L.U. and Christian, R. (1978) *IEEE Transactions on Optical Communications,* **COM-26**, 999
170 STC (1983) *Semiconductor Laser Application Note*, STC Publication 6340/2646E, 2nd edn
* 171 Kressel, H. and Butler, J.K. (1977) *Semiconductor Lasers and Heterojunction LEDs,* Academic Press, London
* 172 RCA (1979) *Optical Communications Products*, RCA Publication OPT-115
* 173 Lexel Co., USA (1977) *Argon and Krypton Ion Lasers*
174 Furlow, W.M. (1969) Laser technology today. *Laser Journal*, Dec. 1969
175 Optical Publication Co., USA (1973) *The Optical Industry and Systems Directory, 1973/4*
176 Muncheryan, H.M. (1983) *Principles and Practices of Laser Technology*, TAB Books, USA
177 Pike, E.R. (ed.) (1976) *High Power Gas Lasers 1975*, Institute of Physics, London
178 Hecht, J. (1982) An introduction to carbon dioxide laser. *Lasers and Applications*, Sept. 1982, 83–90
* 179 Hughes Aircraft Co. Brochure on *Waveguide CO_2 1W laser 30384*
180 Riordan, T. (1983) The free election laser. *Photonic Spectra*, July 1983, 40
181 Svelto, O. (1982) *Principles of Lasers*, 2nd edn, Plenum Press, New York
182 PennWell (1983) *Laser Focus Buyer's Guide 1983*, PennWell, USA
183 Kortz, H.P. (1982) YAG-pumped dye lasers, the proven tunable source. *Laser Focus*, July 1982, 57
184 Grove, R.E. (1982) Copper lasers come of age. *Laser Focus*, July 1982, 45
185 Witherell, C.H.E. (1981) Laser micro-brazing to join parts. *Laser Focus*, Nov. 81, 73
186 Davies, D.E.N. (1984) Opto-electronics – a new dimension in electronics. *The Radio and Electronic Engineer*, **54**, 1–9
187 Charschan, S.S. (1981) The evolution of laser machinery. *Electro Optical System Design*, Aug. 1981, 63
188 Heavens, O.S. (1971) *Lasers*, Duckworth, London

189   Anon. (1982) Scanning laser ophthalmoscope, *Optics News*, Dec. 1982, 10
* 190   Coherent Inc. (1982) *Lasers at Work*
191   High Tech (1984) *Lasers and Applications 84* Directory, High Tech Publications, USA.
192   Wilson, J. and Hawkes, J.F.B. (1983) *Optoelectronics – An introduction,* Prentice Hall, Englewood Cliffs, New Jersey
193   Smith, H.M. (1969) *Principles of Holography*, Wiley Interscience, Chichester
194   SIRA (1967) *New Developments in Optics and their Applications in Industry Conference,* 1967
195   Beiser, L. (1983) The scanner decision. *Laser and Applications*, April 1983, 61
† 196   Marcuse, D. (1974) *Theory of Dielectric Optical Waveguides*, Academic Press, London
197   Grossman, B. (1983) How to select acousto-optic modulators. *Laser Focus*, April 1983, 49
* 198   Coherent (1982) *Laser Beam Modulators*, Coherent®, Connecticut, USA
* 199   *Technical Specification EFLD 250/750*, Interactive Radiation Inc, Northvale, New Jersey
200   Davis, C.M. (1982) An introduction to fiberoptic sensors, *Fiberoptic Technology*, February 1982, 112
201   Davies, D.E.N. and Culshaw, B. (1981) Development and potential of optic fiber sensors. *IFOC 1981–82, Handbook and Buyer's Guide*, p. 120
202   Culshaw, B. (1984) *Optical Fibre Sensing and Signal Processing,* Peter Peregrinus
203   Culshaw, B. (1982) Optical fibre transducers. *The Radio and Electronic Engineer,* **52**, 283
204   IMC (1981) *Optical Sensors and Optical Techniques in Instrumentation*, The Institute of Measurements and Control, London
205   IEE (1984) *Optical Point Sensors for Process Control*, Proceedings of the IEE Colloquium No. 1984/7, London, 1984
206   Davies, D.E.N., Chaimowicz, J.C.A., Economou, G. and Foley, J. (1984) Displacement sensor using a compensated fibre link. In *2nd International Conference on OFS*, Stuttgart, 1984
† 207   Midwinter, J. (1979) *Optical Fibres for Transmission*, John Wiley, Chichester
208   IEE (1984) *Optical Fibre Sensors*, Proceedings of the IEE Conference, 1983, Publication 221, IEE, London, 1984
209   Culshaw, B. (1983) Optical systems and sensors for measurement and control, *Journal of Physics E: Sci. Instrum.*, **16**, 283–289
210   Sincerbox, G. (1982) Holographic scanner. *Optics News* Nov/Dec. 1982, 6
211   Kramer, C.J. (1981) Holographic laser scanners. *Laser Focus*, June 1981, 70
212   Sincerbox, G. and Rosen, G. (1983) Opto-optical light deflection. *Applied Optics,* **22**, 390
213   Siemens Labs (1983) *Optical Communications*, John Wiley, New York
214   Enscoe, R.F. and Kocka, R.J. (1984) Electro-optic modulation, systems and applications. *Laser and Applications*, June 1984, 91–95
* 215   RCA Data Sheet for Laser Type C 806014E
216   Graindorge, P. *et al* (1983) Interferometric sensor using phase conjugate mirrors. Paper Presented at the *Optical Fibre Research Conference*, London, 1983
217   Heitler, W. (1958) *Elementary Wave Mechanics*, 2nd edn, Clarendon Press, Oxford
† 218   Gowar, J. (1983) *Optical Communication Systems*, Prentice Hall, London
219   Van Ruyen, L.J. (1982) *Electronic Components and Applications,* **5**, 42–45
220   Pry, S.M. (1980) *Optical Spectra*, Sept. 1980, 66–67
221   Goodman, J.W. (1971) An introduction to the principles and applications of holography. *Proc. IEE,* **59**, 1292–1304
222   Rallison, R. (1984) Applications of holographic optical elements. *Lasers and Applications*, December 1984, 61–88
223   Tarasov, L.W. (1981) *Laser Age in Optics*, Mir, Moscow
224   Hopper, C. (1984) European prepayment telephone card. *Communications International*, Jan. 1984, 76–80
225   Gaskill, J.D. (1978) *Linear Systems, Fourier Transforms and Optics,* Wiley, Chichester

226   Rosenfeld, A. and Kak, A.C. (1976) *Digital Picture Processing*, Academic Press, London

227   Champeney, D.C. (1973) *Fourier Transforms and their Physical Application*, Academic Press, London

228   Kallard, T. (1977) *Exploring Laser Light*, Optosonic Press Publication, USA

229   IEE (1985) Colloquium on *Optical Techniques in Image and Signal Processing*, IEE, London

230   Holmes, L. (1980) Integration of optoelectronics. *Electronics and Power*, June 1980, 463–464

231   IEE (1985) Colloquium on *Advances in Coherent Optic Devices and Techniques*, IEE, 1985, Digest No. 30

232   Pitt, C.W. and Overbury, A. (1984) In *Annual Review, Department of Electronic and Electrical Engineering*, October 1984, University College London, pp. 20–21

233   Sasaki, H. (1977) Efficient intensity modulation, etc. *Electronic Letters*, **13**, 693–694

234   Pitt, C.W. *et al.* (1984) In *Annual Review, Department of Electronic and Electrical Engineering,* October 1984, University College London, p. 17

235   Pitt, C.W. (1984) Use of thin films in optical waveguiding devices. *Vacuum*, **34**, 399–403

236   Pepper, I.M. (1982) Nonlinear optical phase conjugation, *Laser Focus*, **18**, 71–78

237   Abell, G.O. (1982) *Exploration of the Universe*, Saunders, New York

238   Palais, J.P. (1980) Fiber coupling using graded index rod lenses. *Applied Optics*, **19**, 2011–2018

239   Smith, P.W. and Miller, D.A.B. (1982) Optical bistability. *Laser Focus*, **18**, 77–78

240   Haan, M.R. *et al.* (1979) A system concept for optical data recording. In *IERE Conference Proceedings* 1979, Supplement No. 43

241   Ramaker, J.M. and Vromans, P. (1979) Disc for optical recording. In *IERE Conference Proceedings,* Supplement No. 43

242   Stahlie, T.J. (1985) Principles of optical recording and data integrity. In Proceedings of the Colloquium *Optical Mass Data Storage, IEE Digest*, No. 58

243   Fateh, M.T. *et al* (1984) Optical flip-flops and sequential logic circuits. *Applied Optics*, **23**, 2163–2171

244   Anon. (1985) Laser-levelling puts salt in its place. *Lasers and Applications*, **4,** 42

245   Strohbehn, J.W. (1978) *Laser Beam Propagation in the Atmosphere*, Springer Verlag, New York, p. 10.

246   Chaimowicz, J.C.A. (1979) CRTs for phototypesetting. *Displays*, 1, 111–116

* 247   Skan-A-Matics Corp. (1983) *Bar Code Basics*

248   Chaimowicz, J.C.A. (1971) Optoelectronics in package identification. In *Proceedings of the Conference on Electronic Control of Mechanical Handling*, IERE, 1971

249   Learner, R. (1983) *Astronomy through the Telescope*, Evans Bros, London

250   Sill, W.B. and Hoss, N. (1968) *Encyclopedia of the Sciences*, Popular Science Publishing Co.

† 251   Harger, R.O. (1980) *Optical Communications Theory*, Dowden and Hutch

252   Drain, L.E. (1980) *The Laser Doppler Technique*, Wiley, Chichester

* 253   Polytec. *Laser Velocimeter*, Publication LV-4-810-5000E, Polytec, Waldbronn, Germany

* 254   Polytec. *Laser Doppler Velocimeter*, Publication LV-9-91-3000E, Polytec, Waldbronn, Germany

* 255   Polytec. *Laser Velocimeter*, Publication LD-1-702-2000E, Polytec, Waldbronn, Germany

256   Holmes, L. (1983) Optical bistability. *Electronics and Power*, **33**, 581–584

257   Abraham, E. *et al.* (1983) The optical computer. *Scientific American*, **248**, 85–94

258   Gibbs, H.M. *et al.* (1982) Optical bistability at room temperature? *Optics News*, **8**, 7

259   Jae-Won Song *et al* (1984) Optical bistability etc. *Applied Optics*, **23**, 1521–1524

260   Ashkin, A. (1970) Acceleration and trapping of particles by radiation pressure. *Physical Review Letters,* **24**, 156–159

261   Ashkin, A and Dziedzic, J.M. (1971) Optical levitation by radiation pressure. *Applied Physics Letters,* **19**, 283–285

262   Ashkin, A. and Dziedzic, J.M. (1980) Observation of light scattering using optical levitation. *Applied Physics,* **19**, 660–668

*   263   Spectra Physics. *Laserplane*, Constructional and Agricultural Division, Spectra Physics Inc.

264   Ashkin, A. (1972) Pressure of laser light. *Scientific American,* **226**, 63–71

265   Grimsehl, E. (1933) *Light*, Blackie, London

266   Mallalieu, K. *et al* (1985) An analysis of the photothermal drive of a force sensor. In *Proceedings of the SIRA Conference on Fibre Optics*, 1985, SPIE Publications, Vol. 522

267   Venkatesh, S. and Culshaw, B. (1985) Optically activated vibrations in a micro-machined Si structure. *Electronics Letters,* **21**, 315–317

268   Wherrett, B.S. (1985) Acousto-optical computation. *Applied Optics*, Sept. 1985

269   Devane, M.M. (1984) *Studies of Phase Conjugations etc.* PhD Thesis, University of Dublin

270   Chaimowicz, J.C.A. and Chettle, W.C. (1964) *Proceedings of the IEE Conference on Lasers and Applications*, September 1964, London

271   Chaimowicz, J.C.A. (1975) Designing an optoelectronic communications link. *Video and Audio Visual Review,* January 1975, 1–5

272   Chaimowicz, J.C.A. (1965) Semiconductor light source for industry. *Product Design Engineering*, October 1965

273   Gambling, W.A. *et al.* (1981) Optical fibres for transmission. *The Radio and Electronic Engineer,* **51**, 313–325

274   Pain, H.J. (1983) *The Physics of Vibrations and Waves*, 3rd edn, Wiley, Chichester

275   Doyle, T. (1972) *The Use of RCA Solid State Si Detectors in Small Signal Detection Systems*, RCA Applic. Note AN 4849

276   Schwartz, M. (1982) *Information Transmission, Modulation and Noise*, 3rd edn, McGraw-Hill, Maidenhead

277   Panter, P.F. (1972) *Communication System Design*, McGraw-Hill, Maidenhead

278   Starr, A.M. (1953) *Radio and Radar Technique*, Pitman, London, pp 551–552

279   Lasers and Applics (1985) *Designer's Handbook*, Lasers and Applics, USA

280   Bleaney, B.I. and Bleaney, B. (1959) *Electricity and Magnetism*, Oxford University Press, Oxford

281   Bannister, A. and Raymond, S. *Surveying*, 5th edn, Pitman, London

282   Burnside, C.D. (1983) *Electromagnetic Distance Measurements*, 2nd edn, Collins, London

283   Durst, F., Melling, A. and Whitelaw, J.H. (1976) *Principles and Practice of Laser Doppler Anemometry*, Academic Press, London

284   Aeschylus (1977) *The Oresteia* (trans. R. Fagles), Penguin Books, Harmondsworth

285   Heinzmann *et al.* (1986) Bidirectional full duplex single fibre link with s/c junction transceiver, *Laser Focus*, Nov. 86, 106–119

286   Pepper, D.M. (1986) Applications of optical phase conjugation. *Scientific American,* **254**, 56–66

†   287   Kao, C.K. (1981) *Optical Fibre Technology*. John Wiley, Chichester

†   288   Clarricoats, P. (1980) *Progress in Optical Communications*, Academic Press, London

†   289   Lacey, E.A. (1982) *Fiber Optics*, Prentice-Hall, Englewood Cliffs, NJ

†   290   Marcuse, D. (1981) *Principles of Optical Fibre Measurements*, Academic Press, London

†   291   Palais, J.C. (1984) *Fiber Optic Communications*, Prentice-Hall, Englewood Cliffs, NJ

†   292   Personick, S.D. (1981) *Optical Fiber Transmission Systems*, Plenum Press, New York

†   293   Senior, J.M. (1985) *Optical Fiber Communications*, Prentice-Hall, Englewood Cliffs, NJ

†   294   Taylor, F. (1983) *Fibre Optic Communications*, Artech

†   295   Tingye, L.I. (ed.) (1985) *Optical Fibre Communications*, Vol. 1, Academic Press, London

†   296   Baker, D.G. (1985) *Fiber Optic Design and Applications*, Prentice-Hall, Englewood Cliffs, NJ

Index

Note: illustrations are indicated by *italic page numbers*

Abbreviations, xv–xvii
Acceptor impurities, 25
Acousto-optic (AO) beam deflectors, 182–189, 274
Acousto-optic (AO) modulators, 196–198
Acousto-optic (AO) spectrum analysers, 292
Acronyms, xv–xvii
Agricultural applications, 240
Alignment applications, 239–240
Amplification, optical, 302
Analogue modulation, 196–197
Analogue signals, digital transmission of, 220–221
AND gate, optical version of, 304
Angle of acceptance
  determination of maximum angle, 327–328
  meaning of term, *93*
Angles, three-dimensional geometries considered, 126, 317–320
Antiglare baffles, 86
Associative record identification, 268
Attenuation, fibre optics, 94–95
Autocollimators, 240
Avalanche photodiodes (APDs), 81–82, 105

'Back distance' concept, 331, *332*
Bandwidth, meaning of term, 336
Bar code scanners, 251–253
  in-contact/hand-held readers, 251
  out-of-contact readers, 253
Bar codes, 248–254
  decoding of, 249–251
  reading of, 249, 251–254
Beam deflectors, 179–194
Beam diameter, definition of, 176
Beam divergence
  free-space optical communications, 128
  lasers, 176–177

Beam expanders, 177–179
Beam modulation, 194–203
Beam purification, 205–208
Beam reducers, 197
Beam routing devices, 289, *290–292*, 292
Beam splitters, 286–287, *288*
Beam waist, meaning of term, 176
Bell, Alexander Graham, 119
Biarritz (France), fibre optics communications scheme, 118
Bibliography, 346–353
Bipolar devices, 25
Birefringence, 198, 199
  electrical control of, 201
Bistables, optical, 305, *306*
Bit error rate (BER), 66, 132
  relation to speech intelligibility, 132
Bragg angle, meaning of term, 188
Bragg cell frequency shifters, 229, 230, 257
Bragg cells, 188, 196, 198
  frequency shift by, 198, 229, 230
Bragg reflectors, 284, *286*
Bristol board, 43, 321
British Telecom, fibre optic cable developments, 100, 114, 116, *117*
Builder's chalk line, 237–239
Burrus light-emitting diode, *49*

Calcite crystals, 198
Canada, fibre optics communications scheme, 118
Candela, definition of, 41
Captive ray systems, 89–118
Captive rays, physics explained, 90–94
Carbon dioxide lasers, 153–155
  power of, 155, 158
  TEA type, 155
  waveguide type, 154–155
Carrier energy diagrams, 22
Cassegrain objective, 126

Catadioptric objectives, 126, 127
Cathode ray tube (CRT), light guiding
    assembly for, 243–244
Circularly polarised light, 338–339
COBRA switch, *291*, 292
Coherence, 166–170
    phase coherence, 135
    spatial coherence, 170
    temporal coherence, 166–170
Coherence length
    definition of, 167
    derivation of, 336–337
    typical values quoted, 169
Coherent light guide, *242*, 243
Collimation
    free-space optical communications,
        122–124
    semiconductor laser beams, 208–209
Colour changing, 302
Colour fountains, 19
Colour–voltage relation, 33
Commodore P.50 pocket calculator,
        program for space angle calculation, 319
Computer-generated HOEs (CGHs), 270
Computers
    holographic applications, 269–270
    holography assisted by, 270
Cone channel condensers, 19
Construction industry applications, 237, *238*,
    239
Critical angle of incidence, definition of, 90
Crookes' radiometer, 298
Cube corner (trihedral) retroflectors,
    235–236, 237
Current, optic fibre detection of, 227–228
Current transfer ratio (CTR), meaning of
    term, 214

Dark current, 27, 70
Darlington connection, meaning of term, 84
Decibels
    calculation of, 4, 310–311
    power ratios expressed in, 310–311
Depletion layer, 26–27
Descartes' Law, 7
Diffraction
    frequency sorting by, 274
    physics of, 182–183
Diffraction gratings, *184*, 186
    beams produced by, 186–187
    holograms as, 264
Diffraction-limited lens, 189
Diffuse reflection, 249
Digital optical storage (DOR) systems,
    244–248
    commercial systems, 244–245, 248
    dimensions of, 246–247
    mechanical aspects of, 248
    optics of, 247
    recording principle of, 245–246
    storage capacity of, 248

Digitisation, 220–221
Distance measurements, 237, 240–241
    elapsed-time instruments used, 240–241
    phase comparison instruments used, 241
Divergence half-angle, definition of, 124
Donor impurities, 24
Doppler effect, 254
    physical explanation of, 254–255
Double heterojunction laser diodes
        (DHLDs), 143–146
Dye lasers, 156

Effective apperture (EA), calculation of, 315
Einstein. lasers predicted by, 136–137
Einstein's Equation, 33, 100
Electro-optic (EO) beam deflectors, 190,
    191, 193
Electro-optic (EO) modulators, 201–202
Electronics, comparison with optics, 2, 4
Elliptically polarised light, 338–339
Endoscopy, 17, 163
Evanescent field, 283–284
Excimer lasers, 155
Extraordinary (E) ray, meaning of term,
    199–200
Extrinsic semiconductors, 23–25

f-number, 16
    calculation of, 315
Fabry–Perot etalons (FPEs), 156, *303*, 304
Faraday effect, 227, 228
Fawley–Nursling (Hampshire, UK),
        overhead cable, 100, 115–116
Fermi level concept, 25–26
Fibre diameter, propagating modes affected
        by, 329
Fibre optic beam splitters, 286, *288*
Fibre optic pressure sensors (FOPSs),
    223–225
    transfer function of, 223–224
    University College London version, *224*,
    225
Fibre optics, funnel of acceptance, 93
Fibre optics communications (FOCs),
    89–118
    access point components used, 110–112
    advantages of, 102–103
    applications of, 115–118
    attenuation in, 94–95, 103
    cables used, 99–100
        examples of, *99*, 100
    components of FO link, 103–110
    connection techniques used, 107–110
    disadvantage of, 103
    expanded-beam connectors used, 109–110
    ferrule connectors used, 107–108
    fibre cladding used, 94–95
    glass purity, effects of, 95
    graded index fibres used, *97*, 98–99
    hardware used, 96–101

Fibre optics communications (*cont.*)
  information-carrying capacity of, 103
  jointing techniques used, 107
  jointing/connecting problems, 105, *106*
  kinematic connectors used, 108, *109*
  light-emitting diodes used, 46, 59–60
  light sources used, 104
  loop networking used, 112–113
  modal dispersion in, 95–96
  modulation multiplexing formats used, 114
  monomode fibres used, 97–98
  multimode fibres used, 96, *97*
  networking components in, 112–114
  physics involved, 90–96
  reasons for use of fibres, 102–103
  receiving heads used, 104–105
  reflective star couplers used, 113–114
  slicing techniques used, 107
  speed of light in, 100–101
  star networking used, 113–114
  state-of-art overview, 115–118
    international developments, 117–118
    UK achievements, 115–116
  stepped index fibres used, 96–98
  systems used in networks, 110–114
  transmissive star couplers used, *113*, 114
  transmitting heads used, 103–104
Fibreless optical communications, 119–134
Field flattening lenses, 193
Filter fitted photodiodes, 79
First quadrant mode (FQM), opto-electronic
    P–N junction, 30–32
Flip-flops, optical, 305, 306
Flux, meaning of term, 36–37
Focal length
  calculation of, 314
  meaning of term, 11
Foot-lambert, definition of, 323
Forbidden band, 22–23
Forward-biased P–N junctions, 28–30, *72*
Fourier method, 271–274
  applications of, 274–282
Fourier transform (FT), 272
  optical method for generation of, 274–275
Free electron lasers (FELs), 156–158
Free-space optical communications
    (FSOCs), 119–134
  advantages of, 120–121, 134
  alignment problems, 127
  atmospheric effects on range, 130–131
  beam divergence in, 128
  collimation in, 122–124
  competitors to, 134
  design data required, 57–58
  directivity of, 124–126
  examples of equipment available, 132–134
  fascination of, 121
  flux collection angle in, 124–126
  hardware, 122–127
  history of development, 119–120
  lensless transmitter antennas, 126

Free-space optical communications (*cont.*)
  link as whole, 127–132
  mirror-and-lens systems, 126
  nature of, 120–121
  range of
    achievements so far, 131
    atmospheric effects on, 130–131
    factors affecting, 128–131
  receiver antenna, 127
  transmitter antenna, 122–127
Frequency, meaning of term, 4
Frequency doubling/trebling, 302
Frequency mixing, 302
Frequency shifting, optic fibre sensors used,
    229–230
Frequency sorting, diffraction used for,
    274–276
Frequency spectrum, components removed
    from, 276, 278
Fresnel lenses, 9, *10*, 11
Fresnel zone plates, 261, 292

Galilean beam expanders, 179, 205
Gallium arsenide (GaAs) junctions, 23, 30,
    32, 33, 47–48, 62
Gallium arsenide (GaAs) lasers, 143–150
  as circuit elements, 146–150
  digital optical storage systems use, 244
  energy level diagram for, *145*
  modulation of, 202–203
  power of, 143, 158
  wavelength of radiation, 146
Galvo deflectors, 179–182
Gap detectors, 236
Gaseous flow, monitoring by Doppler
    velocimetry, *256*, 257
Gaussian beam, 174
  diameter of, 175–176
  divergence of, 176–177
  spatial filtering of, 205–207
Gaussian Equation, 12
  electronics equivalent of, *13*, 15
Glass filters, 88
Goldstein–Kreid configuration, 255, *256*
Graded index (GRIN) fibres, *97*, 98–99
Graded index (GRIN) lenses, 296–298
Group velocity, meaning of term, 330
Gyroscope, optic fibre, 225, 227

Half-wave retardation plates, *200*, 340, *344*
Helium–neon (HeNe) lasers, 140–143, 158
  construction of, *141*
  energy transfers in, *142*
  power of, 140, 158
  wavelength of radiation, 143
Heterodyne mixing, Doppler velocimetry,
    255, 256
Heterodyning, optic fibre sensors used,
    229–230
Heterojunctions, 49
  in lasers, 146

Heterojunctions (*cont.*)
  in LEDs, 49
Higashi Ikoma (Japan), fibre optics
    communications scheme, 118
'High-voltage' photovoltaic chips, 80
Holes, concept of, 21–22, 25
Holographic beam deflectors, 189, 264
Holographic interferometry, 268–269
Holographic optical elements (HOEs), 270,
    295
Holographic scanning, 189, *190*
Holography, 259–270
  applications of, 268–269
  compared with photography, 260
  and computer technology, 269–270
  first developed, 259
  interesting aspects of, 266, 268
  lasers in, 261
  physics of, 261–266
  of plane, 264–266
  of point source, 261, *263*, 264
  security aspect of, 266, 268
  of three-dimensional objects, 266, *267*
Hydrophones, optic fibre, 229–230

Iceland spar, 198
Illuminance, 39–40
  units of, 39
Illuminating fibre-optic assemblies, 241–244
Image forming lens, 12–16
Image improvement, spatial filtering
    techniques for, 278–280
Imaging fibre-optic bundles, 241, 243–244
Indispensability of light, 1
Infrared emitting diodes (IREDs), 44, 51
  see also Light-emitting diodes
Insulated signal couplers (ISCs), 212–221
  analogue/digital, 215
  applications of, 216–218
  audio/RF, 215
  characteristics of, 211
  common mode rejection (CMR) units, 216
  as electrical components, 214–218
  gain effects, 214–215
  high sensitivity/high power, 216
  low/high-voltage, 216
  operation of, 213–214
  opto-coupling application, 219–220
  singles/multiples, 216
  structure of, 213
  transfer function of, 215
Integrated optics (IO), 282–295
  aims of, 282
  beam routing devices, 289, *290–292*, 292
  beam splitters, 286–287, *288*
  connections for, 294–295
  devices used, 286–293
  future of, 295
  materials used, 282–283
  modulators, 289
  on/off switches, 289

Integrated optics (*cont.*)
  physics behind devices, 283–285
  substrate-topping methods, 294
  substrate-working techniques, 294
Intensity units, 40–41
Interference, results of, 4
Interference filters, 88
Interference fringes, *185*, 186, *187*
  laser Doppler velocimetry, 257–258
Interference holography, 268–269
Interlaser System, *133*
Intrinsic photoelectric effect, *63*
Intrinsic semiconductors, 22, *23*
Inverse Fourier Transform, 273
Invisibility of light, 1
Ion lasers, 152
Irradiance, 39–40
  units of, 39

Japan, fibre optics communications schemes,
    118
Johnson noise, 68, 70, 325

Keplerian beam expanders, 179–205

Lambert, definition of, 323
Lambertian radiators, 43, 321–323
  meaning of term, 321
Laser beam engineering, 166–211
Laser chip, 143–144
Laser Doppler velocimetry (LDV), 164,
    254–258
  practical equipment, 255–257
Laserdrive 1200 system, 244–245
Lasers
  acousto-optic deflectors used, 182–189
  acousto-optic modulators used, 196–198
  applications of, 157–158, 160–165
  beam diameter of, 175–176
  beam divergence of, 176–177
    reduction of, 177–179
  beam expansion of, 177–179
  beam purification for, 205–208
  beam treatment techniques, 177–211
  brazing of materials by, 162
  carbon dioxide lasers, 153–155
  circular symmetry configurations in, 174
  coherence of light from, 135, *136*, 166–170
  comparison with LEDs, 104
  cutting of materials by, 160, *161*
  deflection of beam, 179–194
  'doughnut' mode of, 174
  drilling of materials by, 160–161
  dye lasers, 156
  electro-optic deflectors used, 190, *191*, 193
  electro-optic modulators used, 198,
    201–202
  engineering industry use of, 164
  entertainment industry use of, 164–165
  excimer lasers, 155
  feedback photodiodes used, 147, *149*

Lasers (*cont.*)
　free electron lasers (FELs), 156–158
　fundamental mode of, 173
　gallium arsenide (GaAs) lasers, 143–150
　heat treatment of materials by, 162–163
　helium–neon (HeNe) lasers, 140–143, 158
　holographic deflectors used, 189–190
　holography in, 261
　holography use of, 165, 261
　ion lasers, 152
　longitudinal modes of, 170–172
　manufacturing applications of, 160–163
　meaning of term, 135
　mechanical deflectors used, 179–182
　mechanical modulators used, 196
　medical applications of, 158, 163–164
　metal vapour lasers, 152
　mode locking techniques, 210–211
　modulation of beam, 194–203
　neodymium yttrium aluminium garnet (Nd-YAG) lasers, 152, 158
　opto-optic modulators used, 202
　opto-optical deflectors used, *192*, 193
　oscillation modes of, 170–174
　physics of, 135–140
　power range of, 150, 158–159
　printing industry use of, 164
　pumping of, 138, 140
　Q switching technique for, 209–210
　safety aspects of, 150, 158, 159
　semiconductor materials purified by, 163
　soldering of materials by, 162
　solid-state lasers, 150–152
　spot-forming techniques for, 204–205
　surgical applications of, 164
　theoretical prediction by Einstein, 136–137
　transverse modes of, 172–174
　wavelength range of, 150
　welding of materials by, 161–162
Le Souffleur à la lampe [Georges de la Tour], 3
Leakage current, 27, 28, 70
Length-dependent lenses, 296–298
Lens, etymology of word, 12
Lens bending, 11–12
Lens ducts, 17
Lens formulae, 313–315
Lens-to-image distance, calculation of, 314
Lens-to-object distance, calculation of, 315
Lenses, 5–16
Liberated-ray systems, 119–134
Library books, bar codes on, 248–254
Lidar (light-radar), 241
Light-emitting diodes (LEDs)
　applications of, 61
　attractiveness as light sources, 60–61
　bi-colour LED, 58, *59*
　as circuit elements, 50–52
　collimation of radiation from, 61

Light-emitting diodes (*cont.*)
　comparison with lasers, 104
　cooling of, 58
　data sheets for, 55–58
　driver circuits used, 52–55
　edge emitters, 50
　exotic types, 46–47, 58–60
　geometries of, 47–50
　heterojunction used, 49
　manufacturers' information on, 55–58
　modulatability of, 60
　nature of, 44
　packaging of, 45–47
　pigtailed emitters, 46, 59–60, 104
　radiation flux calculated from polar diagram, 316–320
　of radiation from, 61
　solid-state nature of, 61
　special devices, 46–47, 58–60
　spectral compactness of, 61
　surface emitters, 47–49
　temperature effects on, 52, *53*
　wavelength varieties, 44
Light funnel, 19
Light guides, 17–20
Light pipe, 19
Light receiver (RX), components of, 62, 85
Light transmitter (TX), components of, 44, 103–104
Light trap, 19
Light-activated silicon-controlled rectifier (LASCR), 85
Lightwave communications, frequency possible, 102
Linewidth, meaning of term, 336
Liquid flow, monitoring by Doppler velocimetry, *256*, *257*
Lithium niobate, 282, 283
Loch Fyne (UK), fibre optics cable, 100, 116, *117*
London (UK), fibre optics network, 116
Long-wave light emitting diodes (LWLEDs), 44, 51
　see also light-emitting diodes
Long-wave photodiodes, 80
Low-pass filtering, electronic compared with optical, *207*
Lumens, relation to watts, 38
Luminance, 42–43
　units of, 42
Luminous flux, 36–38
Luminous intensity, 40–41, *42*
Lux, definition of, 39

Mach–Zehnder interferometer configuration, *229*, *230*, 289
'Magic wand' readers, 251
Magnification, calculation of, 14, 314
Maser (Microwave Amplification by Stimulated Emission of Radiation), 150

Matched filters, 280–282
Mechanical [beam] deflectors, 179–182
Mechanical [beam] modulators, 196
Metal vapour lasers, 152
Microwave (MW) links, features of, 134
MIE scattering, 300
Milton Keynes, fibre optics scheme, 115
Mirages, 98
Mirror-based beam deflectors, 179–182
Mirrors, as light-guides, 17–18
Modal dispersion, fibre optics, 95–96
Mode locking, 210–211
    active mode locking, 211
    passive mode locking, 211
    results of, 210–211
Modulators, integrated optics, 289
Moon–Earth distance measurement, 237
Moonbeam Telegraph, 119
Multifacet polygonal rotating deflectors, 181

Neodymium [doped] glass (Nd-glass) lasers,
    150–152, 158
Neodymium yttrium aluminium garnet (Nd-
    YAG) lasers, 152, 158
Newtonian lens formula, 15
Noise
    load resistance effects, 326
    meaning of term, 66
    total, 326
    types of, 324–325
Notation, xi–xiii
Nova laser, 151, 152, 158
Numerical apperture (NA), 16
    fibre optics, 93, 94, 328

Operating modes
    P–N junctions, 31, 34–35, 70
    photodiodes, 62, 63, 70
Optic fibre gyroscope, 225, 227
Optic fibre hydrophones, 229–230
Optic fibre sensors (OFSs), 222–231
    applications of, 223–231
    categorisation of, 222
    electric measurement by, 227–229
    pressure sensors, 223–225
    rotation sensors, 225–227
Optical barriers, 231–237
    range considerations for, 236–237
    reflection barriers used, 234–236
    transmission barriers used, 232–234
Optical bistability, 305, 306
Optical data storage, 244–248
Optical density
    calculation of, 4, 311–312
    transmittance expressed as, 311–312
Optical eight arrangement, 233
Optical fibres, 20
    see also Fibre optics. . .; Optic fibre. . .
Optical levitation, 298–300
Optical phase conjugation, 306–308

Optical retardation, 200, 201
    electrical control of, 201
Optical retardation plates, 340
Optical signal amplifiers, 305–306
Optical switches, 304–305
Optically activated mechanical vibrations,
    300–301
Optics, comparison with electronics, 2, 4
Opto-couplers, 212
    see also Insulated signal couplers (ISCs)
Opto-optic (OO) beam modulators, 202
Opto-optical beam deflectors, 192, 193
Optofollowers, 219–220
Optoschmitt, 85
OR gate, optical version of, 305
Ordinary (O) ray, meaning of term, 199–200
Overfilling (of lens), meaning of term, 204

P–N junction
    forward biasing of, 28–30
    operating modes of, 31, 34–35, 70
    physics of, 21–27
    quadrants of, 30–34, 70
    reverse biasing of, 27–28
Pantheon (Rome), 37
Pattern identification techniques, 268, 280
Pattern recognition, 280
Peltier cooler, 203
Periodicals listed, 345
Periscopes, 17–18
Perspex (Lucite, Plexiglass) light guides, 20
Phase coherence, 135
Phase conjugate mirrors (PCMs), 306–308
Phase velocity, definition of, 101, 330
Phase-locked loop techniques, optic fibre
    sensors used, 230
Phase-preserving spatial filter masks, 280,
    281
Photodarlingtons, 84
Photodiodes
    amplifying light-receiving devices, 81–85
    characterisation of, 64–69
    as circiut elements, 70–73
    detectivity of, 66–67
    incidence response of, 66
    light (reverse) current for, 65
    linear-output amplifiers used, 73–75
    logarithmic-output amplifiers used, 73–75
    luminous sensitivity of, 65
    manufacturers' data on, 64
    noise equivalent power of, 66
    noise limits of usefulness for, 66–68
    non-amplifying light-receiving junctions,
        62–81
    operating modes of, 62, 63
    preamplifying circuits used, 73–76
    quantum efficiency of, 65
    response speed of, 68
    responsivity of, 65
    specific detectivity of, 67–68
    spectral response of, 68–69

Photodiodes (*cont.*)
  varieties of, 76–80
Photoelectric field effect transistors
    (photofets), 84–85
Photometric quantites, 36–43
Photon, meaning of term, 2
Photophone transmitter, 119, *120*
Photoreceivers, 85–88
  energy management in, 86–87
  optical filters used, 88
  spectral flux management in, 87–88
Photosensitive integrated circuits, 85
Photosynthesis, 1–2
Phototransistors, 82–84
Pigtailed LEDs, 46, 59–60, 104
Pigtailed photodiodes, 80, 105
PIN photodiodes, 77, *78*, 105
Pinholes, spatial filtering by means of,
    205–207
Plane polarised light, 198–201, 340
Plane-wave approach, 5–11
Plano-convex lenses, 9, 11–12
Pockel's (electro-optic) effect, 201–202, 284
Polarised light, 198–201, 338–343
Population inversion (in lasers), 138
Potential barrier, 26–27
Power generators, photodiodes used in, 72,
    73, *76*, 77
Power levels, optics compared with
    electronics, 4
Presence detectors, 232–233
Pressure sensors, 223–225
Prismatic refraction, *6*, *8*, 9
Prisms, as light guides, 18–19
Propagating modes, fibre diameter effects
    on, 329
Pseudoscopic image, 266
Pyramidal deflectors, 181, *182*

Q switching, 209–210
  active Q switching, 209
  passive Q switching, 210
  results of, 210
Quadrant detectors, 239
Quadrants, 30–34, *70*
Quarter-wave plates, 340, *343*, *344*

Radiance, 43
Radiant intensity, 41, *42*
Radiation flux, calculation from polar
    diagram, 316–320
Radiometric quantities, 36–43
Random access spot-positioning techniques,
    193–194
Range
  calculation of, 331–335
  free-space optical communications
    achievements so far, 131
    calculation of, 130
    factors affecting, 128–131
Rangefinders, 240–241

Raster scanning, 193–194
Reach-through avalanche photodiode
    (RAPD), 81–82
Real images, 14
Rectification, optical, 302
References listed, 346–353
Refraction, schematic diagram of, *7*
Refraction dispersion, meaning of term, 330
Refractive index, 5, 7
  electrical control of, 284
  raising of, 294
Retardation plates, 340
Retroflective targets, *235*, 236
Retroflectors, 234–236, 237
Reverse saturation current, meaning of
    term, 27
Reverse-biased P–N junctions, 27–28, *72*
Rotating mirror deflectors, 181–182
Rotation sensors, 225–227
Ruby lasers, 150, 158

Sagnac interferometer, *227*
Sandwich ribbon coupling, 294, *295*
Schottky noise, 70, 324–325
Schottky photodiodes, 77, *78*
Second quadrant mode (SQM)
  opto-electronic P–N junction, 30, 31,
    32–33
  photodiodes operating in, *63*, 71, *72*, 73
Selective positive feedback, lasers, 140, 143
Selfoc elements, 296–298
Self adhesive retroreflective tape, 236
Semiconductor diode detectors, noise in,
    324–326
Semiconductor lasers
  collimation of beams, 208–209
  modulation of, 202–203
  see also Gallium arsenide lasers
Semiconductors, impurities in, 23–25
Shiva laser, 152, 158
'Shot' noise, 68, 70, 324–325
Siemens cordless telephone, *132*
Sign convention, 14
Signal, meaning of term, 66
Signal-to-noise ratio (SNR)
  free-space optical communications, 132
  photodiodes, 66
Silicon P–N junctions, *22*, 23
Single heterojunction laser diode (SHLD),
    147
Snell's Law, 7, 9
Solar cells, photodiodes as, *72*, 73, *76*, 77
Space angle, calculation from flat half-angle,
    317–319
Spatial coherence, 170
Spatial filtering, 205–208, 276, 278
  applications of, 278–282
  image improvement by, 278–280
Spatial frequency, meaning of term, 271
Spectrum analysers, integrated optics, 292

Spectrum width, meaning of term, 336
Speech intelligibility, parameters affecting, 132
Speed of light
  value in vacuum, 100
  velocity greater than, 100–101
Spot-forming techniques, 204–205
Spot-positioning techniques, 193–194
Steradian, definition of, 40, 41
Stimulated photon emission, 137
Stone oil lamp, 3
Strathclyde University, resonant bridge, 300–301
Supercouplers, 219–220
Supermarket bar codes, 248–254
Surface acoustic wave (SAW) generators, 285, 287
Surface acoustic waves (SAWs), 285
Surface barrier (Schottky) photodiodes, 77, 78
Surveying applications, 237, 238, 239
Switches, integrated optics, 289, 290–291, 292

Temporal coherence, 166–170
Thermal noise, 70, 325
Third quadrant mode (TQM)
  opto-electronic P–N junction, 30, 31, 33–34
  photodiodes operating in, 63, 71, 72, 73
Total conjugate length (TCL), 14, 15
  calculation of, 314
Total internal reflection (TIR), 19, 90, 91, 327
Transatlantic Automatic Telephone (TAT-8) link, 116
Transphasors, 305–306
Transreceivers, 80–81
Transverse electric and magnetic (TEM) modes, 172–173

Transverse gratings, 284–285, 286
Truncated shaft deflectors, 181, 182
Two-way electro-optic junctions, 80–81

UK, fibre optics communications developments, 115–116
Underfilling (of lens), meaning of term, 205
Universal Product Code (UPC), 249–251
University College London
  fibre optic pressure sensor, 224, 225
  force-measuring resonant level bar, 300, 301
  free-space optical communications equipment, 133, 134
  integrated optics spectrum analyser, 292, 293
USA, fibre optics communications schemes, 117–118

Virtual images, 14
Visibility curve, 38
Visibility factor, definition of, 169
Visible light-emitting diodes (VLEDs), 44, 51
  see also Light-emitting diodes
Voltage
  optic fibre detection of, 228–229
  relation with colour, 33
Volume holograms, 269

Wave nature of light, 5–11
Wavelength dedicated devices, 340
Wratten gelatine filters, 88

Y splitters, 287, 288
Young's Experiment, 185, 186